"*Phenomenology and QBism* joins the forces of the leading proponents of Quantum Bayesianism and phenomenology of physics to appraise the promise of phenomenology for the foundations of physics. The outcome is an exciting and inspiring collection of essays."

Mirja Hartimo, *University of Helsinki, Finland*

Phenomenology and QBism

This volume brings together philosophers and physicists to explore the parallels between Quantum Bayesianism, or QBism, and the phenomenological tradition. It is the first book exclusively devoted to phenomenology and quantum mechanics.

By emphasizing the role of the subject's experiences and expectations, and by explicitly rejecting the idea that the notion of physical reality could ever be reduced to a purely third-person perspective, QBism exhibits several interesting parallels with phenomenology. The central message of QBism is that quantum probabilities must be interpreted as the experiencing agent's personal Bayesian degrees of belief – degrees of belief for the consequences of their actions on a quantum system. The chapters in this volume elaborate on whether and specify how phenomenology could serve as the philosophical foundation of QBism. This objective is pursued from the perspective of QBists engaging with phenomenology as well as the perspective of phenomenologists engaging with QBism. These approaches enable us to realize a better understanding of quantum mechanics and the world we live in, achieve a better understanding of QBsim, and introduce the phenomenological foundations of quantum mechanics.

Phenomenology and QBism is an essential resource for researchers and graduate students working in the philosophy of physics, philosophy of science, quantum mechanics, and phenomenology.

Philipp Berghofer is a Post-Doc researcher and lecturer at the Philosophy Department of the University of Graz, Austria, and a former visiting fellow at the Center for Philosophy of Science at the University of Pittsburgh, USA (Spring Term 2023). His research focus centers around epistemology, phenomenology, and the philosophy of physics.

Harald A. Wiltsche is Full Professor in Theoretical Philosophy at Linköping University, Sweden. Prior to that, he held positions at Stanford University, the University of Toronto, and the University of Graz. His main areas of research are the general philosophy of science, the philosophy of physics, and phenomenology.

Routledge Studies in the Philosophy of Mathematics and Physics

Edited by Elaine Landry, University of California, Davis, USA and Dean Rickles, University of Sydney, Australia

The Emergence of Spacetime in String Theory
Tiziana Vistarini

Origins and Varieties of Logicism
On the Logico-Philosophical Foundations of Mathematics
Edited by Francesca Boccuni and Andrea Sereni

The Foundations of Spacetime Physics
Philosophical Perspectives
Edited by Antonio Vassallo

Scientific Understanding and Representation
Modeling in the Physical Sciences
Edited by Kareem Khalifa, Insa Lawler, and Elay Shech

The Non-Fundamentality of Spacetime
General Relativity, Quantum Gravity, and Metaphysics
Kian Salimkhani

Bertrand's Paradox and the Principle of Indifference
Nicholas Shackel

Phenomenology and QBism
New Approaches to Quantum Mechanics
Edited by Philipp Berghofer and Harald A. Wiltsche

For more information about this series, please visit: https://www.routledge.com/Routledge-Studies-in-the-Philosophy-of-Mathematics-and-Physics/book-series/PMP

Phenomenology and QBism

New Approaches to Quantum Mechanics

Edited by Philipp Berghofer
and Harald A. Wiltsche

NEW YORK AND LONDON

First published 2024
by Routledge
605 Third Avenue, New York, NY 10158

and by Routledge
4 Park Square, Milton Park, Abingdon, Oxon, OX14 4RN

Routledge is an imprint of the Taylor & Francis Group, an informa business

Library of Congress Cataloging-in-Publication Data
Names: Berghofer, Philipp, editor. | Wiltsche, Harald A., editor.
Title: Phenomenology and QBism : new approaches to quantum mechanics / edited by Philipp Berghofer, Harald A. Wiltsche.
Description: New York, NY : Routledge, 2024. | Series: Routledge studies in the philosophy of math and physics | Includes bibliographical references and index.
Identifiers: LCCN 2023035892 (print) | LCCN 2023035893 (ebook) | ISBN 9781032191812 (hardback) | ISBN 9781032194059 (paperback) | ISBN 9781003259008 (ebook)
Subjects: LCSH: Quantum Bayesianism. | Phenomenological theory (Physics) | Physics--Philosophy.
Classification: LCC QC174.17.Q29 P44 2024 (print) | LCC QC174.17.Q29 (ebook) | DDC 530.13/3--dc23/eng/20231107
LC record available at https://lccn.loc.gov/2023035892
LC ebook record available at https://lccn.loc.gov/2023035893

ISBN: 978-1-032-19181-2 (hbk)
ISBN: 978-1-032-19405-9 (pbk)
ISBN: 978-1-003-25900-8 (ebk)

DOI: 10.4324/9781003259008

Typeset in Sabon
by KnowledgeWorks Global Ltd.

Contents

List of Contributors

Michel Bitbol, Archives Husserl, CNRS/ENS, France

Florian J. Boge, Wuppertal University, Germany

Laura de La Tremblaye, Archives Husserl, CNRS/ENS, France

Steven French, University of Leeds, UK

Christopher A. Fuchs, University of Massachusetts, Boston, USA and University of Colorado, Boulder, USA

Philip Goyal, University at Albany (SUNY), USA

Jacques Pienaar, University of Massachusetts, Boston, USA

Thomas Ryckman, Stanford University, USA

Rüdiger Schack, Royal Holloway, University of London, Egham, UK

Blake C. Stacey, University of Massachusetts, Boston, USA

Hans Christian von Baeyer, College of William and Mary, USA

List of Contributors

1 Introducing Phenomenology to QBism and Vice Versa

Phenomenological Approaches to Quantum Mechanics

Philipp Berghofer and Harald A. Wiltsche

1.1 Introduction

Books on philosophical implications of quantum mechanics typically start by pointing out that although quantum mechanics is the most successful theory in the history of science, we still do not (agree on how to) understand it. As paradoxical as this might seem, this is undoubtedly true. In terms of accuracy and range of applications, quantum mechanics is unrivaled. And yet, there has never been a widely accepted scientific theory that has caused so much perplexity and disagreement. What does quantum mechanics tell us about the nature of reality? Although disputes about interpretational issues are as old as the theory itself, the situation is still that over a dozen different interpretations give diverging answers to this question. For some, quantum mechanics implies that there exist infinitely many worlds and that every time a quantum event is observed, reality branches. Others have argued that quantum mechanics must be seen against the backdrop of a strict mind-body dualism, and that consciousness causes the wave function to collapse. Quite generally, said wave function, a central notion in quantum mechanics, is a particular bone of contention. Many philosophers in the analytic tradition tend to consider the wave function a physically real entity, suggesting that the mathematical space it populates is real and ontologically fundamental. Physicists, by contrast, typically relegate the wave function to the status of a mere mathematical tool. And although the collapse postulate is accepted since the 1930s, most working physicists prefer to ignore the question of what it exactly means or why the process of measuring seems to bring about the collapse in the first place. For philosophers, on the other hand, questions surrounding the apparent collapse of the wave function—and the infamous measurement problem more generally—lie at the heart of any serious philosophical attempt to come to terms with modern physics.

There are two main reasons why the present volume occupies a special place in the vast philosophical literature on quantum mechanics. First,

DOI: 10.4324/9781003259008-1

being the first collection that explicitly focuses on the relationship between phenomenology and quantum physics, it advances the somewhat unusual thesis that the phenomenological tradition has much to offer to advance our understanding of the latter. Section 2 of this introductory chapter thus aims at identifying, clarifying, and motivating some of the cornerstones of phenomenological approaches to quantum mechanics. Here we will already see that phenomenology shares crucial systematic similarities with a recent interpretation of quantum mechanics that goes by name of QBism. This, then, is the second unique feature of this volume: Being the first collection on the philosophical implications of QBism, it advances the thesis that phenomenology and QBism offer rich potentials for mutual enlightenment. While phenomenology provides several building blocks that could support QBists in their attempts to explicate the philosophical underpinnings of their own position, QBism is attractive for phenomenologists because it renders quantum mechanics close in spirit to some of the main characteristics of phenomenological philosophy.

There are several additional points of contact between QBism and phenomenology: For one thing, both projects consider the sphere of *lived experience* the ineluctable starting point and epistemological foundation of any scientific investigation. In the eyes of many, this emphasis on experience has put QBism and phenomenology on a direct collision course with the picture of science that is still dominant in large parts of contemporary philosophy of science. For instance, "mainstream" philosophers of science have accused QBism of rendering quantum physics explanatorily impotent because the reduction to experience and subjective degrees of belief seem to cut all ties to the external world which, according to the critics, is the only explanandum in physics (Hagar 2003; Timpson 2008; Brown 2019; Earman 2019). However, instead of rejecting QBism because of its incompatibility with the received view about science, one could also consider the radical alternative: What if quantum mechanics can be consistently interpreted as revealing that our picture of science and reality, as inherited from classical mechanics, is fundamentally ill-headed? And what if, furthermore, QBists are right that the mathematical formalism of quantum theory is no representational vehicle, but rather a tool for embodied agents to manage their expectations about future experiences? Since, arguably, phenomenology is the most thoroughly developed experience-first approach in modern philosophy, it would not be altogether surprising under these assumptions that the phenomenological tradition—and not "mainstream" analytic philosophy of science—offers the most suitable framework for the understanding and interpretation of science in general and quantum mechanics in particular.

Conversely, interpretational disputes about the nature of quantum physics could also have an impact on the reception of phenomenology within the wider scientific arena: It is a common criticism that phenomenology's emphasis on first-person experience puts it at odds with a purely objective third-person methodology that is usually associated with the sciences. In the face of this charge, phenomenologists typically concede, insisting that philosophy and science are indeed very different projects. But if QBism is successful in showing that a third-person methodology is by no means a universal characteristic of science, this would have obvious consequences for questions regarding the "scientificality" of phenomenology. We will come back to these and similar other points in Section 3.

1.2 Phenomenological Approaches to Quantum Mechanics

Phenomenology, the philosophical tradition that has been inaugurated by Edmund Husserl at the beginning of the 20th century, is a movement that requires the experiencing subject to focus on how she experiences the world. Its basic objective is to unveil the structures of consciousness. What does it mean to undergo an experience? What distinguishes, for instance, perceptual experiences from other types of experiences such as introspective experiences, mathematical intuitions, evaluative experiences, etc.? On a more basic level: What distinguishes mental states that in a sense to be specified "present" their objects (such as when a perceptual experience presents a tree as bodily present) from mental states that do not possess this kind of presentiveness or givenness (such as beliefs)? Regarding methodology, phenomenology constitutes a *descriptive* and *eidetic* approach to investigating consciousness. It is descriptive in the sense that it is a first-person analysis. The aim is to clarify what it is like for the subject to undergo a specific experience, not, for instance, how the brain behaves when having the experience. The ambition, however, is not simply to describe how it feels to undergo a specific experience, but to specify necessary phenomenal features that distinguish different types of experiences and mental states. In this sense, phenomenology pursues an eidetic methodology, using tools like conceptual analysis to unveil structural moments of consciousness. One such structural moment typically identified by phenomenologists as a mark of the mental is *intentionality*. Here intentionality denotes the "aboutness" or "directedness" of mental states. Experiences, wishes, or desires are essentially characterized by their being directed at something *beyond themselves*. Importantly, intentionality comes in many different flavors. One can be intentionally directed toward the same object in many different ways, such as when one first believes that one's bike is in the office, and then sees that one's bike is in the office. For Husserl, these different modes of "givenness" are of utmost *epistemological* importance. Experiences in which the object is given in a presentive manner are contrasted with empty (or signitive) acts in which what is given is not the object in its actual presence, but the object as something that is only meant. While believing that one's bike is in the office is an empty act, the presentive act of seeing the bike *fulfills* the empty act of believing. For many phenomenologists, fulfillment, i.e., the congruence between the object as it is emptily intended and the object as it is given in a presentive experience, is what distinguishes knowledge from mere belief (see, e.g., Hopp 2020, Section 5.1). Phenomenology, in Husserl's tradition, also has the ambition to be the First Philosophy, i.e., the most basic science. This is precisely due to the central epistemic role that phenomenology ascribes to experience: Every science, every piece of knowledge, can be traced back to epistemically foundational experiences (see Berghofer 2022). Since experiences are our basic justifiers, and since phenomenology can be viewed as the study of experience that clarifies how experiences justify, phenomenology enjoys epistemological priority over the individual sciences. In what follows, we focus on features, methods, and teachings of phenomenology that we consider important when it comes to developing phenomenological approaches to quantum mechanics. For more details on phenomenological key concepts and how they are relevant to (philosophy of) physics, see Berghofer and Wiltsche (2020).

1.2.1 *Life-World First! (Against Wave Function Realism)*

Of all works of classical phenomenology, we consider Husserl's last major publication, *The Crisis of European Sciences and Transcendental Phenomenology*, particularly relevant for phenomenological interpretations of science. In our view, there are two key concepts that deserve special attention: First, the notion of the *life-world* which is not only crucial for a more general understanding of Husserl's late philosophy but which was also highly influential in areas such as sociology or anthropology. Second, the *Crisis* is the locus classicus for Husserl's critique of the *mathematization of nature*. In this subsection we briefly discuss both notions and focus specifically on their relevance for the interpretation of quantum mechanics.

As mentioned above, in phenomenology experiences play the central epistemological role. Phenomenology considers experience to be the ineluctable starting point and epistemological foundation of any scientific investigation. The life-world, in turn, can be understood as the world of our everyday experiences. It is the world of ordinary objects, the world of tables and chairs, the world as it is immediately perceivable and familiar to us. However, the life-world is not only the pre-scientific world in which we all live. According to Husserl, it is also the "meaning-fundament of natural science" (Husserl 1970, 48) and the "realm of original evidences" to which "[a]ll conceivable verification leads back" (Husserl 1970, 127 f.; translation slightly modified). This is to say that the life-world is both the meaning fundament and the epistemic basis of all scientific endeavors.

In light of this initial characterization, it becomes immediately clear that the life-world thesis puts certain restrictions on our understanding of science and scientific theories. Most importantly, once the thesis of the priority of the life-world is accepted, all interpretations that relegate the life-world to the status of a mere illusion are ruled out from the outset. Such a view was popularized, for instance, by Wilfrid Sellars:

> [S]*peaking as a philosopher*, I am quite prepared to say that the common sense world of physical objects in Space and Time is unreal-that is, that there are no such things. Or, to put it less paradoxically, that in the dimension of describing and explaining the world, science is the measure of all things, of what is that it is, and of what is not that it is not.
>
> (Sellars 1963, 173)

The view expressed here is still common in contemporary analytic philosophy of science (see, e.g., Ladyman & Ross 2007). From our phenomenological vantage point, however, any attempt to demote the life-world to the status of an illusion amounts to an empirically incoherent, self-defeating line of reasoning. To put it crudely, declaring our experiences to be illusory is like sawing off the branch on which we are sitting. This is, of course, because our perceptual experiences and the life-world are the *epistemic foundation* of the sciences. As Husserl puts it: "Straightforward experience, in which the life-world is given, is the ultimate foundation of all objective knowledge" (Husserl 1970, 226). If the sciences reveal the illusory character of our experiences, they cast doubt on their own epistemic foundation. Following Jeffrey Barrett (1999, 116), Huggett and Wüthrich define "a theory to be *empirically incoherent* in case the truth of the theory undermines our empirical justification for believing it to be true" (Huggett & Wüthrich 2013, 277). Hence, if our everyday experiences and the life-world are the epistemic foundation of modern physics, but modern physics is interpreted as revealing that the life-world is mere illusion, this interpretation of physics is in danger of being empirically incoherent.[1] One strategy for reconciling the life-world and the world of science could be to regard ordinary experiences and mathematical models as two distinct ways of being intentionally directed toward one and the same world.

This discussion regarding the life-world/science relationship brings us directly to Husserl's critique of the mathematization of nature which, in slogan form, amounts to the warning that we must not "take for *true being* what is actually a *method*" (Husserl 1970, 51). Quite generally, the Husserlian term "mathematization" refers to the cognitive process through which nature is turned into a mathematical manifold. What this means, concretely, is best understood through Husserl's interpretation of the works of the "father of modern science," Galileo Galilei. On Husserl's view, Galileo marks a watershed in the history of physics not primarily because of any of his individual theoretical or experimental accomplishments. What sets Galileo apart from the tradition before him is rather the larger methodological vision "that trying to deal with physical problems without geometry is attempting the impossible" (Galilei 1967, 203). On Husserl's reading, however, the purpose of Galileo's introduction of mathematical models into physics was not merely to "save the appearances" in individual sub-segments or reality. Rather, Galileo linked his methodological innovation to the much more radical ontological thesis that his mathematical-geometrical models are direct representations of the one true reality *which is mathematical in nature*. It is this metaphysical view that forms the background of Galileo's famous book-metaphor:

> [T]his all-encompassing book that is constantly open before our eyes, that is the universe, [...] cannot be understood unless one first learns to understand the language and knows the characters in which it is written. It is written in mathematical language, and its characters are triangles, circles, and other geometrical figures; without these it is humanly impossible to understand a word of it, and one wanders around pointlessly in a dark labyrinth.
>
> (Galilei 2008, 183)

Following Husserl's account, then, Galileo's contribution to the history of modern science cannot be reduced to the insight that mathematical models are highly suitable tools for the representation of empirical reality. On Husserl's reading, Galileo's radicality lends itself to the much more radical view that reality literally consists of and is exhausted by geometrical-mathematical structures and entities. "[T]hrough Galileo's *mathematization of nature*, *nature itself* is idealized under the guidance of the new mathematics; nature itself becomes [...] a mathematical manifold" (Husserl 1970, 23). Accordingly, concerning the formal and technical apparatus of the mathematical sciences, Husserl warns us not to be "misled into taking these formulae and their formula-meaning for the true being of nature itself" (Husserl 1970, 44).

We note that Galileo, according to this reading, is even more radical than Sellars. Both demote the world of our everyday experiences to some form of illusion but while Sellars holds that reality consists of the objects of our fundamental physical theories, Galileo believes that reality *is* a mathematical manifold. Of course, this appears highly counter-intuitive at first glance, and one might argue that while Husserl's criticism of Galileo is sound, it is only of historical interest. But views very similar to Galileo's are prominently championed in contemporary philosophy of quantum mechanics. This brings us to the doctrine of *wave function realism*.

In quantum mechanics, the quantum state is represented by the so-called wave function. Mathematically speaking, *wave functions are vectors in a Hilbert space*. This is often expressed by saying that "[w]ave functions live in Hilbert space" (Griffiths 2018, 94). A Hilbert space is an abstract mathematical concept, namely a complete vector space on which an inner product is defined. But if one thinks of the wave function as something real, doesn't this mean that the mathematical Hilbert space must be granted physical existence too? This question is directly connected to the more general issue regarding the relationship between abstract mathematical spaces and the space we actually live in. In light of this issue, one possible reaction would consist in the straightforward reification of Hilbert space. And indeed, one can find prominent voices championing Hilbert space realism (e.g., Carroll & Singh 2019). However, most consider this an implausible and unwarranted hypostatization of mathematical objects and it has been pointed out that only "[v]ery few people are willing to defend Hilbert space realism in print" (Wallace 2013, 216).

A similar but more subtle form of mathematization takes place in configuration space realism, i.e., the project of reifying the 3*N*-dimensional configuration space, *N* being the number of the particles in the universe. The main proponent of this view is David Albert.

> [I]t has been essential […] to the project of quantum-mechanical *realism* to learn to think of wave functions as physical objects *in and of themselves*. And of course the space those sorts of objects *live* in, and (therefore) the space *we* live in, the space in which any realistic understanding of quantum mechanics is necessarily going to depict the history of the world as *playing itself out* (if space is the right name for it – of which more later) is *configuration-space*. And whatever impression we have to the contrary (whatever impression we have, say, of living in a three-dimensional space, or in a four-dimensional space-time) is somehow flatly illusory.
>
> (Albert 1996, 277)

This configuration space realism, often referred to as wave-function realism,[2] has been quite popular and has sparked much controversy (see particularly the contributions in Ney & Albert 2013). In fact, "[t]his view of the ontology of (no hidden variable) quantum mechanics has probably been the most commonly assumed in the recent literature" (Wallace 2013, 217). Note that Albert explicitly says that our impression to live in a three-dimensional space is "flatly illusory." Obviously, from a Husserlian perspective, such a claim is highly suspect, to say the least. Accordingly, many objections to this claim have a phenomenological touch.[3]

David Wallace objects that configuration space realism "makes the same unmotivated conceptual move as Hilbert space realism: it reifies a mathematical space without any particular justification" (Wallace 2013, 217). Bradley Monton argues "that our everyday commonsense constant experience is such that we're living in three spatial dimensions, and nothing from our experience provides powerful enough reason to give up that prima facie obvious epistemic starting point" (Monton 2013, 154). Peter Lewis has even argued "for the converse of Albert's initial position; the world really is three-dimensional, and the 3*N*-dimensional appearance of quantum phenomena is the theoretical analog of an illusion" (Lewis 2013, 124).

It would go beyond the scope of this introductory chapter to discuss these arguments in detail, but we note the Husserlian idea that no matter how abstract our scientific theories are, their justification, ultimately, lies in ordinary experiences, in what is immediately given. In Husserl's words, "the inductive scientific judging" of the "exact objective sciences that by going beyond the immediately experienced infers the non-experienced is always dependent on its ultimate legitimizing basis, on the immediate data of experience" (Husserl 1973, 121; our translation). For phenomenologists, arguing based on scientific theories that the life-world is mere illusion is empirically incoherent (Wiltsche forthcoming), and, indeed, this worry has been raised against Albert's configuration space realism (Chen 2019, 6).

In sum, since the wave function is the central concept of quantum mechanics, reifying the wave function is the natural move for interpretations that are in the spirit of standard scientific realism. It is rather unsurprising, then, that wave function realism is widely held in contemporary philosophy of quantum mechanics.[4] Reifying the wave function, however, is similar to the reification of geometrical concepts that Husserl ascribed to Galileo. It runs the risk of constituting a counter-intuitive mathematization of nature that confuses reality with what is a method to describe or represent reality. Phenomenologists should thus be cautious to subscribe to any interpretation that is in danger of implying such a mathematization. Here phenomenologists are in agreement with the QBist worry that in standard realist interpretations "the strategy has been to reify or objectify all the mathematical symbols of the theory and then explore whatever comes of the move" (Fuchs & Stacey 2019, 136). As a critical reaction to this argumentative strategy, an idea that has recently emerged in the foundations of quantum mechanics is that the whole project of interpreting quantum mechanics is problematic—if this project is understood as taking the quantum formalism as a given and attempting to read off the nature of physical reality from the mathematical structure of the formalism. Instead, it has been argued that one first needs to identify intuitive physical principles from which the formalism can be *reconstructed*. What is then interpreted are the physical principles and not the mathematical formalism. We return to this project of reconstruction below when discussing Philip Goyal's contribution to this volume.

1.2.2 *Physics First! (Against Modificatory Interpretations)*

Quite generally, a driving motivation for phenomenology is a deep respect for the phenomena. This is to say that phenomenologists strive to be attentive to how intentional objects are given to them, aiming at a study of the given that is as unprejudiced as possible. This idea has been expressed by Husserl in various ways, most prominently in his call to "go back to the 'things themselves'" and his verdict that phenomenologists "are the genuine positivists" (Husserl 1982, 39). One implication of this is, according to Husserl, that logical truths should not be reduced to psychological facts. Logical truths are given to us as being independent of our encountering them, which is why we should be cautious in our naturalist ambitions to reduce them to psychological or physical laws. In this spirit, Mirja Hartimo has recently argued that Husserl's phenomenological approach to mathematics can be described as a mathematics-first approach.

> While philosophers' and mathematicians' activities are complementary, each should also respect the others' autonomy. [...] In general, his approach is 'mathematics first.' [Husserl] *does not develop a philosophical view of what mathematics should be like, but aims rather to describe mathematics as it is for mathematicians* [our emphasis].
> (Hartimo 2021, 27)

The relationship between mathematics and philosophy of mathematics is summarized by Hartimo as follows: "For Husserl mathematics comes first – in accordance to the slogan 'back to the phenomena themselves' – yet at the same time, philosophy comes first in the sense that it seeks to give reflective foundations to the other disciplines" (Hartimo 2021, 31). Edith Stein clarifies the foundational role of phenomenology as follows:

> Being a foundational science, indeed, does not mean that phenomenology generates presuppositional statements for all other sciences, from which the latter would be able to logically derive their own theorems. Rather, by removing the 'self-forgetfulness' of the dogmatic scientist, phenomenology reveals the dimension of unclarity

> that attaches to *every* dogmatic science, and transforms 'naive' science, which does not inquire into the meaning and justification of its methodology, into a science that has been clarified by critical reason.
> (Stein 2018, 312; Husserl 1987, 263f.)

On our view, these comments can be summarized as follows: Phenomenology aspires to be the most basic science not because it claims to deliver the axioms or theorems of every or even any individual science, but because it addresses the epistemic foundation of any given individual science, identifies justification-conferring experiences as foundational justifiers, and seeks to clarify how we legitimately get from these experiences to scientific theories.

Regarding physics, one of our main interests as phenomenologists lies in the question of how the respective theory and mathematical formalism *emerges from its experiential and life-worldly foundation*. It should be noted, however, that phenomenologists should be *very* cautious to demand modifications of the formalism on philosophical grounds.[5] One should be particularly cautious if the urge for modification is based on ontological principles or intuitions (that might turn out to be ontological prejudices). Consider, for instance, Bohmian mechanics. In contemporary philosophy of quantum mechanics, Bohmian mechanics is a highly popular choice. In the physics community, however, it remains disliked and largely ignored. This is because Bohmian mechanics actually *changes* the formalism of quantum mechanics, adding so-called hidden variables that are, in principle, unobservable. Importantly, it remains technically challenging to square Bohmian mechanics with special relativity and, as of yet, we do not have a relativistic extension of Bohmian mechanics that rivals the predictive scope of well-established relativistic extensions of textbook quantum mechanics (see Goldstein 2021, Section 1.4; Kofler & Zeilinger 2010; Wallace 2022). Bohmian mechanics, strictly speaking, is thus not an interpretation of our most successful scientific theory (i.e., quantum mechanics), but a *rival* theory that is clearly inferior in its predictive power. But why, then, is Bohmian mechanics so popular in philosophy? One main reason is that Bohmian mechanics promises a *clear ontology*. This desideratum has been expressed as follows:

> In fact, the lack of a clear ontology in orthodox quantum mechanics is the real root of the measurement problem (and many other problems). If the ontology is clear—if it is clear what the fundamental entities in nature are that the theory seeks to describe—there can't be any paradoxes.
> (Dürr & Lazarovici 2020, 47)

In Bohmian mechanics, the fundamental objects of the theory are point particles that have a definite position at each time, their dynamics being governed by deterministic equations. Accordingly, Bohmian mechanics is the interpretation whose ontology is closest to classical mechanics. Of course, there are some important dissimilarities between classical and Bohmian mechanics. Although both theories are deterministic, it is impossible in Bohmian mechanics to know all initial conditions. As a consequence, Bohmian mechanics—just like textbook quantum mechanics—is limited to predictions that are probabilistic in nature. Furthermore, in Bohmian mechanics the wave function plays an important role in determining the dynamics of the point particles. Accordingly, one might argue that its ontology is not as clear as its proponents would like it to be since the ontological status of the wave function remains contentious. As mentioned above, Bohmians disagree on whether the wave function is physically real, or more like a nature of law, or a new kind of entity that is ontologically sui generis. Also, it should be mentioned that Bohmian mechanics implies non-locality which has often been understood, most notably by Einstein, as a counter-intuitive "spooky action at a distance."

However, the main worry addressed in this section is that Bohmian mechanics is not in accordance with the "physics first" idea. As we have seen, this is because Bohmian mechanics does not interpret the formalism that is actually used by the majority of quantum physicists, but *modifies* the quantum formalism and thereby introduces a rival theory. This is true for Bohmian mechanics as well as for objective collapse theories such as the Ghirardi-Rimini-Weber (GRW) theory. Proponents of the many-worlds interpretation (MWI) typically regard this as a major advantage of their interpretation, arguing that on their account no modification of the quantum formalism is required (Wallace 2022). Although we do not contest this claim, we wish to note that the ontological status of the wave function is a problem for MWI as well, as is the postulation of infinitely many, in principle unobservable worlds. We shall turn to the topic of observability in the next subsection.

The "physics-first" idea can be broken down to the following guideline which should be taken seriously by every phenomenological approach to quantum mechanics: We should be cautious to modify the formalism of our most successful scientific theory, particularly if the rival modified theory is (i) predictively less successful and (ii) less parsimonious due to surplus mathematical structure. Furthermore, as will become clearer in the course of this chapter, we may add the following guideline: We should not consider the properties of classical mechanics as properties that must be preserved in quantum theory. This is to say that if quantum phenomena—such as the phenomenon that quantum objects apparently do not have pre-determined and pre-existing values—suggest a worldview that is different from our classical intuitions, we should be open to that possibility, respecting the quantum phenomena.

1.2.3 *Experience First! (Against Objectivist Interpretations)*

Phenomenology is an "experience-first" project in an epistemological and a methodological sense. Epistemologically, because it acknowledges that all epistemic justification, every piece of knowledge, and any successful scientific endeavor can be traced back to epistemically foundational experiences. One of its aims, then, is to analyze which experiences are involved and which role they play in the practice and reasoning of the respective science (Berghofer 2022). In a methodological sense, phenomenology qualifies as an "experience-first" project because it relies on a descriptive first-person analysis of consciousness. Applied to phenomenological approaches to science, this latter aspect leads to questions such as whether we should accept the existence of scientific entities that are in principle unobservable, whether science can successfully abstract away from the subject and her experiences, or whether, to the contrary, science should actively seek to incorporate the first-person perspective into science. In Section 2.3.1, we briefly discuss the criterion of *observability* which took center stage in the reasoning of various phenomenologists and physicists. If scientific theories must conform to this criterion, this alone would be enough to undermine the idea that science can be purged of all subjective or operational notions. In Section 2.3.2, we move on to the stronger claim that physics needs to incorporate the physicist and that quantum mechanics in particular should be understood in exactly this way. Interestingly enough, this is a claim that unites classical phenomenologists such as Merleau-Ponty and contemporary QBists. Finally, in Section 2.3.3, we discuss a concrete interpretation of quantum mechanics that exemplifies such a phenomenological approach.

An additional goal in this section is to make explicit the tensions between some of the basic tenets of phenomenology on the one hand and objectivist interpretations of quantum mechanics on the other. In our terminology, interpretations qualify as "objectivist" if they (i) reify the wave function, i.e., assign it the status of a physically real entity, and/or (ii) claim to deliver a purely objective third-person perspective that does not contain any irreducibly subjective or perspectival moments. While (i) was in the center of attention in Section 2.1, the focus in this section is on (ii).

1.2.3.1 Observability

For Husserl, the most fundamental question in epistemology is how subjectivity can be the source of objective knowledge (see Melle in Husserl 1984, page XXXI). In his view, the kind of acts that play the role of justifiers for all sorts of beliefs are a particular type of experiences, namely *originary presentive intuitions*. What makes this particular category of acts special is the fact that they present their objects as "bodily present," "actually present," or simply "self-given" (Husserl 1997, 12). Since all mediate justification leads back to immediate justification, and since originary presentive intuitions are the source of this kind of justification, originary presentive intuitions also play the role of *ultimate* (albeit fallible) justifiers. The epistemic significance of these acts is most firmly stated in the famous *principle of all principles*:

> No conceivable theory can make us err with respect to *the principle of all principles: that every originary presentive intuition is a legitimizing source of cognition, that everything originarily* (so to speak, in its "personal" actuality) *offered to us in 'intuition' is to be accepted simply as what it is presented as being*, but also *only within the limits in which it is presented there.*
>
> (Husserl 1982, 44)

Particularly the last part of this principle is sometimes interpreted as suggesting that one cannot be justified in believing in the existence of an entity that cannot, in principle, be originally given. This, of course, would have far-reaching consequences for the interpretation of scientific theories. For instance, Wiltsche argued for a phenomenologically motivated anti-realism about unobservable scientific entities (Wiltsche 2012). On an exegetical level, this anti-realism can be backed up by passages in which Husserl explicitly demands that being a physical object implies being perceptually experienceable (Husserl 2003, 74). Although it remains controversial whether phenomenologists should feel committed to such a form of anti-realism (see the response to Wiltsche in Berghofer 2018; Hardy 2020, and also Wiltsche forthcoming), we note that from a phenomenological perspective it is prima facie suspect if a scientific theory implies the existence of entities that are in principle unobservable.

Another way for the concept of observability to enter the discussion is the idea that "a theory shouldn't make distinctions that it cannot empirically honor" (Carrier 2012, 28; our translation). To put it differently, if A and B are two distinct physical states, it must at least in principle be possible to empirically distinguish between them, there must be an observable difference. This was the driving idea behind Einstein's development of relativity theory as well as Heisenberg's quantum mechanics (see, e.g., Carrier 2012; Rovelli 2021). Consider how Heisenberg opens his 1925 article that marks the beginning of modern quantum mechanics:

> The objective of this work is to lay the foundations for a theory of quantum mechanics based exclusively on relations between quantities that are in principle observable.
>
> (Heisenberg, as cited in Rovelli 2021, 20)

Arguably, hidden-variable theories like Bohmian mechanics violate this idea, as does the MWI.

1.2.3.2 *Incorporating the First-Person Perspective into Science*

Considering Husserl's aforementioned critique of Galileo, it cannot be emphasized enough that Husserl nowhere questions the tremendous success of Galileo's amalgamation of mathematics and physics. In fact, the technological and predictive success of modern mathematized science is so tremendous that it poses questions on its own: One can wonder, as Eugene Wigner famously did (1960), why mathematical models can be so successfully used to represent reality at all. In answering this question, phenomenologists typically do not follow mathematical monists such as Galileo or Max Tegmark in claiming that "our successful theories are [...] mathematics approximating mathematics" (Tegmark 2008, 125). Instead, phenomenologists like to point out that the application of mathematical models involves a *process of idealization* such that our mathematical framework is not applied to nature itself but an idealization, a model of nature (see Islami & Wiltsche 2020).

The success of mathematical physics has indirectly led also to some further problems. For instance, it has been argued that the undeniable success of quantum physics in general and of relativistic quantum field theory in particular has shrouded and overshadowed a lack of conceptual clarity in physics (Berghofer et al. 2023). It seems that a new generation of physicists is learning how to apply certain concepts, methods, and theories in order to solve suitable problems, but that the meaning of these concepts, their historical embeddedness, and their relationship to our real world is more and more lost. One can suspect that this alienation between physical theory and physical reality is not only a philosophical problem but actually hinders progress in physics. This problem has been explicitly mentioned by Einstein:

> Concepts that have proven useful in ordering things easily achieve such an authority over us that we forget their earthly origins and accept them as unalterable givens. Thus they come to be stamped as 'necessities of thought,' 'a priori givens,' etc. The path of scientific advance is often made impassable for a long time through such errors. For that reason, it is by no means an idle game if we become practiced in analyzing the long commonplace concepts and exhibiting those circumstances upon which their justification and usefulness depend, *how they have grown up*, *individually*, *out of the givens of experience* [our emphasis]. By this means, their all-too-great authority will be broken.
>
> (Einstein 1916, 102; cited in Howard 2014, 358)

A similar point has been made by Husserl:

> And it is precisely for this reason that a theoretical task and achievement like that of a natural science [...] can only be and remain meaningful in a true and original sense *if* the scientist has developed in himself the ability to *inquire back* into the *original meaning* of all his meaning-structures and methods, i.e., into the *historical meaning of their primal establishment*, and especially into the meaning of all the *inherited meanings* taken over unnoticed in this primal establishment, as well as those taken over later on.
>
> (Husserl 1970, 56)

A further problem that indirectly arose from the success of mathematical physics is the naturalistic attitude according to which a third-person mathematical description can describe everything there is. In this context Stein said: "What physics [...] reveals pertains to the real nature but it never *exhausts* nature. And what evades the web of mathematical formulas is not less 'real' than what is captured by mathematics" (Stein 2004, 62). Here we find two motifs that are typical for a phenomenology of physics. First, although phenomenology does not, of course, dispute the success of physics or object to the implementation of mathematics, a claim often seen in phenomenology is that the mathematical picture delivered by physics only constitutes one perspective on nature, and that what we gain from the successful application of mathematical tools can never be an exhaustive picture of nature. "We have seen that the methods of the exact natural sciences do not capture reality in its totality, instead they are only concerned with certain sides of nature" (Stein 2004, 73).[6] Second, and closely connected, most phenomenologists reject the view according to which mathematizability is a criterion for existence. Typically, this not only holds true in the case of physical existence but also for entities such as values, essences, or consciousness. Ultimately, this amounts to the overall view that, although mathematics can be an extremely useful tool in many areas of science, we must not mistake mathematizability for scientificality.

It is a commonplace in phenomenology that a purely objective third-person perspective is unreachable (see, e.g., Berghofer 2020 and Khalili 2022). Here is how Zahavi puts it: "There is no pure third-person perspective, just as there is no view from nowhere. This is, of course, not to say that there is no third-person perspective, but merely that such a perspective is, precisely, a perspective from somewhere. It is a view that *we* can adopt on the world" (Zahavi 2019, 54). However, instead of merely discussing whether the natural sciences can reach a pure third-person perspective, some phenomenologists went a step further and considered the considerably *stronger* view that the natural sciences should actively try to incorporate the *first-person* perspective. Merleau-Ponty is perhaps the most prominent phenomenologist who explicitly championed this claim.

> But a physics that has learned to situate the physicist physically, a psychology that has learned to situate the psychologist in the socio-historical world, have lost the illusion of the absolute view from above: they do not only tolerate, they enjoin a radical examination of our belongingness to the world before all science.
>
> (Merleau-Ponty 1968, 27)

The upshot here is that physics in its most sophisticated form not only abandons the project of delivering a completely objective picture of the world but instead actively incorporates the cognizing subject and thus accounts for the fact that the life-world predates all scientific endeavors. It is only by doing so that we can hope to unveil the most fundamental relation, namely the one between the observer and the observed.[7] According to the late Merleau-Ponty, quantum mechanics represents the closest approximation to such a new kind of physics, which—unlike classical physics—not only "posits nature as an object spread out in front of us, [but rather] places its own object *and its relation to this object in question*" (Merleau-Ponty 2003, 85; our emphasis). Adopting the terminology from the French physicist and logician Paulette Destouches-Février, Merleau-Ponty calls the worldview that emerges from quantum mechanics a "participationist conception" and the kind of realism he subscribes to a "partial realism" (Merleau-Ponty 2003, 97f.). As we will see below, this is a striking similarity both in content and in terminology to the QBist notion of a "participatory realism" (Fuchs 2017).

It should also be noted that Merleau-Ponty's understanding of quantum mechanics was influenced by the interpretation offered by Fritz London and Edmond Bauer in their *La Théorie de l'Observation en Mécanique Quantique* (1939). The London and Bauer interpretation constitutes the first genuinely phenomenological approach to quantum mechanics. One of its main ideas is that quantum mechanics should be interpreted from the perspective of a phenomenological theory of knowledge that seeks to clarify the relationship between the observer and the observed. We return to this below. First, we briefly point out what it is that makes quantum mechanics so interesting for phenomenology and how this connects to the main question of this sub-section, namely whether physics can or should incorporate the first-person perspective.

In textbook quantum mechanics "measurement" is a central and irreducible notion. This finds expression in the so-called collapse postulate. According to this postulate, when a measurement takes place, the wave function collapses such that the quantum state is not in a state of superposition anymore, which in turn results in us observing a definite value. This raises the question, of course, as to why it actually is that the wave function collapses upon measurement. Yet even more importantly, one distinctive feature of (textbook) quantum mechanics is that it includes the subjective-operational term "measurement" as a primitive notion. It is "primitive" in the sense that it cannot be reduced to mathematical terms, which also explains why—according to several scholars—the way quantum mechanics is taught and understood in some physics textbooks is not only misleading but plainly unscientific. The background of this verdict is the conviction that the aim of a sensible approach to quantum mechanics must be to "develop an objective description of nature in which 'measurements' are subject to the same laws of nature as all other physical processes," thus resulting in a situation in which "any form of interpretation [...] is superfluous" (Dürr & Lazarovici 2020, viii). The intention behind such purely objectivistic interpretations—which tend to be referred to as "quantum theories without observers" (Goldstein 1998; Dürr & Lazarovici 2020, viii)—is to purge scientific theories of all subjective, experiential, and operational notions such as "consciousness," "experience," or "measurement." In the words of Tim Maudlin: "A precisely defined physical theory [...] would never use terms like 'observation,' 'measurement,' 'system,' or 'apparatus' in its fundamental postulates. It would instead say precisely *what exists and how it behaves*" (Maudlin 2019, 5). As the reader will find, the mindset underlying this volume goes in the opposite direction. Virtually all authors agree that we should welcome the operational flavor of quantum mechanics, and that we should consider the central, irreducible role of measurement or experience as a virtue instead of a vice. One way to cash this out is to view the wave function not as a physically real entity but as a mathematical tool that encodes the subject's probabilistic expectations about her future experiences. The notorious "collapse" of the wave function, then, is not a physical process but simply corresponds to the updating of the subject's information that results from the measurement process. As we shall see in more detail in Section 3, this is precisely how QBists approach the infamous measurement problem in quantum mechanics.

1.2.3.3 *Phenomenological Approaches to the Measurement Problem*

As mentioned in the previous subsection, the notion of measurement plays a central and irreducible role in quantum mechanics. According to textbook presentations, a unique feature of the theory is that the wave function always collapses upon measurement. But why? Understanding the apparent collapse of the wave function is the central theme of the infamous measurement problem. An early "solution" to this problem was to argue that consciousness causes the wave function to collapse. In fact, this view goes back to none other than John von Neumann, who provided quantum mechanics with its rigorous mathematical foundation. In his monumental *The Mathematical Foundations of Quantum Mechanics* von Neumann approached the process of quantum measurement by making the following distinction: "[L]et us divide the world into three parts: I, II, III. Let I be the system actually observed, II the measuring instrument, and III the actual observer" (von Neumann 2018, 273). Now, the distinctive feature of III is that, in contrast to I and II, "III itself remains outside of the calculation" (von Neumann 2018, 273). This was usually taken to amount to the view that non-material consciousness is responsible for the collapse of the wave function, a view that is of course diametrically opposed to the objectivist interpretations mentioned before. Although the "consciousness causes collapse" idea has also been endorsed by other prominent physicists such as Eugene Wigner, it plays virtually no role in current debates. This has to do with what is considered to be its main defect, its inability to make clear how non-material consciousness could have a causal physical effect on material reality (Shimony 1963; Putnam 1979).

As Steven French has pointed out in a series of papers and a forthcoming book (French 2002, 2020, forthcoming), there exists an alternative to the von Neumann/Wigner approach which is relevant in our context for at least three reasons: First, it avoids the main problems usually associated with the "consciousness causes collapse" idea; second, it restores the intuition that consciousness is crucial to our understanding of quantum mechanics; third and last, it explicitly relies on the framework of Husserlian phenomenology. The account we are referring to is laid out in Fritz London's and Edmond Bauer's short 1939 monograph *La Theorie de l'Observation en Mecanique Quantique*. In essence, the book had two aims, namely to provide a "concise and simple" (London & Bauer 1983, 219) account of the measurement problem in the spirit of von Neumann's groundbreaking work and to shed more light on the relationship between the observed and the observer. Although London and Bauer's book was generally well-known in the community, its genuinely phenomenological dimensions had been overlooked. This is unfortunate because, arguably, if "interpreted correctly, it offers a much more sophisticated account of measurement which, being grounded in the tradition of Husserlian phenomenology, is capable of responding to" the objections that have been raised against the von Neumann-Wigner interpretation that consciousness causes collapse (French 2020, 208).[8]

Considering the specifics of the London and Bauer approach, the first thing to note is that, in their view, "a measurement is achieved only when the" outcome "has been *observed*" (London & Bauer 1983, 251).[9] This is to say, according to them, a measurement is only complete when the observer has observed the outcome of the measurement procedure. What is more, London and Bauer "consider the ensemble of three systems, (*object* x) + (*apparatus* y) + (*observer* z), as a combined and unique system" that is described "by a global wave function" (London & Bauer 1983, 251). Here object x is the quantum system upon which a measurement is conducted, apparatus y is the measuring apparatus, and the observer z observes the measurement outcome. The global wave function is expressed as follows: $\Psi(x, y, z) = \sum \Psi_k u_k(x) v_k(y) w_k(z)$. In their unusual terminology, Ψ_k are the coefficients. The states of the quantum system are represented by u_k, v_k represent the states of the apparatus, and w_k the states of the observer.

What this all means, in essence, is that the consciousness of the observer is not something non-physical that impinges on the quantum system from the outside, magically causing the wave function to collapse. Instead, the wave function represents an *interrelated* system of object, apparatus, and observer. According to London and Bauer, however, there is something special about an observer who relates back to her own consciousness. Such an observer

> possesses a characteristic and quite familiar faculty which we can call the 'faculty of introspection.' He can keep track from moment to moment of his own state. By virtue of this 'immanent knowledge' he attributes to himself the right to create his own objectivity – that is, to cut the chain of statistical correlations summarized in $\sum \Psi_k u_k(x) v_k(y) w_k(z)$ by declaring, 'I am in the state w_k' or more simply, 'I see $G = g_k$.'
>
> (London & Bauer 1983, 252)

It is through the faculty of introspection that conscious observers have privileged access to their own states. This is what makes the observer special, and what distinguishes her from the object and the apparatus. By introspecting her own state, "the observer establishes his own framework of objectivity and acquires a new piece of information about the object in question" (London & Bauer 1983, 252). Consequently, we should not think of the relationship between consciousness and the wave function in terms of consciousness causing the wave function to collapse but rather in terms of a conscious observer who is able to *separate* herself from the wave function.

> Thus it is not a mysterious interaction between the apparatus and the object that produces a new Ψ for the system during the measurement. It is only the consciousness of an 'I' who can separate himself from the former function $\Psi(x, y, z)$ and, by virtue of his observation, *set*

> *up a new objectivity* in attributing to the object henceforward a new function $\Psi(x) = u_k(x)$.
>
> (London & Bauer 1983, 252)

As Steven French rightly points out, many of the concepts employed by London and Bauer "clearly demand a phenomenological reading" (French 2002, 484). French also emphasizes that, furthermore, the original term for what is here translated as "set up" is the French "constituer." So, basically, what London and Bauer are saying here is that the consciousness of an I *constitutes* objectivity, which, again, prompts a phenomenological reading. Hence, according to French, London and Bauer's view of the "separation" between the ego and the wave function can be understood "not [...] in terms of consciousness 'causing', in whatever sense, the wave function to collapse, but rather in Husserlian terms, as that of a *mutual separation* of both an Ego-pole and an object-pole through a characteristic act of reflection" (French 2002, 484).

Another clear testament of London and Bauer's commitment to Husserlian phenomenology can be found toward the end of their booklet.[10] Here, London and Bauer point out that the discussion surrounding the concept of a quantum measurement relates to a broader philosophical problem, namely "the determination of the necessary and sufficient conditions for an object of thought to possess objectivity and to be an object of science" (London & Bauer 1983, 259). They continue by adding that "[m]ore recently Husserl [...] has systematically studied such questions and has thus created a new method of investigation called 'Phenomenology'" (London & Bauer 1983, 259). Summarizing the philosophical implications of their understanding of the quantum formalism, London and Bauer write:

> [T]he discussion of this formalism taught us that the apparent philosophical point of departure of the theory, the idea of an observable world, totally independent of the observer was a vacuous idea. Without intending to set up a theory of knowledge, although they were guided by a rather questionable philosophy, physicists were so to speak trapped in spite of themselves into discovering that the formalism of quantum mechanics already implies a well-defined theory of the relationship between the object and the observer, a relation quite different from that implicit in naïve realism, which had seemed, until then, one of the indispensable foundation stones of every science.
>
> (London & Bauer 1983, 220)

Similar ideas have been articulated by several of the founding figures of quantum mechanics, most notably by Niels Bohr who believed quantum mechanics to reveal that "physics is to be regarded not so much as the study of something *a priori* given, but rather as the development of methods for ordering and surveying human experience" (Bohr 1963, 10). Or, to put it differently: The main lesson of quantum mechanics is that science, at a fundamental level, is not supposed to provide "a description of *reality in itself* [but] a description of *reality as experienced by an agent*" (Goyal 2012, 584).[11]

In sharp contradistinction to the MWI—according to which the quantum formalism provides us with an objective description of the deterministic evolution of the universal wave function—and the Bohmian interpretation—according to which the formalism is about the deterministic evolution of point particles—London and Bauer take the quantum formalism to be a "well-defined theory of the relationship between the object and the observer." One consequence of this view is that, on London and Bauer's reading, quantum mechanics is clearly at odds with naïve realism. In fact, London and Bauer go so far as to say that "the idea of an observable world, totally independent of the observer was a vacuous idea" (London & Bauer 1983, 220). Yet, despite this anti-realist flavor, it would still be wrong to see in London and Bauer straightforward instrumentalists according to whom quantum mechanics does not tell us anything about reality. It tells us something very important namely precisely that naïve realism is wrong and that, in the words of the QBists, "reality is *more* than any third-person perspective can capture" (Fuchs 2017, 113).

1.3 QBism

The distinctive idea of QBism is to apply a personalist Bayesian account of probability, as it has been developed by Bruno de Finetti, to quantum probabilities (Fuchs et al. 2014). This means that probabilities in quantum mechanics are interpreted not as objective but as subjective probabilities. Another way to put this is that, according to QBism, quantum states do not represent objective reality but instead *represent an agent's subjective degrees of beliefs about her future experiences*. Consequently, instead of being construed as (the representation) of something physically real, the wave function is considered to be a mathematical tool that encodes one's expectations about one's future experiences. In short, QBism argues that quantum states do "not represent an element of physical reality but an agent's personal probability assignments, reflecting his subjective degrees of belief about the future content of his experience" (Fuchs & Schack 2015, 1). A measurement is understood as an act of the subject on the world and the outcome of a measurement is the very experience that results from this process (see DeBrota & Stacey 2019). Instead of subscribing to a worldview according to which the world is objectively "out there," waiting to be discovered, QBists think of the relationship between world and subject in terms of a reciprocal one. Building on the insight that "reality is *more* than any third-person perspective can capture" (Fuchs 2017, 113), they propose a kind of realism that has been aptly labeled a "participatory realism" (Fuchs 2017). Accordingly, one of the QBists' objectives is to put the scientist back into science (Mermin 2014). In this section, we highlight some of the similarities between QBism and phenomenology and discuss possible tensions. The overall objective is to bring both into a fruitful dialogue, expecting a better understanding of quantum mechanics to emerge from this engagement.

1.3.1 QBism and Phenomenology: Points of Contact

Perhaps the most obvious and systematically most significant similarity between QBism and phenomenology is their commitment to an experience-first approach. In phenomenology, this commitment finds its expression in the thesis that all epistemic justification and every piece of knowledge can be traced back to epistemically foundational experiences. In QBism, it manifests itself in the interpretation of quantum mechanics as a tool that allows the experiencing subject to predict future experiences, and in the construal of measurement outcomes as the very experiences of the subject. Furthermore, both QBism and phenomenology agree that experiences constitute our main points of contact with the world and that a purely objective third-person perspective that abstracts away from the subject and her experiences is in principle impossible. Accordingly, in their contribution to this volume, Michel Bitbol and Laura de la Tremblaye write:

> And since Phenomenology is the only contemporary philosophical research program that does not turn lived experience into some ghostly epiphenomenon, and that takes instead experience as its absolute starting point, we claim it is the only unified framework suitable for making sense of QBism.

This, of course, is in stark opposition to the received view in "mainstream" analytic philosophy of science according to which scientific theories purport to describe a reality that is assumed to be completely independent from the observer. Here, the underlying picture is that physical objects have a number of intrinsic properties (such as position and momentum), that the states of these properties are objectively fixed, and that the aim of science is to offer an exhaustive third-person description of what these states are. To be sure, our point is not to deny that this picture has some initial plausibility and that it is in agreement with a straightforward interpretation of many successful scientific theories. This is true, in particular, of classical mechanics which has crucially shaped the way we think about the nature of science and reality.[12] Yet, its tremendous successes notwithstanding, the fact of the matter is that classical mechanics is dead. It now has been dead for over a century and it is not coming back. To put it provocatively, then, the impression is that while physicists tend to accept the classical picture to be undermined by quantum mechanics, large parts of the philosophical community do not seem ready to move on. The three main interpretations in contemporary philosophy of quantum mechanics—Bohmian mechanics, the MWI, and objective collapse theories—still cling to the idea that physics is in the business of providing a description that is completely free from all irreducibly perspectival or subjective moments. But the price to be paid is either a modification of the quantum formalism or the introduction of a rather baroque ontology of infinitely many unobservable worlds.

Although there is no universal agreement regarding phenomenology's stance in the metaphysical realism debate, it seems safe to say that phenomenologists tent to be more open than standard analytic philosophers to some of the stronger implications of QBism. For instance, when QBists insist that "there is no such thing as the universe in any completed and waiting-to-be-discovered sense" and that "nature is being hammered out as we speak" (Fuchs in Schlosshauer 2011, 285), this is indeed reminiscent of Husserl's description of the constituting role of transcendental subjectivity.

> Every imaginable sense, every imaginable being, whether the latter is called immanent or transcendent, falls within the domain of transcendental subjectivity, as the subjectivity that constitutes sense and being. The attempt to conceive the universe of true being as something lying outside the universe of possible consciousness, possible knowledge, possible evidence, the two being related to one another merely externally by a rigid law, is nonsensical. They belong together essentially; and, as belonging together essentially, they are also concretely one, one in the only absolute concretion: transcendental subjectivity. If transcendental subjectivity is the universe of possible sense, then an outside is precisely – nonsense.
>
> (Husserl 1960, 84)

In light of this, it does not come as a surprise that phenomenologists typically embrace the idea that "[t]he reality of the object is not hidden behind the phenomenon, but unfolds itself in the phenomenon" (Zahavi 2003, 16). There are obvious similarities here with John Archibald Wheeler's position which, employing Bohr's concept "phenomenon," comes to expression when he asks "what other kind of universe can we expect to see than one built as 'phenomenon' is built, upon query of observation and reply of chance, a *participatory* universe?" (Wheeler 1980, 359). Wheeler elaborates on the crucial role of the participator in the following way:

> More generally, we would seem forced to say that no phenomenon is a phenomenon until – by observation, or some proper combination of theory and observation – it is an observed phenomenon. The universe does not 'exist, out there,' independent of all acts of observation. Instead, it is in some strange sense a participatory universe.
>
> (Wheeler 1978, 41)

The terminological similarity between Wheeler's "participatory universe" and the QBist "participatory realism" is no coincidence. Wheeler was the main inspiration for this terminology and had a crucial impact on Fuchs' intellectual development (see Fuchs 2017 and Crease & Sares 2021). Wheeler not only articulated ideas that resemble phenomenological teachings but was in fact influenced by phenomenological thinkers (see Berghofer 2022, Section 15.4).

We also note that Wheeler's slogan "no phenomenon is a phenomenon until it is an observed phenomenon" is similar to the QBist principle that "[an] experiment has no outcome until I experience one" (Fuchs et al. 2014, 751). This is the QBist personalized version of Asher Peres' saying that "[u]nperformed experiments have no results" (Peres 1978). As noted in the previous section, this resembles London and Bauer's claim that "a measurement is achieved only when the [respective outcome] has been *observed*" (London & Bauer 1983, 251). As trivial as these statements may initially sound, they encapsulate one of the central tenets of QBism because they directly connect to the question "of whether quantum measurements reveal some pre-existing value for something that's unknown, or whether in some sense they go toward creating that very value, from the process of measurement" (Fuchs & Stacey 2016, 289). QBists, of course, subscribe to the latter view, claiming that measurements do *not* reveal pre-existing values.

> This leads us down the path of participatory realism in which experiences are as real and as fundamental as what we hope to uncover behind the normative part of quantum theory. When I said that for participatory realism, reality is more than any third-person perspective can capture, I view that as a positive statement—that we grasp some feature of reality that says it resists representation. Experience becomes a fundamental and irreducible element in the universe. The

world is such that we cannot give a block universe representation of it. There is no view from nowhere, and I view that as an ontological statement.

(Fuchs in Crease & Sares 2021, 555)

The claim that measurements do not reveal pre-existing values but that the very act of measurement is responsible for the object in question having the observed value is in agreement with the "orthodox position" of textbook quantum mechanics. According to orthodoxy, *before* the measurement takes place, "[t]he particle wasn't really anywhere. It was the act of measurement that forced it to 'take a stand'" (Griffiths 2018, 17). It is needless to say that this is hard to reconcile with the prevailing objectivist interpretations commonly found in contemporary philosophy of quantum mechanics.

What distinguishes QBism from textbook quantum mechanics is that QBism constitutes a consistent approach that specifies the role of the experiencing subject, the nature of the wave function, the nature of measurement, and crucial implications for the nature of reality. Textbooks and the original Copenhagen approaches are either silent on (some of) these topics or inconsistent, for instance when it comes to the nature of the wave function. Also, sometimes the main objectives of science are spelled out in terms of the subject's experience (Bohr 1963, 10), other times in terms of factive notions such as knowledge (Heisenberg 1958, 15) or information (Zeilinger 1999). QBism, by contrast, is the best-developed interpretation in which experience plays a fundamental role. Phenomenology, analogously, is the most thoroughly developed experience-first project in philosophy. In light of this similarity, bringing QBism and phenomenology into mutual dialogue is the obvious move. The question is not whether QBists and phenomenologists should attempt to join forces but what has taken us so long. Here is a list of exemplary ways in which, on our view, phenomenologists and QBists can benefit from each other.

What phenomenology offers QBism

1 Even opponents of QBism tend to agree that QBism delivers a consistent interpretation of quantum mechanics that avoids problems surrounding the apparent collapse of the wave function and non-locality (Vaidman 2014, 17f.). However, the main objection is that there is a lack of a clear philosophical foundation (Timpson 2008, 580). One of the main objectives of this project of engaging QBism and phenomenology is to introduce phenomenology as a suitable philosophical-conceptual framework for QBism.

2 QBists argue that (quantum) measurement outcomes are the very experiences of the observing subject. However, QBists are physicists with no formal training in phenomenology or epistemology. So when asked what precisely the experience looks like that is supposed to correspond, for instance, to the outcome of a spin-up/spin-down measurement or how exactly an instrument-mediated experience gains its justificatory force, answers remain vague. It is precisely here that phenomenologists could come to the rescue.

What QBism offers phenomenology

1 Phenomenology specifies experiences as our ultimate evidence/justifiers, but, of course, it would go beyond the scope of phenomenology to provide an answer to the question: Based on my *actual* experiences, what should I believe (to experience next)? According to QBism, quantum mechanics should be understood as delivering this formalism. This would imply an intimate connection between philosophy and science.
2 Contemporary analytic epistemology is dominated by anti-phenomenological externalist accounts according to which evidence is not constituted by our experiences but by facts and the epistemic status of our beliefs is not determined by what is internally accessible to us but by external factors such as reliability. A crucial rationale for externalism is the idea that philosophy should strive to be methodologically similar to the natural sciences. Here the natural sciences are typically understood as adopting a third-person perspective that successfully abstracts away from the subject and her personal experiences. If it turns out that this requirement does not even work for the most fundamental physical theory, this motivation vanishes. This would open the door for a phenomenological experience-first epistemology according to which epistemology clarifies how experiences justify and science clarifies what a subject should believe (to experience next) based on her actual experiences.

Summarizing the above, here are some of the main claims that unite phenomenologists and QBists.

QP1: We must be careful with the project of mathematizing nature and should abstain from reifying mathematical quantities such as the wave function. Since such reifications/objectifications "take for *true being* what is actually a *method*" (Husserl 1970, 51), we should be highly critical of "the strategy [...] to reify or objectify all the mathematical symbols of the theory and then explore whatever comes of the move" (Fuchs & Stacey 2019, 136).

QP2: We should not modify the quantum formalism—the formalism of the most successful theory in the history of humanity—simply to make our fundamental scientific theory a better fit for our ontological intuitions. Instead, we must acknowledge that these intuitions and our scientific worldview have historically sedimented inherited meanings. If theory A (classical mechanics) dominates for hundreds of years, significantly shaping our view of science and reality, but then turns out to be empirically unacceptable and is superseded by B (quantum theory), it is problematic to require of B to fit with the intuitions we inherited from A. Instead, we should take the quantum phenomena seriously as well as the implications that "[q]uantum theory itself threw [...] before us!" (Fuchs 2017, 115).

QP3: The notion of experience is an irreducible primitive that is the ineluctable starting point of any encounter with and knowledge of the external world.

QP4: The life-world is the "meaning-fundament of natural science" and thus we should not expect science to deliver a purely objective view on the world. "There is no pure third-person perspective, just as there is no view from nowhere" (Zahavi 2019, 54). Accordingly, we should not be surprised if quantum mechanics can be understood as suggesting that "reality is *more* than any third-person perspective can capture" (Fuchs 2017, 113).

QP5: Instead of striving for the unreachable goal of a purely objective science that offers a comprehensive third-person description of reality, we may look for "a physics that has learned to situate the physicist physically" and has "lost the illusion of the absolute view from above" (Merleau-Ponty 1968, 27). Accordingly, it should be appreciated if "QBism puts the scientist back into science" (Mermin 2014).

QP6: "[A] measurement is achieved only when the [respective outcome] has been *observed*" (London & Bauer 1983, 251). Or in the QBist version: "This experiment has no outcome until I experience one" (Fuchs et al. 2014, 751). This is to say that the quantum formalism does not offer an objective description of the evolution of some external entities such as wave functions or point particles. Instead, it tells us something about the interaction between the observed and the observer.

QP7: The most fundamental aim of science is to deliver the formalism that allows the experiencing subject to predict what she should expect to experience next.

We believe that QP1-7 should be universally accepted by all QBists. Concerning the phenomenological tradition, QP1-4 should be uncontroversial among Husserlian phenomenologists. QP5 is a stronger claim endorsed by Merleau-Ponty, probably anticipated by Husserl. QP6 has been endorsed by the phenomenologically minded physicist Fritz London, regarding Husserlian phenomenology as providing the broader philosophical framework to address the problems that arise in the context of quantum measurements. Obviously, if this were true, this would be good news for the phenomenological movement. QP7 is true if the QBist interpretation of quantum mechanics is correct and if quantum mechanics truly is fundamental. Phenomenologists could regard QP7 as revealing the close connection between philosophy and science.

1.3.2 QBism and Phenomenology: Possible Points of Conflict

Although we do believe that phenomenology and QBism are natural bedfellows, we also want to address some possible points of conflict. As we have repeatedly noted, the core claim of phenomenological epistemology is that all knowledge/justification leads back to epistemically foundational experiences. If this is true, it would be of great significance to develop a formalism that allows the experiencing subject to answer the question of what, based on her previous actual experiences, she should expect to experience next. Assuming that quantum mechanics is a good candidate in this regard, QBism is the currently best-developed interpretation that embraces the idea that "*experience* is fundamental to an understanding of science" in the sense that "quantum mechanics is a tool anyone can use to evaluate, on the basis of one's past experience, one's probabilistic expectations for one's subsequent experience" (Fuchs et al. 2014, 749). More precisely, in QBism quantum states are doxastically interpreted as representing the subject's beliefs about her future experiences (DeBrota & Stacey 2019, 10). QBists rightly emphasize that this is a *doxastic* and not an *epistemic* interpretation of the quantum state/wave function. This is because knowledge is a factive notion. If one knows that p, then p is the case. Belief is non-factive. From a phenomenological perspective, it is a clear advantage of QBism to be formulated in non-factive terms such as experience and belief. Importantly, this is also why QBism avoids the PBR no-go theorem (Pusey et al. 2012). According to QBism, the wave function neither represents an underlying ontic state, nor is it about our knowledge/uncertainty of an underlying ontic state. In the terminology introduced by Harrigan and Spekkens (2010), it is neither *ψ-ontic* nor *ψ-epistemic*. While the PBR theorem rules out *ψ-epistemic* interpretations, it is silent on the QBist claim that wave functions represent degrees of beliefs about one's future experiences (DeBrota & Stacey 2019; Glick 2021; Hance et al. 2022).[13]

Importantly, however, QBists explicitly deny that the quantum state represents what the subject *should* believe.[14] For QBists, the quantum state is a set of probability assignments. These probabilities are personalist Bayesian probabilities. As long as your assignments are consistent, you cannot be wrong about them. This is to say that for an event X, the probability P(X) represents the subject's degree of belief that X will occur. But perhaps epistemologically minded phenomenologists should disagree. Maybe quantum mechanics has a more straightforward epistemically normative dimension in the sense that quantum states represent what a subject *should* believe to experience next. And perhaps it would even be more accurate to say that quantum states assign *degrees of justification* to beliefs about possible future experiences.[15] This is to say that for the phenomenologist it remains to be seen whether QBism succeeds in developing the methodology of taking experience as the starting point of science or whether we need an approach in which quantum states tell us what we objectively should believe. If the latter is the way to go, then phenomenological approaches to quantum mechanics might share crucial similarities with Richard Healey's pragmatist interpretation according to which quantum mechanics "is a source of objectively good advice about *how* to describe the world and what to believe about it as so described. This advice is tailored to meet the needs of physically situated, and hence informationally-deprived, agents like us" (Healey 2022, Section 4.3).

A further potential point of conflict concerns the QBist account of scientific instrumentation. As we have seen, QBists identify the outcome of (quantum) measurements with the *experiences* of the measuring subject. However, this is understood by the QBists as implying that the instruments used in quantum mechanical measurements must be regarded as *bodily extensions* of the measuring subject such that the respective objects (e.g., electrons) can be observed *directly* (Pienaar 2020). In his contribution to this volume, Pienaar puts this in the following way:

> Prolongation Thesis: Those measuring instruments which the agent regards as being the source of their experiences (i.e. of their measurement outcomes) are to be regarded as prolongations of the agent's body, and thus having the same metaphysical status as the bodily sense organs of the agent.

As Pienaar elaborates, the prolongation thesis goes back to a remark of Wolfgang Pauli, but while it remains unclear what role this idea played in Pauli's considerations, the QBists take it "deadly seriously" (Fuchs 2017). This raises at least two questions. First, does the prolongation thesis really follow from the QBist postulate that measurement outcomes are experiences? This is the question of whether and to what degree QBists are committed to the prolongation thesis. Second, is the prolongation thesis plausible, particularly from a phenomenological perspective? Concerning the latter question, one thing to note is that phenomenologists typically do employ a broader notion of perception (more precisely: originary presentive intuition). For Husserl, not only bodily objects can be originally given in perceptual experiences but also eidetic truths in eidetic intuitions, one's own mental states in introspective experiences, and perhaps even values in evaluative experiences (see Husserl 1996, 286, 290).

Most notably, the prolongation thesis seems to resonate well with Merleau-Ponty's famous example of the blind man's stick.

> The blind man's stick has ceased to be an object for him, and is no longer perceived for itself; its point has become an area of sensitivity, extending the scope and active radius of touch, and providing a parallel to sight.
>
> (Merleau-Ponty 2002, 165)

However, while the prolongation thesis seems to make sense in this example, it is far from clear that it also holds with respect to quantum measurements. Merleau-Ponty seems to reject the thesis in the latter context. Contrasting the role of the measuring apparatus in classical physics and quantum mechanics, Merleau-Ponty states that while classically "the apparatus is the prolongation of our senses" in quantum mechanics "[t]he apparatus does not present the object to us." Instead, "[i]t realizes a sampling of this phenomenon as well as a fixation. [...] Known nature is artificial nature" (Merleau-Ponty 2003, 93). Although, unfortunately, Merleau-Ponty does not offer a detailed analysis of what is "artificial" about quantum measurements, it is prima facie plausible to assume that indeed there is a fundamental difference between looking through telescopes or microscopes on the one hand and using measuring devices in modern particle physics on the other hand. In the case of telescopes and microscopes, there is a rather straightforward sense in which we directly observe the object in question. But it would be quite a stretch to say that we directly observe particles when looking at the *photographs* gained by cloud chambers and bubble chambers that visualize the *tracks* of charged particles. What is more, while cloud chambers and bubble chambers have a photographic readout, the devices that are now common, such as particle colliders like the LHC, have a purely electronic readout. What we gain from LHC experiments is data—big data. "Data pours out of the LHC detectors at a blistering rate. Even after filtering out 99% of it, in 2018 we gathered 88 petabytes of data."[16] To say, for instance, that the Higgs boson can be originally given in LHC experiments is highly implausible to us.

This discussion, of course, concerns the question of whether we should believe in the existence of so-called "unobservable" scientific entities such as atoms and electrons. As mentioned in Section 2.3.1, this is hotly debated in analytic philosophy as well as in phenomenology. Here, we only note that Husserl's conception of *horizontal intentionality* may prove useful in this context. It is commonly accepted in phenomenology that we can distinguish between what is originally given in experience and what is co-given in the horizon of the experience. When you look at the table in front of you, what is originally given to you is the frontside of the table, while its backside is co-given. This table-experience is rich in anticipations of what the table looks like from different angles. Importantly, co-givenness is *not* just another term for background beliefs. Co-givenness—just like originary givenness—is a name for how experiences present (parts of) their objects/contents. This means it denotes a distinctive kind of phenomenal character. Joel Smith has convincingly argued that the phenomenal character of co-givenness is "belief independent" (Smith 2010, 736). We understand this as implying that if an object is co-given in the horizon of an experience, the object can be regarded as experienced (although not as originally given). Perhaps QBists might want to soften their claim, saying that in quantum measurements scientific entities like atoms and electrons are co-given in instrumentally mediated experiences.

1.4 Contributions

This volume is structured in three parts. Roughly speaking, Part I constitutes the QBist perspective on phenomenology, Part II the phenomenologist perspective on QBism, and Part III sheds light on complementary ideas and supplementary approaches.

Part I comprises four chapters, authored by leading QBists. Blake Stacey gets the ball rolling by addressing the history of QBism and by elaborating on some of the ideas that have been dropped and renounced over the years. This is a service of extreme value to the readers of this volume because QBism must not be understood as a project that saw the light of day in a completed fashion. Instead, it is an ongoing research program whose history "is a series of hard-fought battles for self-consistency" (Stacey, this volume). This is to say that some views that are still often ascribed to QBists have already been dismissed, and that instead of trying to synthesize older and newer QBist texts, one should rather turn to the latter. In fact, Stacey suggests that "nothing posted on the arXiv before 2009 should be cited as an example of QBism." Stacey depicts the QBist struggle for consistency as a process "of becoming consistently Jamesian." When addressing the future of QBism, Stacey points out that William James' pragmatism and Husserl's phenomenology are evolutionary cousins. Building on this insight, Stacey sketches "ways in which phenomenological turns may be inspirational for future developments in the QBist project." In Stacey's view, one particular avenue for phenomenologists who wish to contribute to the development of QBism is the ongoing project of reconstructing quantum theory. This project of theoretical reconstruction is discussed in detail in Philip Goyal's contribution, marking the final chapter of this volume.

Chris Fuchs' chapter "QBism, Where Next?" complements Stacey's foregoing contribution. While Stacey focused on the historical development of QBism, Fuchs discusses the present and future of QBism. Fuchs' chapter stands out as the most comprehensive and conceptually/philosophically clearest depiction of the QBist research program to date. Fuchs begins by introducing two concepts central to QBism, "agent" and "user of quantum mechanics," and proceeds by clarifying eight key tenets of QBism. Fuchs refers to this as "QBism's eightfold path." What is still missing is the "noble truth" to which this eightfold path is supposed to lead. In Fuchs' words: "What is the precise ontology that compels the eightfold path?" As the reader will see in the course of this volume, which ontological conclusions to draw from the theory is one of the most pressing open questions QBists have to face. As Fuchs notes, this is also a consequence of the QBist mindset that we must be careful not to read our ontological preconceptions into the quantum formalism. Instead, "it became clear that the pertinent way to move forward was to get the 'epistemics' of the theory right before anything else: Getting reality right would follow for those who had patience enough to pass the marshmallow test." From a phenomenological perspective, this seems the most reasonable way to proceed, although one might be cautious regarding the optimism expressed here. Concerning the quest of uncovering the ontological implications of quantum theory, Fuchs speculates that the best way to move forward is by analyzing the famous Wigner's friend thought experiment precisely in view of the tenets constituting QBism's eightfold path. Accordingly, the penultimate section of Fuchs' contribution is devoted to Wigner's friend, pointing out that both agents—Wigner and his friend—are agent as well as system and thus must be treated symmetrically. Fuchs concludes his chapter by elaborating on parallels between the QBist approach to Wigner's friend and how Merleau-Ponty illustrates his ontology of chiasm and flesh by discussing the example of a person touching one hand with the other.

In Chapter 3, Rüdiger Schack sheds light on points of contact between QBism and Merleau-Ponty. More precisely, he discusses the QBist claim that "[t]he world does not admit a third-person description and is fundamentally indeterministic" in the context of Merleau-Ponty's essay "The intertwining—The chiasm." Schack's chapter begins with an excellent exposition of the normative character of QBism. According to QBists, quantum states represent an agent's degrees of beliefs and the Born rule is a *normative* constraint that "functions as a consistency criterion which puts constraints on the agent's decision-theoretic beliefs." Schack stresses that measurement outcomes do not reveal pre-existing properties. Instead, they are viewed as "personal consequences for the agent taking the measurement action and come into existence only through the measurement action itself." In this QBist picture, the measurement apparatus is a bodily extension of the measuring subject that gives direct access to the physical system. While this constitutes a consistent interpretation of quantum mechanics that, as mentioned above, avoids the PBR theorem and other no-go theorems, it leads to several philosophical challenges. Here is how Schack approaches one such challenge:

> What remains a big challenge is to develop an explicit ontology for QBism, an ontology that [is] based on first-person experience but which accounts for a real world beyond any particular agent's experience and thus avoids the trap of idealism. It turns out that there is a significant overlap between the project of finding such a QBist ontology and the philosophy of Maurice Merleau-Ponty and other phenomenologists.

In this context, Schack discusses the "Copernican principle" of QBism according to which we human beings are both experiencing agents *and* physical systems. This is contrasted with Merleau-Ponty's notion of the flesh and his remarks about the relationship between the seeing subject and the visible world. Throughout his chapter Schack hints at discussions surrounding the issue of free will, concluding that the scientific world view of QBism is the one best suited for free agents.

In his chapter "Unobservable Entities in QBism and Phenomenology," Jacques Pienaar also addresses the issue "that reality in QBism must be somehow founded upon an agent's subjective experiences." Contrasting and comparing QBist and phenomenological approaches, Pienaar discusses the view according to which in-principle observability is a criterion for physical existence. What implications does this criterion have for atoms, fields, quantum states, and probabilities? In this context, Pienaar introduces the above-mentioned prolongation thesis, stating that certain measurement instruments in modern physics "are to be regarded as prolongations of the agent's body, and thus having the same metaphysical status as the bodily sense organs of the agent." Making a distinction between probability-like entities on the one hand and system-like entities, such as atoms and fields, on the other hand, Pienaar seeks to establish a QBist approach as a middle-position between two phenomenological proposals:

> We found that QBism distinguishes between system-like entities and probability-like entities, and adheres to the Prolongation Thesis, which renders unobservable system-like entities potentially observable, contrary to Wiltsche's objection (Wiltsche 2012). On the other

hand, in order to maintain its claim that probability-like entities cannot be said to physically exist, QBism should embrace Wiltsche's 'Originality Thesis of Justification' over Berghofer's counter-proposed 'Criterion of Justification', as the latter opens the door to the possible physical existence of quantum states and hidden variables.

Hans Christian von Baeyer's contribution marks the beginning of Part II. The chapter begins by assessing that the field of interpreting quantum mechanics is about to enter a new epoch, one that centers on the notion and fundamental role of *perception*. Due to this focus, von Baeyer elaborates, it is natural to assume that phenomenology is the ideal philosophical framework researchers working on quantum foundations should consult. This is illustrated by building upon the phenomenological work of the little-known American philosopher Samuel Todes. Von Baeyer is particularly interested in Todes' claim that the active human body must not be reduced to passive matter but should be recognized as the "material thing whose capacity to move itself generates and defines the whole world of human experience in which any material thing, including itself, can be found" (Todes 2001, 88). This brings us back to the question of whether science should be viewed as a completely de-humanized endeavor. Von Baeyer stresses similarities between Todes and the QBist David Mermin. Mermin has recently addressed the "problem of the now," i.e., the issue that worried Einstein so deeply, namely the fact that physics seems to be unable to single out the present moment as something special. Why is there such a gap between how we experience time and how time is treated in physics? Von Baeyer writes:

> Mermin showed that the principal difficulty was not the relativity of simultaneity, but the compulsive exclusion of subjectivity from science. [...] With the emphasis of both QBism and phenomenology on a first-person perspective, the problem of the NOW melts away. NOW, according to both Mermin and Todes, is simply the moment in time at which the anticipation and prediction of a future event turns into memory of the past experience of that event. Since each of us is capable of distinguishing between those two modes of cognition, that moment is not only well defined, but surely of surpassing importance.

Above we mentioned that the descriptive methodology of phenomenology is typically considered to be in tension with the third-person approach of science. However, if von Baeyer and Mermin are on the right track, science and phenomenology may turn out to be just two sides of the same coin.

Like several other contributions to this volume, Thomas Ryckman's chapter, entitled "QBism: Realism about What?," focusses on one of the most apparent worries regarding QBism, namely its relationship to realism and anti-realism. While several critics have expressed the worry "that lurking under QBism's philosophical hood is an ogre of solipsism," QBists themselves have traditionally been busy distancing themselves from subjectivism and instrumentalism, and—following Wheeler—to present "participatory realism" as a counterproposal. However, even if one deems the QBists' distancing from all sorts of anti-realism as successful, the question still remains what participatory realism actually amounts to and how it relates to other positions in the field. Ryckman addresses this issue by discussing a historical figure whose philosophical outlook has been intensely discussed in the literature and who is considered by QBists as a "*sine qua non*" of their own position, Niels Bohr. Following Ryckman's reading, what makes Bohr's thought attractive to QBists is fruitful connection between two ideas that are seen as incompatible by many, epistemological anti-representationalism on the one hand and a rather minimal empirical realism about the quantum world on the other. In this context, Ryckman summarizes one of his central insights as follows:

> [Bohr's] rejection of representationalism stems entirely from the implications of the quantum postulate for the subject-object relation concerning the objective description of quantum phenomena. Many interpreters of Bohr regard this non-representationalism to imply an anti-realist or instrumentalist stance towards quantum theory. This is not correct. Non-representationalism, occasionally termed 'renunciation of visualization', is a necessary epistemological step when fashioning objective description of the physical world beyond the objectifying methods employed in classical physics. Such knowledge as we have of the microphysical world is empirically based and fallible, it can only be projected from the interpretation of our experiences and is therefore indeterminate, but it is nonetheless knowledge of the real.

Engaging QBism with phenomenology is a field of research that originated in the contributions of Michel Bitbol (2020) and Laura de la Tremblaye (2020) in Wiltsche and Berghofer (2020). Their joint paper "QBism: An Eco-Phenomenology of Quantum Physics" published in the present volume may turn out to be similarly groundbreaking. Here they focus on a tension within QBism that has been worrying many researchers sympathetic to this program, namely the "discrepancy between the instrumentalist and realist inclinations of QBism." On the one hand, QBism is anti-realist regarding the wave function and views quantum mechanics as a tool anyone can use to make better predictions about one's future experiences. This is the instrumentalist dimension of QBism. On the other hand, QBists view themselves as being part of a realist program that seeks to uncover the deep structure of reality. But what can quantum mechanics tell us about reality if it is only a tool? The typical move of QBists is to argue that the simple fact that the quantum formalism is so successful tells us much about reality. The idea is that by analyzing the tool we use to systematize our experiences, we learn significant lessons about the world we experience. Still, it remains a widespread suspicion that the QBists can't have it both ways and that their "realism" does not deserve this label. In their contribution, Bitbol and de la Tremblaye resolve this discrepancy by "unambiguously choosing the first-person standpoint as a radical origin of knowledge, and by ascending from the situated lived experience of a knowing and acting subject, to the structure and use of the quantum formalism." In this context, they suggest to replace "the concept of a somehow extrinsic relation between subject and world, with the immersive experience of a subject partaking of the world." This leads them to the proposal that QBists should replace their "participatory realism" with a form of "*radical participatory empiricism.*" We have seen that QBism has a history of becoming more consistent by sacrificing views that are sacred to standard realists. The proposal of Bitbol and de la Tremblaye might be the next step in this evolution.

Part III begins with Steven French's chapter "Putting Some Flesh on the Participant in Participatory Realism." One apparent feature of French's contribution is that it resonates nicely with topics that are also in the center of several other chapters, most notably those by Ryckman, Boge and Bitbol and de la Tremblaye. Most importantly, French focuses on the question of whether QBism is successful in its attempts to navigate between the extremes of subjectivist idealism and naïve realism. In line with the general topic of this volume, French considers whether phenomenology provides a suitable middle ground and thus can serve as a useful philosophical framework for QBism. In particular, French takes a closer look at a recent suggestion by Laura de la Tremblaye who has argued that QBism's insistence on subjective expectations squares nicely with Husserl's understanding of perception as based on a horizontal structure of intentionality. In a nutshell, de la Tremblaye's proposal is that the Husserlian perceptual horizon corresponds to the QBist's quantum state, the content of the perceptual act parallels the measurement outcome, and the modification of the horizon corresponds to the post-measurement modification of the state vector. However, French sees several problems with this comparison. Apart from circularity concerns, French pays special attention to the dangers of drawing parallels between everyday perception and quantum cases. As French argues, one way to overcome these problems is to move beyond the strictures of Husserlian phenomenology and to consider the later works of Merleau-Ponty, and the notion of the *flesh* in particular.

According to French, however, Merleau-Ponty's later philosophy is harder to mesh with QBism than it might initially appear:

> It is [the] perspectival aspect of Merleau-Ponty's thought that encourages a positive comparison with QBism. However, [...] he also drew on London and Bauer's analysis, with its explicit incorporation of a correlationist aspect, both phenomenologically and physically, as manifested via quantum entanglement. This is anathema to the QBist, of course, as is Merleau-Ponty's centering of the relations represented by the theory more generally. If, then, the QBist wants to draw on phenomenology to philosophically underpin her position, she is going to have to either modify the latter or exclude the correlationist understanding of the former.

Florian Boge's chapter "Back to Kant! QBism, Phenomenology, and Reality of Invariants!" can be read as a continuation but also a significant refinement of some of the main objections that were levelled against QBism in the philosophy of physics literature. Among these criticisms, QBism's alleged inability to distance itself from extreme forms of subjectivism and solipsism as well as problems associated with its anti-representationalism appear particularly relevant. While some of the earlier critics were arguably guilty of misconstruing the basic tenets of QBism (Earman 2019; see, for a rejoinder, Fuchs & Stacey 2020), Boge is very cautious in critically engaging with the actual QBists and not with a strawman version of their views. After formulating what he sees as the main problems with QBism, Boge goes a step further and considers the question whether phenomenology could help QBism to overcome its philosophical shortcomings. The assumption that it could is not only in line with the overall topic of this volume. Since subjectivism, solipsism, and anti-representationalism are topics that are traditionally high on the phenomenological agenda, the suspicion that phenomenological arguments might be of use for QBists is indeed not unreasonable. But ultimately, Boge's conclusion is negative. Accordingly, the rest of Boge's chapter is devoted to show that

> [...] QBism [...] could profit from a Neo-Kantian philosophy of science. The reason is that this allows for a solid, comprehensible abduction-basis and a solid framework for a non-reductionist semantics, thus doing justice to actual physical practice. The suggestion should actually not come as a big surprise, since a connection has been made before [...], and since not only phenomenology, the philosophy currently 'flirted with' by QBism, has its roots in Kant, but also QBism's 'old love' *pragmatism*.

The volume concludes with Philip Goyal's contribution which elucidates the virtues and prospects of the quantum reconstruction program. The program of reconstruction constitutes a novel field of research that already enjoys popularity among physicists working on quantum foundations but remains largely ignored in the philosophy community. Goyal's chapter is perhaps the philosophically richest and conceptually most stringent account of reconstruction that exists to date. The main idea of reconstruction is introduced as follows:

> The methodology of quantum reconstruction seeks to remove the interpretative bottleneck by systematically deriving the quantum formalism in an operational framework from postulates that are, ideally, physically well-motivated, thereby *distilling* the full mathematical content of the theory into precise natural-language statements that—unlike the abstract mathematical postulates of quantum theory—are amenable to philosophical reflection.

Goyal specifies the method of reconstruction as a two-step procedure: Reconstruct the quantum formalism, and then interpret the reconstruction. This is to say that we should not treat the quantum formalism as a given but instead derive it from precise physical principles and then interpret these principles. The project of reconstruction emerged at the turn of the millennium as a consequence of the booming interest in quantum information. By now, there exist several successful reconstructions. As Goyal elaborates, typically they are formulated in an *operational* framework, deriving the quantum formalism from *information-theoretic* principles. In our view, the operational dimension coheres perfectly with phenomenological and QBist ideas. This is because in operational frameworks the notion of a measuring subject is considered an irreducible part of the theory. Since "information" is a factive concept, it is less clear to us how the informational dimension squares with phenomenological and QBist approaches. Perhaps future reconstructions can be spelled out in terms of non-factive mental states such as experience and belief. Goyal's thorough and rigorous chapter will play an important role in advancing the project of reconstruction and increasing its popularity among philosophers.

Acknowledgments

The majority of the chapters in this volume were presented at the conference "Phenomenological Approaches to Physics: QBism and Phenomenology, Compared and Contrasted" which finally, after two pandemic-related postponements, took place at Linköping University, Sweden, in June 2022. We would like to thank our co-organizer Jan-Åke Larsson as well as all participants who helped to make this conference a memorable event. Financial support for organizing this conference was provided by the Faculty of Arts and Science at Linköping University and Riksbankens Jubileumsfond (project number F19-1529:1). Many thanks to Chris Fuchs and Mahdi Khalili for their feedback on earlier versions of this chapter. Furthermore, we would like to thank our families for emotional support, and the Austrian Science Fund for a generous grant to carry out research on the intersection between physics and phenomenology (project number: P31758). Philipp Berghofer's research was funded in part by the research project "Quantum Mechanics and Phenomenology: Specifying the Philosophical Foundations of QBism" granted by the Styrian Government. Finally, parts of this publication were made possible through the support of Grant 62424 from the John Templeton Foundation. The opinions expressed in this publication are those of the author(s) and do not necessarily reflect the views of the John Templeton Foundation.

Notes

1 For similar discussions in the philosophy of quantum gravity of what it would mean for the very endeavor of physics if it turned out that space and time are not fundamental, cf. Huggett and Wüthrich (2013) and Oriti (2014). We must not forget that "[a] central concern of philosophy of science is understanding how the theoretical connects to the empirical, the nature and significance of 'saving the phenomena'" (Huggett & Wüthrich 2013, 276). Cf. for a phenomenological rendering of this argument, Wiltsche (forthcoming).

2 It should be noted that this common terminology is misleading since there are other positions that can be viewed as realist positions concerning the wave function that do not subscribe to configurations space realism (see Chen 2019).

3 For an early anticipation of and objection to this kind of wave function realism, see the comments of the phenomenologically minded physicist Hermann Weyl in Weyl (2009, 147f.).

4 However, while the most common realist interpretations of quantum mechanics—the many-worlds interpretation, Bohmian mechanics, and Ghirardi-Rimini-Weber theory—are all in danger of leading to a problematic mathematization of nature, not all proponents of these interpretations argue for reifying the wave function. Even the main proponents of the respective interpretation do not agree on the ontological status of the wave function. In the case of the many-worlds interpretation, one of its main proponents, David Wallace, is also one of the strongest critics of wave function realism (Wallace 2021). In Bohmian mechanics, we find versions of wave function realism that consider ordinary three-dimensional space as illusory (Albert 1996), or emergent and non-fundamental (Ney 2013), as well as the idea that there is a fundamental three-dimensional ontology and that the wave function might not be physically real like particles or fields but ontologically sui generis (Maudlin 2013).

5 However, there are noteworthy exceptions: Following Ryckman's interpretation (2005), the motivation for Weyl's criticism of Einstein's choice of Riemannian geometry as the mathematical backbone of General Relativity Theory was philosophical and not scientific in nature.

6 "Phenomenology is not out to dispute the value of science and is not denying that scientific investigations can lead to new insights and expand our understanding of reality. But phenomenologists do reject the idea that natural science can provide an exhaustive account of reality. Importantly, this does not entail that phenomenology is, as such, opposed to quantitative methods and studies. The latter are excellent, but only when addressing quantitative questions" (Zahavi 2019, 52).

7 Consider in this context also Weyl's *Mind and Nature* where it is argued that "the structure of our scientific cognition of the world is decisively determined by the fact that this world does not exist in itself, but is merely encountered by us as an object in the correlative variance of subject and object. The world exists only as that met with by an ego, as one appearing to a consciousness; the consciousness in this function does not belong to the world, but stands out against the being as the sphere of vision, of meaning, of image, or however else one may call it" (Weyl 2009, 83).

8 From a phenomenological perspective, it is Fritz London who deserves particular historical attention: The fact that London was nominated four times for the Nobel Prize in chemistry and one time for the Nobel Prize in physics speaks volumes about the breadth and significance of his oeuvre. However, London's academic career had started in philosophy where he completed his first doctoral dissertation *Über die Bedingungen der Möglichkeit einer deduktiven Theorie* under the supervision of the Munich phenomenologist Alexander Pfänder. London's dissertation, which appeared in Husserl's *Jahrbuch für Philosophie und Phänomenologische Forschung* in 1923, was described as "a set theoretic concretization of Husserl's largely programmatic account of a *macrological philosophy of science*" (Mormann 1991, 70). After graduating in philosophy at the age of 21, London's focus shifted toward quantum physics where he worked under the likes of Arnold Sommerfeld and Erwin Schrödinger. However, as we will see, London's phenomenological training remained the background of his approach to quantum mechanics.

9 Compare this with the central QBist thesis: "This experiment has no outcome until I experience one" (Fuchs et al. 2014, 751).

10 However, it should be noted that French's phenomenological reading of the London and Bauer approach has recently been challenged by Otávio Bueno (2019).

11 Consider in this context also Weyl's claim that modern physics reveals that science "does not state and describe states of affairs—'Things are so and so'—but that it constructs symbols by means of which it 'represents' the world of appearances" (Weyl 2009, 83).

12 See, e.g., Rovelli (2006). Husserl would describe this as a process of historical sedimentation in which we (unwittingly) inherit meanings from previous generations.

13 Unfortunately, in the literature the PBR theorem is often misunderstood as ruling out any interpretation that is not *ψ-ontic* (e.g. Maudlin 2019, 83–89). This means that large parts of the community falsely believe that a successful approach to quantum mechanics must be *ψ-ontic* and are ignorant of how the QBist escapes the PBR theorem. In fact, the implications of PBR should be

understood as revealing that QBism is one of the main alternatives to *ψ-ontic* interpretations.

14 Personal email conversation between Chris Fuchs, Jacques Pienaar, and Philipp Berghofer.

15 In more technical terms, this relates to the question of whether quantum probabilities should be understood as subjective Bayesian probabilities along the lines of Bruno de Finetti or rather as objective Bayesian probabilities in the Coxian sense according to which probabilities represent reasonable expectations (Cox 1946). For a discussion from the QBist perspective, see DeBrota and Stacey (2019).

16 https://wlcg-public.web.cern.ch/about. Retrieved on February 13, 2020. In this context (see, e.g., Karaca 2017, 344).

References

Albert, David (1996): "Elementary Quantum Metaphysics," in J. Cushing, A. Fine, and S. Goldstein (eds.): *Bohmian Mechanics and Quantum Theory: An Appraisal*, Dordrecht: Springer, 277–284.

Barrett, Jeffrey (1999): *The Quantum Mechanics of Minds and Worlds*, Oxford, UK: Oxford University Press.

Berghofer, Philipp (2018): "Transcendental Phenomenology and Unobservable Entities," *Perspectives* 7, 1, 1–13.

Berghofer, Philipp (2020): "Scientific Perspectivism in the Phenomenological Tradition," *European Journal for Philosophy of Science* 10, 3, 1–27.

Berghofer, Philipp (2022): *The Justificatory Force of Experiences: From a Phenomenological Epistemology to the Foundations of Mathematics and Physics*, Synthese Library, Cham: Springer.

Berghofer, Philipp & Wiltsche, Harald (2020): "Phenomenological Approaches to Physics: Mapping the Field," in H. Wiltsche & P. Berghofer (eds.): *Phenomenological Approaches to Physics*, Synthese Library, Cham: Springer, 1–47.

Berghofer, Philipp, François, Jordan, Friederich, Simon, Gomes, Henrique, Hetzroni, Guy, Maas, Axel & Sondenheimer, René (2023): "Gauge Symmetries, Symmetry Breaking, and Gauge-Invariant Approaches," *Cambridge Elements*, Cambridge: Cambridge University Press.

Bitbol, Michel (2020): "A Phenomenological Ontology for Physics: Merleau-Ponty and QBism," in H. Wiltsche & P. Berghofer (eds.): *Phenomenological Approaches to Physics*, Synthese Library, Cham: Springer, 227–242.

Bohr, Niels (1963): *Essays 1958–1962 on Atomic Physics and Human Knowledge*, New York, NY: Interscience Publishers.

Brown, Harvey (2019): "The Reality of the Wavefunction: Old Arguments and New", in: Cordero, Alberto (ed.): Philosophers Look at Quantum Mechanics, Cham: Springer, 63–86.

Bueno, Otávio (2019): "Is There a Place for Consciousness in Quantum Mechanics?" in J. Acacio de Barros & C. Montemayor (eds.): *Quanta and Mind: Essays on the Connection between Quantum Mechanics and the Consciousness*, Cham: Springer, 129–139.

Carrier, Martin (2012): "Die Struktur der Raumzeit in der klassischen Physik und der allgemeinen Relativitätstheorie," in M. Esfeld (ed.): *Philosophie Der Physik*, Berlin: Suhrkamp, 13–31.

Carroll, Sean & Singh, Ashmeet (2019): "Mad-Dog Everettianism: Quantum Mechanics at Its Most Minimal," in A. Aguirre, B. Foster, and Z. Merali (eds.): *What Is Fundamental?* Cham: Springer, 95–104.

Chen, Eddy (2019): "Realism about the Wave Function," *Philosophy Compass* 14, 1–15.

Cox, Richard (1946): "Probability, Frequency and Reasonable Expectation," *American Journal of Physics* 14, 1–13.

Crease, Robert & Sares, James (2021): "Interview with Physicist Christopher Fuchs," *Continental Philosophy Review* 54, 541–561.

de la Tremblaye, Laura (2020): "QBism from a Phenomenological Point of View: Husserl and QBism," in H. Wiltsche & P. Berghofer (eds.): *Phenomenological Approaches to Physics*, Synthese Library, Cham: Springer, 243–260.

DeBrota, John & Stacey, Blake (2019): "FAQBism," https://doi.org/10.48550/arXiv.1810.13401.

Dürr, Detlef & Lazarovici, Dustin (2020): *Understanding Quantum Mechanics*, Cham: Springer.

Earman, John (2019): "Quantum Bayesianism Assessed", *The Monist* 102/4, 403–423.

French, Steven (forthcoming): *Cutting the Chain of Correlations. Reviving a Phenomenological Approach to Quantum Mechanics.* Oxford: Oxford University Press.

French, Steven (2002): "A Phenomenological Solution to the Measurement Problem? Husserl and the Foundations of Quantum Mechanics," *Studies in the History and Philosophy of Modern Physics* 33, 467–491.

French, Steven (2020): "From a Lost History to a New Future: Is a Phenomenological Approach to Quantum Physics Viable?" in P. Berghofer & H. Wiltsche (eds.): *Phenomenological Approaches to Physics*, Synthese Library, Cham: Springer, 205–225.

Fuchs, Christopher (2017): "On Participatory Realism," in Ian Durham & Dean Rickles (eds.): *Information and Interaction*, Cham: Springer, 113–134.

Fuchs, Christopher, Mermin, David & Schack, Rüdiger (2014): "An Introduction to QBism with an Application to the Locality of Quantum Mechanics," *American Journal of Physics* 82, 749–754.

Fuchs, Christopher & Schack, Rüdiger (2015): "QBism and the Greeks: Why a Quantum State Does Not Represent an Element of Physical Reality," *Physica Scripta* 90, 1–6.

Fuchs, Christopher & Stacey, Blake (2016): "Some Negative Remarks on Operational Approaches to Quantum Theory," in: G. Chiribella & R. Spekkens (eds.): *Quantum Theory: Informational Foundations and Foils*, Dordrecht: Springer, 283–305.

Fuchs, Christopher & Stacey, Blake (2019): "QBism: Quantum Theory as a Hero's Handbook," in E. Rasel, W. Schleich, & S. Wölk (eds.): *Foundations of Quantum Theory*, Amsterdam: IOS Press, 133–202.

Fuchs, Christopher & Stacey, Blake (2020): *Qbians do not exist. arXiv*, [quantph] (1911.07386).

Galilei, Galileo (1967): *Dialogue Concerning the Two Chief World Systems-Ptolemaic and Copernican*, Berkeley and Los Angeles, CA: University of California Press.

Galilei, Galileo (2008): *The Essential Galileo*, edited and translated by M. Finocchiaro, Indianapolis, IN: Hackett.

Glick, David (2021): "QBism and the Limits of Scientific Realism," *European Journal for Philosophy of Science* 11, 1–19.

Goldstein, Sheldon (1998): "Quantum Theory without Observers – Part One," *Physics Today* 51, 42–46.

Goldstein, Sheldon (2021): "Bohmian Mechanics," in *The Stanford Encyclopedia of Philosophy* (Fall 2021 Edition), Edward N. Zalta (ed.), https://plato.stanford.edu/archives/fall2021/entries/qm-bohm/.

Goyal, Philip (2012): "Information Physics – Towards a New Conception of Physical Reality," *Information* 3, 567–594.

Griffiths, David (2018): *Introduction to Quantum Mechanics*, 3rd edition, Cambridge, UK: Cambridge University Press.

Hagar, Amit (2003): "A Philosopher Looks at Quantum Information Theory", in: *Philosophy of Science* 70, 752–775.

Hance, Jonte, Rarity, John & Ladyman, James (2022): "Could Wavefunctions Simultaneously Represent Knowledge and Reality?" *Quantum Studies: Mathematics and Foundations* 9, 333–341.

Hardy, Lee (2020): "Physical Things, Ideal Objects, and Theoretical Entities," in H. Wiltsche & P. Berghofer (eds.): *Phenomenological Approaches to Physics*, Synthese Library, Cham: Springer.

Harrigan, Nicholas & Spekkens, Robert (2010): "Einstein, Incompleteness, and the Epistemic View of Quantum States," *Foundations of Physics* 40, 125–157.

Hartimo, Mirja (2021): *Husserl and Mathematics*, Cambridge, UK: Cambridge University Press.

Healey, Richard (2022): "Quantum-Bayesian and Pragmatist Views of Quantum Theory," *The Stanford Encyclopedia of Philosophy* (Summer 2022 Edition), Edward N. Zalta (ed.), https://plato.stanford.edu/archives/sum2022/entries/quantum-bayesian/.

Heisenberg, Werner (1958): *Physics and Philosophy*, New York, NY: Harper & Brothers Publisher.

Hopp, Walter (2020): *Phenomenology: A Contemporary Introduction*, New York, NY: Routledge.

Howard, Don (2014): "Einstein and the Development of Twentieth-Century Philosophy of Science," in M. Janssen & C. Lehner (eds.): *The Cambridge Companion to Einstein*, Cambridge, UK: Cambridge University Press, 354–376.

Huggett, Nick & Wüthrich, Christian (2013): "Emergent Spacetime and Empirical (in)Coherence," *Studies in History and Philosophy of Modern Physics* 44, 276–285.

Husserl, Edmund (1960): *Cartesian Meditations*, transl. by Dorion Cairns, The Hague: Martinus Nijhoff.

Husserl, Edmund (1970): *The Crisis of European Sciences and Transcendental Phenomenology*, transl. by David Carr, Evanston: Northwestern University Press.

Husserl, Edmund (1973): *Zur Phänomenologie der Intersubjektivität, Texte aus dem Nachlass, Erster Teil: 1905–1920*, Den Haag: Martinus Nijhoff.

Husserl, Edmund (1982): *Ideas Pertaining to a Pure Phenomenology and to a Phenomenological Philosophy, First Book*, transl. by Fred Kersten, The Hague: Martinus Nijhoff.

Husserl, Edmund (1984): *Einleitung in die Logik und Erkenntnistheorie, Vorlesungen 1906/07*, Den Haag: Martinus Nijhoff.

Husserl, Edmund (1987): *Aufsätze und Vorträge (1911–1921)*, Den Haag: Martinus Nijhoff.

Husserl, Edmund (1996): *Logik und allgemeine Wissenschaftstheorie, Vorlesungen Wintersemester 1917/18*, Den Haag: Martinus Nijhoff.

Husserl, Edmund (1997): *Thing and Space: Lectures of 1907*, transl. by Richard Rojcewicz, Dordrecht: Springer.

Husserl, Edmund (2003): *Transzendentaler Idealismus, Texte aus dem Nachlass (1908–1921)*, Den Haag: Martinus Nijhoff.

Islami, Arezoo & Wiltsche, Harald (2020): "A Match Made on Earth: On the Applicability of Mathematics in Physics," in H. Wiltsche & P. Berghofer (eds.): *Phenomenological Approaches to Physics*, Synthese Library, Cham: Springer, 157–177.

Karaca, Koray (2017): "A Case Study in Experimental Exploration: Exploratory Data Selection at the Large Hadron Collider," *Synthese* 194, 333–354.

Khalili, Mahdi (2022): "From Phenomenological-Hermeneutical Approaches to Realist Perspectivism," *European Journal for Philosophy of Science* 12, 1–26.

Kofler, Johannes & Zeilinger, Anton (2010): "Quantum Information and Randomness," *European Review* 18, 469–480.

Ladyman, James & Ross, Don (2007): *Every Thing Must Go: Metaphysics Naturalized*, Oxford, UK: Oxford University Press.

Lewis, Peter (2013): "Dimension and Illusion," in A. Ney & D. Albert (eds.): *The Wave Function*, New York, NY: Oxford University Press, 110–125.

London, Fritz & Bauer, Edmond (1983): "The Theory of Observation in Quantum Mechanics," in J. A. Wheeler and W. H. Zurek (eds.): *Quantum Theory and Measurement*, Princeton, NJ: Princeton University Press, 217–259.

Maudlin, Tim (2013): "The Nature of the Quantum State," in A. Ney & D. Albert (eds.): *The Wave Function*, New York, NY: Oxford University Press, 126–153.

Maudlin, Tim (2019): *Philosophy of Physics: Quantum Theory*, Princeton, NJ: Princeton University Press.

Merleau-Ponty, Maurice (1968): *The Visible and the Invisible*, Evanston: Northwestern University Press.

Merleau-Ponty, Maurice (2002): *Phenomenology of Perception*, London, UK: Routledge.

Merleau-Ponty, Maurice (2003): *Nature: Course Notes from the Collège de France*, Evanston: Northern University Press.

Mermin, David (2014): "QBism Puts the Scientist Back into Science, *Nature* 507, 421–423.

Monton, Bradley (2013): "Against 3n-Dimensional Space," in A. Ney & D. Albert (eds.): *The Wave Function*, New York, UK: Oxford University Press, 154–167.

Mormann, Thomas (1991): "Husserl's Philosophy of Science and the Semantic Approach," *Philosophy of Science* 58, 1, 61–83.

Ney, Alyssa (2013): "Ontological Reduction and the Wave Function Ontology," in A. Ney & D. Albert (eds.): The Wave Function, New York, NY: Oxford University Press, 168–183.

Ney, Alyssa & Albert, David (eds.) (2013): *The Wave Function*, Oxford, UK: Oxford University Press.

Oriti, Daniele (2014): "Disappearance and Emergence of Space and Time in Quantum Gravity," *Studies in History of Philosophy of Modern Physics* 46, 186–199.

Peres, Asher (1978): "Unperformed Experiments Have No Results," *American Journal of Physics* 46, 745–747.

Pienaar, Jacques (2020): "Extending the Agent in QBism," *Foundations of Physics* 50, 1894–1920.

Pusey, Matthew, Barrett, Jonathan & Rudolph, Terry (2012): "On the Reality of the Quantum State," *Nature Physics* 8, 475–478.

Putnam, Hilary (1979): "A Philosopher Looks at Quantum Mechanics", in *Mathematics, Matter and Method: Philosophical Papers, Volume 1*, Cambridge, UK: Cambridge University Press, 215–227.

Rovelli, Carlo (2006): "The Disappearance of Space and Time," in D. Dieks (ed.): *The Ontology of Spacetime*, Amsterdam: Elsevier, 25–36.

Rovelli, Carlo (2021): *Helgoland*, Dublin: Penguin Books.

Ryckman, Thomas (2005): *The Reign of Relativity: Philosophy in Physics 1915–1925*, Oxford, UK: Oxford University Press.

Schlosshauer, Maximilian (ed.) (2011): *Elegance and Enigma: The Quantum Interviews*, Dordrecht: Springer.

Sellars, Wilfrid (1963): *Science, Perception and Reality*, Atascadero, CA: Ridgeview Publishing Company.

Shimony, Abner (1963): "Role of the Observer in Quantum Theory," *American Journal of Physics* 31, 755–777.

Smith, Joel (2010): "Seeing Other People," *Philosophy and Phenomenological Research* 81, 3, 731–748.

Stein, Edith (2004): *Einführung in die Philosophie*, Edith Stein Gesamtausgabe Band 8, Freiburg: Herder.

Stein, Edith (2018): "Concerning Heinrich Gustav Steinmann's Paper 'On the Systematic Position of Phenomenology,'" transl. by Evan Clarke, in Andrea Staiti and Evan Clarke (eds.): *The Sources of Husserl's 'Ideas I'*, Berlin: De Gruyter, 301–315.

Tegmark, Max (2008): "The Mathematical Universe," *Foundations of Physics* 38/2, 101–150.

Timpson, Christopher (2008): "Quantum Bayesianism: A Study," *Studies in History and Philosophy of Modern Physics* 39, 579–609.

Todes, Samuel (2001): *Body and World.* Cambridge, MA: MIT Press.

Vaidman, Lev (2014): "Protective Measurement of the Wave Function of a Single System," in Shan Gao (ed.): *Protective Measurement and Quantum Reality*, Cambridge, UK: Cambridge University Press, 15–27.

von Neumann, John (2018): *Mathematical Foundations of Quantum Mechanics*, Princeton, NJ: Princeton University Press.

Wallace, David (2013): "A Prolegomenon to the Ontology of the Everett Interpretation," in A. Ney & D. Albert (eds.): *The Wave Function*, New York, NY: Oxford University Press, 203–222.

Wallace, David (2021): "Against Wavefunction Realism," in S. Dasgupta, R. Dotan, and B. Weslake (eds.): *Current Controversies in Philosophy of Science*, New York, NY: Routledge, 63–74.

Wallace, David (2022): "The Sky is Blue and other Reasons Quantum Mechanics is not Underdetermined by Evidence," arXiv:2205.00568.

Weyl, Hermann (2009): *Mind and Nature*, Princeton, NJ: Princeton University Press.

Wheeler, J. A. (1978): "The 'Past' and the 'Delayed-Choice' Double-Slit Experiment," in A. R. Marlow (ed.): *Mathematical Foundations of Quantum Theory*, New York, NY: Academic Press, 9–48.

Wheeler, J. A. (1980): "Beyond the Black Hole," in H. Woolf (ed.): *Some Strangeness in Proprotion. Einstein Centennary Volume*, 341–375.

Wigner, Eugene (1960): "The Unreasonable Effectiveness of Mathematics in the Natural Sciences," *Communications in Pure and Applied Mathematics* 13, 1–14.

Wiltsche, Harald (forthcoming): "The Coordination Problem: A Challenge for Transcendental Phenomenology of Science," *New Yearbook for Phenomenology and Phenomenological Philosophy*.

Wiltsche, Harald (2012): "What Is Wrong with Husserl's Scientific Anti-Realism?" *Inquiry* 55, 2, 105–130.

Wiltsche, Harald & Berghofer, Philipp (eds.) (2020): *Phenomenological Approaches to Physics*, Synthese Library, Cham: Springer.

Zahavi, Dan (2003): "Phenomenology and Metaphysics," in Dan Zahavi, Sara Heinämaa, and Hans Ruin (eds.): *Metaphysics, Facticity, Interpretation*, Dordrecht: Kluwer, 3–22.

Zahavi, Dan (2019): *Phenomenology: The Basics*, London, UK: Routledge.

Zeilinger, Anton (1999): "A Foundational Principle for Quantum Mechanics," *Foundations of Physics* 29, 631–643.

Part I

From QBism to Phenomenology

2 Toward a World Game-Flavored as a Hawk's Wing

Blake C. Stacey

Discourse about quantum physics tends to become separated from its original subject matter. In extreme cases, the profitability of sounding profound completely overwhelms the content. In this regime we find "quantum international relations" and other schemes where people read someone who once read *New Scientist* and then stand around saying "entanglement" to each other in expense-accounted hotels. But even setting aside the assorted types of crystal healing that assume the mantle of physics, it is a genre convention in quantum foundations to rely upon hearsay and oversimplification. The volume with which a physicist will expound upon Bohr, for example, grows with their distance from anything that Bohr himself wrote. And though the genesis of QBism is very much within living memory, and the written record is almost comical in its abundance, QBism faces the problem that to many, that history simply has not occurred. Conversations and citations alike indicate that the frankly radical changes during the first decade of this century, by which "Quantum Bayesianism" became QBism, have gone unacknowledged when recognizing them was most essential.

The history of QBism is a series of hard-fought battles for self-consistency. Of course, I am phrasing it this way to heighten the drama. The amount of blood that has spilled is actually rather small.

The first goal of this chapter will be to trace those changes and find the clearest examples in the fossil record to illustrate the conceptual evolution. The overall theme will be a motion into a Jamesian pragmatism: Empiricism becomes radical, and correspondence theories of truth wither under scrutiny. In the course of this historical expedition, we will meet technical topics that come with a philosophical charge. Consequently, by the time we have completed our survey, we will be positioned to speculate about the *future* of QBism. In particular, we will see possibilities for cross-pollination between pragmatism and phenomenology that may have tangible effects upon everyday physics.

DOI: 10.4324/9781003259008-3

QBism is a research program that can briefly be defined as

> an interpretation of quantum mechanics in which the ideas of *agent* and *experience* are fundamental. A "quantum measurement" is an act that an agent performs on the external world. A "quantum state" is an agent's encoding of her own personal expectations for what she might experience as a consequence of her actions. Moreover, each measurement outcome is a personal event, an experience specific to the agent who incites it. Subjective judgments thus comprise much of the quantum machinery, but the formalism of the theory establishes the standard to which agents should strive to hold their expectations, and that standard for the relations among beliefs is as objective as any other physical theory.
>
> (DeBrota & Stacey 2018)

The first use of the term *QBism* itself in the literature was by Fuchs and Schack in June 2009 (Fuchs & Schack 2009a). Prior to this, they had employed it in talks and correspondence (Fuchs 2014, p. 1707), illustrating the lexicographer's principle that words predate their preservation in books. Introducing a collection of correspondence, Fuchs wrote that three characteristics of the QBist research program distinguish it from existing interpretations (Fuchs 2014, p. ix).

> First is its crucial reliance on the mathematical tools of quantum information theory to reshape the look and feel of quantum theory's formal structure. Second is its stance that two levels of radical "personalism" are required to break the interpretational conundrums plaguing the theory. Third is its recognition that with the solution of the theory's conundrums, quantum theory does not reach an end, but is the start of a great journey.

QBism grew out of Quantum Bayesianism, a loosely defined school of thought typified by a paper by Caves, Fuchs and Schack, "Quantum probabilities as Bayesian probabilities" (Caves, Fuchs & Schack 2002a, hereinafter CFS 2002). The most prominent thread within Quantum Bayesianism may be the one that, in the hands of Fuchs and Schack, later developed into QBism, but the term is applicable much more broadly. It encompasses some writings of Bub and Pitowsky (Pitowsky 2002, Bub & Pitowsky 2010, Bub 2019), for example, and could easily include earlier suggestions of Youssef (1994) and Baez (2003), work by Leifer and Spekkens (2013), the "entropic dynamics" of Caticha (2007) and so forth. These views overlap to the extent that they all advocate interpreting the probabilities in quantum physics according to some variety of Bayesianism. They overlap, but they do not coincide. What else should one expect, given that even before quantum physics was brought into the game, the jest that there were "46,656 varieties of Bayesianism" (Good 1983) was only a mild exaggeration?

Perusing the quantum-foundational literature and partaking in conversations with quantum foundationers, I have been surprised by the inclination to cite CFS 2002 as defining QBism. Sometimes, it is invoked by itself as the canonical QBist document, and on other occasions, it is mixed together indiscriminately with genuinely QBist sources. Without commenting in depth on the merits of these varied works, I find this situation puzzling. All three authors disavow the perspective of this paper (Caves, Fuchs & Schack 2007, Fuchs & Schack 2009a), whether they have continued along the path of developing QBism (Fuchs and Schack) or not (Caves).

The philosophy of science, we fondly imagine, should be the discipline that exposes the distinctions which physicists coarsely gloss over. Perhaps QBism has generated too many expositions to pay attention to just one (Fuchs, Mermin & Schack 2014, Fuchs 2017a, 2017b, Fuchs & Stacey 2019), but reading is in the scholar's job description, and fortunately, not all QBist writings are as long as some of them are. Indeed, some authors have gotten the coordinates right for genuine QBist expositions (Cabello 2017, Frauchiger & Renner 2018, Koberinski & Müller 2018, Schaffer & Barreto Lemos 2021).

Space constraints and the mostly ahistorical writing style of physics journals have prevented earlier QBist articles from delineating which Quantum Bayesian papers still have valuable portions, which are nearly obsolete, which are by authors who may sympathize with QBism without fully subscribing to it, and so forth. And among the more leisurely portrayals of QBism, the book by von Baeyer (2016) was pitched to the general pop-science audience, making it ill-suited to address the "inside baseball" matters like different schools of Bayesianism. Apart from a brief note in the context of a trendy but confined discussion (Stacey 2019a), I myself have not drawn a hard line between QBism and the more amorphous Quantum Bayesianism that came before it, thinking in my innocence that the progression of thought was dramatic enough that it did not need pointing out. Correcting this deficit — and atoning, in part, for my blithe naïveté — turns out to be an educational exercise. This is the second time I have marshalled historical evidence to show that a well-cited work was not in fact QBist, despite third-party claims to that effect (Stacey 2016a). From the viewpoint of QBism's *h*-index, this must appear a quixotic or even self-destructive effort, but integrity is never easy.

2.1 Locating QBism

Statements about probability have been given many different interpretations over the years. One sect would read an equation like "$p = 0.7$" as a claim about relative frequency in a large ensemble, while another would like to take it as concerning the extent to which a proposition follows logically from evidence. The Bayesian tradition regards probabilities as quantities asserted by gamblers, and it is within this tradition that QBism situates itself. Each of these intellectual genera contains many species. Within Bayesianism, one might mandate that in principle, all probabilities should reduce to 0 or 1 — maximal and complete information must resolve all uncertainty. This is the spirit we find, for example, in Jaynes or Garrett (Garrett 1993). Another aspiration has it that within each physical situation, there dwells something like a chance density or a ratio of up to down probabilitons, so that any gambler aware of the value of that "objective chance" must set her odds to the exact figure it implies. This is difficult, perhaps impossible, to make logically self-consistent or to integrate with known quantities in physics; in day-to-day work, the postulation of such properties seems extraneous to scientific practice (Fuchs & Stacey 2019). QBism instead follows the lead of *personalist Bayesianism,* a view historically associated with Ramsey and de Finetti (Ramsey 1931, Jeffrey 1989). It eschews the probabilitons and finds objectivity at a different level. For the QBist, no intrinsic attribute of a physical system can itself compel the outcome of a quantum measurement upon that system, nor even the probabilities that an agent should ascribe to the potential outcomes of that measurement before she performs it.

The research program of QBism does not content itself with providing a story for the familiar mathematical formalism of quantum theory. Nor is it satisfied with detailing how QBism differs from previous attempts to interpret the quantum — a harmless pastime for those who treasure the peace of library basements. Rather, the goal is to understand why that formalism is useful: Why quantum theory, as opposed to any alternative we might envision? QBism finds the exhortation to "shut up and calculate!" unstable against perturbations by curiosity: "Were the world a different way, would we not, after we shut up, calculate in a different fashion?" (Stacey 2019d).

Three Greek-derived words are helpful in discussing interpretations of quantum mechanics. *Ontic* refers to entities and quantities that exist, in their own right, in blunt reality — in a Newtonian worldview, the mass of a rock is ontic. *Epistemic* quantities have the character of knowledge, while *doxastic,* from the Greek for "belief", captures the personalist Bayesian view of probabilities, and thus the QBist interpretation of quantum states. Writing a wavefunction $|\psi\rangle$ is staking out a doxastic claim, though the fact that it has proven useful to use vectors in complex Hilbert spaces to express our doxastic statements has an ontological lesson subtly coded within it.

QBism is largely orthogonal to matters of "Bayesian inference" as understood in statistics or big-data science (DeBrota & Stacey 2018, §§9). Attempting to grasp what QBism is about by extrapolating a Google University education in those subjects has led more than a few poor souls into confusion, whether they recognize it or not.

Our first task will be identifying the differences between QBism and what we might call "proto-QBism", the views articulated by Fuchs, Schack and coauthors in the 1990s and early 2000s. We will begin with the CFS 2002 paper mentioned above, which in my informal experience has been most commonly confused with QBism proper, and which provides a rather nice contrast with it. We will follow that with a close study of a follow-up article that Caves, Fuchs and Schack wrote a few years later, which as we will see still is not QBism. We will then backtrack and examine older writings of Fuchs himself that are less widely invoked, and whose distance from QBism is in certain aspects almost shocking. Our final exhibit will be a position statement that Fuchs and Asher Peres wrote for *Physics Today*. My hope is that revealing this history may help explicate why QBism developed as it did, and that it may aid those displeased with QBism to be unhappy with QBism itself instead of a confabulation. By tracing the history of QBism in this manner, we will then be positioned to speculate about the future, considering potential cross-currents of inspiration between schools of philosophy.

2.2 Caves–Fuchs–Schack (2002)

CFS 2002 makes a case that quantum probabilities should be regarded as Bayesian probabilities, but it doesn't do much more than that — at least, not very well, and from a QBist perspective, not convincingly.

The ceaseless and uncritical invocation of "maximal information" in CFS 2002 is legitimately grating to a QBist ear. At best, one can wincingly try to find a reading where it is tautological, treating the statements *Alice has maximal information about a system S* and *Alice ascribes a pure quantum state to system S* as wholly synonymous. But one would be cruelly paid back for such generosity. The years have taught us that it is just not possible to shake the connotations of "maximal information" — connotations of pre-existing properties, of the "ontic states" that a more timid view would want to underlie quantum theory. As Fuchs and Schack would write much later (Fuchs & Schack 2009a).

> The trouble with the phrase "maximal information is not complete" and the imagery it entails is that, try as one might to portray it otherwise (by adding "cannot be completed," say), it hints of hidden variables. What else could the "not complete" refer to?

Therefore, we must read CFS 2002 through the prism of this later repudiation of it. Section IV begins,

> Our concern now is to show that if a scientist has maximal information about a quantum system, Dutch-book consistency forces him to assign a unique pure state. Maximal information in the classical case means knowing the outcome of all questions with certainty. Gleason's theorem forbids such all-encompassing certainty in quantum theory. Maximal information in quantum theory instead corresponds to knowing the answer to a maximal number of questions (i.e., measurements described by one-dimensional orthogonal projectors).

The language about "knowing the answer to a maximal number of questions" is redolent of ideas that QBism learned after much scrutiny to leave behind. These ideas go by names like "the eigenstate-eigenvalue link" and "the EPR criterion of reality"; their underlying, unexamined premise is that a *probability-1 prediction* equates to an *objective, agent-independent physical truth*.

How could we ever reconcile this language with the QBist insistence that a quantum measurement does not simply read off a pre-existing physical quantity? It just doesn't work. Every instance of "maximal information" in CFS 2002 is a moment when, to adapt the closing line of so many scientific papers, *further research was needed.*

"Knowing the answer to a maximal number of questions" applies to the Spekkens toy model, by construction (Spekkens 2007). In this model, the fundamental atom is a system with four possible physical states. The observer is restricted never to know more than one bit of information about an atom that would require two bits to describe fully. It follows that there are three possible binary-valued tests that the observer can perform on an atom, but no state of knowledge can allow the answer to more than one of them to be foreseen. Consequently, the posit that "Maximal information ... corresponds to knowing the answer to a maximal number of questions" does not really get at anything uniquely quantum at all (Spekkens 2007).

QBists have argued that the restriction in certainty is not fundamental, but rather derived. In order to explore this point, we need to develop some mathematical machinery. The crucial idea is the concept of a *reference measurement* (Appleby et al. 2017). Now that the kilogram is no longer defined as a particular lump of platinum-iridium alloy (Conover 2018), there is room in the vault for a Bureau of Standards quantum measurement device. Consider a physicist Alice who has a qubit system in her possession. She can go to the Bureau of Standards and drop her system into the standard measurement device for qubits. Such a device must have at least four possible outputs that it can generate; that is, Alice's mathematical representation of it must be a positive operator valued measure (POVM) with at least four elements. Let $p(H_i)$ denote Alice's probability for obtaining outcome number i. Suppose that she intends to perform some *other* measurement $\{E_j\}$: Perhaps a von Neumann test corresponding to an orthonormal basis, or perhaps a trine POVM (Fuchs 1997), or any other POVM that Alice might find of interest. Let $r(E_j \mid H_i)$ denote her probability for eliciting outcome j in this other experiment *given* that she has performed the Bureau of Standards reference POVM and obtained outcome i in it first. Quantum theory then furnishes the tools for Alice to compute her probability $q(E_j)$ for eliciting outcome j in this other measurement *without* her carrying out the reference experiment first. The vector of these probabilities will be

$$q = \mu(p, r), \tag{2.1}$$

where μ is some function that depends upon the details of the reference measurement. A simple and illuminating choice is to make the reference measurement a POVM that corresponds to a regular tetrahedron inscribed in the Bloch sphere. For example, letting a and b take the values ± 1, then the four positive semidefinite operators

$$H_{ab} = \frac{1}{4}\left(I + \frac{1}{\sqrt{3}}(a\sigma_x + b\sigma_y + ab\sigma_z)\right) \tag{2.2}$$

sum to the identity and thus constitute a POVM. With $p(H_{ab}) = \mathrm{tr}(\rho H_{ab})$ by the Born Rule and $r(E_j \mid H_{ab}) = 2\mathrm{tr}(E_j H_{ab})$ by the Lüders rule, we have

$$q(E_j) = \mathrm{tr}(\rho E_j) = \sum_{ab}\left[3p(H_{ab}) - \frac{1}{2}\right] r(E_j \mid H_{ab}). \tag{2.3}$$

In this case, the function μ takes the form of the classical Law of Total Probability but with an elementwise deformation of the probability vector p. The reference measurement establishes a mapping from density matrices into probability vectors, thereby yielding a wholly probabilistic representation of the quantum theory of a qubit. Not all probability vectors p correspond to valid quantum states in this representation. In fact, with any *minimal informationally complete* (MIC) experiment as the reference POVM, the state space is mapped into a proper subset of the four-outcome probability simplex. No more than *one* entry in a valid probability vector p can be equal to zero. Or, geometrically speaking, the vertices of the reference probability simplex are unavailable. This has the character of an uncertainty principle: Alice's state of expectation can only be so sharp. But this is just a consequence of a deeper truth, namely that because intrinsic "hidden variables" do not exist, one should not use the Law of

Total Probability to intermediate between different experiments (Stacey 2019a, DeBrota, Fuchs & Stacey 2020). This is a central point in the QBist research program of *reconstructing quantum theory* from physical principles, and so we will be returning to it later.

CFS 2002 includes other problematic passages, like the following:

> In the classical case an i.i.d. assignment is often the starting point of a probabilistic argument. Yet in Bayesian probability theory, an i.i.d. can never be strictly justified except in the case of maximal information, which in the classical case implies certainty and hence trivial probabilities. The reason is that the only way to be sure all the trials are identical in the classical case is to know everything about them, which implies that the results of all trials can be predicted with certainty.
>
> (Jaynes 2003)

Note that the citation supporting this argument is to E. T. Jaynes' unfinished textbook (Jaynes 2003). A QBist naturally asks, "If my probabilities really are *mine*, then who's to stop me from choosing an i.i.d. prior? Experience may lead me to revise my beliefs away from that prior, but I have every right to assert it in the first place". One could square the argument against the legitimacy of classical i.i.d. priors with a Jaynesian view, but ultimately not with a Ramseyan one. Saying "to be sure all the trials are identical" amounts to saying "to be sure the probabilities are physically equal". The argument in CFS 2002 is a relic of an objective-Bayesian interpretation. One way to express this shift of interpretation is to say that in CFS 2002, probabilities are epistemic (about knowledge), while in QBism proper, they are doxastic (about belief). CFS 2002 is saying that an i.i.d. prior is only justified when the ratio of up to down probabilitons is constant across all the trials, and that is just not a kind of Bayesianism that QBism can endorse.

CFS 2002 declares (italics in original),

> Since one of the chief challenges of Bayesianism is the search for methods to translate information into probability assignments, *Gleason's theorem can be regarded as the greatest triumph of Bayesian reasoning.*

From a QBist perspective, this is peculiar. Gleason's theorem proves that it is possible to chop off part of the standard formalism of quantum theory and then re-grow it from the remainder (Gleason 1957). More specifically, Gleason showed that if measurements correspond to orthonormal bases on a Hilbert space, and if the probability of a measurement outcome does not depend upon which basis the corresponding vector is embedded in, then any consistent way of assigning probabilities to measurement outcomes has to take the form of the Born rule. Thus, if Π is a projection operator and $p(\Pi)$ is the probability ascribed to obtaining the outcome corresponding to Π, then we must have

$$p(\Pi) = \mathrm{tr}(\rho\Pi) \tag{2.4}$$

for some density matrix ρ. Both the set of valid ρ and the rule for what to do with a ρ come tumbling out of Gleason's insight. This is of course pertinent to the project of reconstructing quantum theory, a task to which much QBist and QBist-adjacent effort has been devoted — but Gleason's theorem itself has barely figured in that effort. Why? One reason is that the premises of Gleason's theorem are themselves rather late in the game: Gleason's starting point is a Hilbert space and orthonormal bases upon it. The natural question is thus how to arrive at Hilbert space — out of all the mental contrivances that the mathematicians have conjured, why that very particular class of structure? The stated goal of the reconstruction project in which Fuchs and others have participated is to derive complex Hilbert space, linear operators, the space of valid quantum states and all the rest of the formalism from principles that are more deeply rooted. In that light, Gleason's theorem is more a proof of principle, a historically significant demonstration that the machinery *can* be taken apart and rebuilt, rather than "the greatest triumph" of anything.

Busch (2003) — and, later, independently Caves et al. (2004) — proved an analogue of Gleason's theorem in which POVMs are the basic notion of measurement.[1] Gleason's original theorem fails for two-dimensional Hilbert spaces. The essential reason is that in two dimensions, one cannot hold one vector of a basis in place and twirl the rest of the basis around to generate multiple distinct measurements. This difficulty does not apply to the POVM version of the theorem. (The question of which classes of measurements allow the proof of a Gleason-type theorem continues to be studied (Granström 2006, Wright & Weigert 2019a, 2019b).) But the move from orthonormal bases to general POVMs is much more than a mathematical convenience: It represents a fundamental shift in the meaning of the measurement concept.

The underlying intuition that motivates giving a fundamental role to orthonormal bases is that quantum measurements discover the truth values of propositions. Working in the background of the mind, there's always the idea that what a measurement does is uncover the value of a property. Maybe an eigenvalue is a property that has to be reasoned about using a "quantum logic", but it's a property regardless. All of quantum logic's efforts at organizing measurement outcomes into lattices of propositions ultimately rely upon this feeling. The "Quantum Bayesianism" of CFS

2002 had not yet shed this temperament, but QBism spurns it. And when outcomes are consequences, the need to think in terms of "quantum propositions" drops away, like the obligation to please a domineering teacher after graduation day (Fuchs & Stacey 2020).

Next, we consider the interpretation that CFS 2002 places on the *quantum de Finetti theorem* (Hudson & Moody 1976, Caves, Fuchs & Schack 2002b, Fuchs, Schack & Scudo 2004, Brandão & Harrow 2017). This is a quantum analogue of the de Finetti theorem in classical probability theory, which provides a viable meaning for the term "unknown probability" in a subjectivist, or personalist, form of Bayesianism. Consider a scenario in which an agent wishes to conduct a long experiment, made up of many successive trials. We can represent the outcome of each trial by a random variable x_j, and Alice assigns a joint probability distribution $p(x_1,x_2,\ldots,x_N)$ over the possible outcomes of an N-trial experiment. Imposing two conditions on this joint distribution turns out to simplify its form dramatically. First, we require that it be *finitely exchangeable*: Its value is invariant under permutations of its arguments. If π is any permutation of the indices $\{1,\ldots,N\}$, then

$$p(x_1,\ldots,x_N) = p(x_{\pi(1)},\ldots,x_{\pi(N)}). \tag{2.5}$$

Second, we require that Alice's $p(x_1,\ldots,x_N)$ be *extendable,* in the following manner. For any integer $M > 0$, there must be a finitely exchangeable distribution with more arguments, p_{N+M}, such that

$$p(x_1,\ldots,x_N) = \sum_{x_{N+1},\ldots,x_{N+M}} p_{N+M}(x_1,\ldots,x_N,x_{N+1},\ldots,x_{N+M}). \tag{2.6}$$

These two requirements make precise the idea that Alice's probability assignment p derives from an arbitrarily long sequence of random variables, the order of which is, in Alice's judgment, inconsequential. We say that a p which satisfies both conditions, finite exchangeability and extendable, is *exchangeable.*

Let Δ_k denote the space of valid probability assignments over k outcomes. Then, the classical de Finetti theorem shows that exchangeability implies that

$$p(x_1,\ldots,x_N) = \int_{\Delta_k} d\vec{p}\,P(\vec{p})p_{x_1}\cdots p_{x_N} = \int_{\Delta_k} d\vec{p}\,P(\vec{p})p_1^{n_1}\cdots p_k^{n_k}, \tag{2.7}$$

where $P(\vec{p})$ is properly normalized over Δ_k:

$$\int_{\Delta_k} d\vec{p}\,P(\vec{p}) = 1. \tag{2.8}$$

The quantum version replaces an integral over the probability simplex Δ_k with an integral over quantum state space, furnishing a representation of exchangeable quantum-state ascriptions:

$$\rho^{(N)} = \int d\rho P(\rho)\rho^{\otimes N}. \tag{2.9}$$

Just as the classical de Finetti theorem revealed how the term "unknown probability" is merely a convenient shorthand, so does the quantum de Finetti theorem for "unknown quantum state".

Having established this background, we turn to the following passage from CFS 2002:

> Exchangeability permits us to describe what is going on in quantum-state tomography. Suppose two scientists make different exchangeable state assignments and then jointly collect data from repeated measurements. Suppose further that the measurements are "tomographically complete"; i.e., the measurement probabilities for any density operator are sufficient to determine that density operator. The two scientists can use the data D from an initial set of measurements to update their state assignments for further systems. In the limit of a large number of initial measurements, they will come to agreement on a particular product state $\hat{\rho}_D \otimes \hat{\rho}_D \otimes \cdots$ for further systems, where $\hat{\rho}_D$ is determined by the data. *This is what quantum-state tomography is all about.* The updating can be cast as an application of Bayes's rule to updating the generating function in light of the data [Schack, Brun and Caves (2001)]. The only requirement for "coming to agreement" is that both scientists should have allowed for the possibility of $\hat{\rho}_D$ by giving it nonzero support in their initial generating functions.

QBism refuses to go down this path. Indeed, it balks at the first step. "Suppose two scientists make different exchangeable state assignments and then jointly collect data from repeated measurements" — no, we're stopping right there. QBism insists that *measurement outcomes are personal to the agent who elicits them.*

A QBist take on the quantum de Finetti theorem puts all the state assignments into a single user's internal mesh of beliefs. Alice supposes that tomorrow, she will make a multipartite, exchangeable state ascription, perhaps $\rho_1^{(N)}$ or perhaps $\rho_2^{(N)}$. Using the quantum de Finetti theorem, she represents the first joint state as a "meta-probability density" $P_1(\rho)$:

$$\rho_1^{(N)} = \int d\rho P_1(\rho)\rho^{\otimes N}, \tag{2.10}$$

and likewise for $\rho_2^{(N)}$. If the density functions $P_1(\rho)$ and $P_2(\rho)$ have at least a little agreement, Alice can *expect* that her initial choice will wash out. That is, her expectation *now* for her *future mesh of beliefs* is that the choice between the ascriptions $P_1(\rho)$ and $P_2(\rho)$ will eventually become inconsequential. So, there is definitely a story to be told about the mathematics, perhaps a rather important one, but it is not the story given in CFS 2002.[2]

Recalling Fuchs' three distinguishing characteristics of QBism, it is difficult to argue that they are present in CFS 2002. The "formal structure" is not reshaped in look or in feel, merely propped up by a couple extant theorems, one of them known since 1957. Two levels of radical personalism are not present — there is only one, and it is but tepidly embraced, while the other is flatly contradicted. The third characteristic, the "start of a great journey", receives an endorsement of sorts in the paper's send-off, but one so disconnected from everything that went before, it reads more like an afterthought than an outlook.

2.3 Caves–Fuchs–Schack (2007)

Caves, Fuchs and Schack moved away from some positions held in their 2002 paper just a few years later. "Subjective probability and quantum certainty" (hereinafter CFS 2007) is the last work coauthored by all three of them, since as that paper was being written and published, it became clear that Caves disagreed with Fuchs and Schack on various points that are essential to Fuchs and Schack's further development of QBism.[3] In their introduction, they state the following:

> In a previous publication [CFS 2002], the authors were confused about the status of certainty and pure-state assignments in quantum mechanics and thus made statements about state preparation that we would now regard as misleading or even wrong.

Discussions leading up to this change of heart can be found in Fuchs (2014, pp. 193 ff.).

There is less material in CFS 2007 that is overtly contra-QBist than there is in CFS 2002, yet Fuchs' later warning covers it as well (Fuchs 2010):

> The present work, however, goes far beyond those statements in the metaphysical conclusions it draws—so much so that the author cannot comfortably attribute the thoughts herein to the triumvirate as a whole. Thus, the term QBism to mark some distinction from the known common ground of Quantum Bayesianism.

One would think that this hazard sign would prompt a degree of caution to be taken before treating all the citations in a list of "Quantum Bayesian" papers on equal footing.[4] We can see a trace of a divergence already manifesting in this aside:

> Bayesian updating is consistent, as it should be, with logical deduction of facts from other facts, as when the observed data d logically imply a particular hypothesis h_0, i.e., when $\Pr(d \mid h) = 0$ for $h \neq h_0$, thus making $\Pr(h_0 \mid d) = 1$. Since the authors disagree on the implications of this consistency, it is fortunate that it is irrelevant to the point of this paper. That point concerns the status of quantum measurement outcomes and their probabilities, and quantum measurement outcomes are not related by logical implication.

CFS 2007 is evasive on the question of whether measurement outcomes are personal to the agent who elicits them. They write of "the facts an agent acquires about the preparation procedure", and they say, "The occurrence or nonoccurrence of an event is a *fact* for the agent". But if the occurrence or nonoccurrence of an event is a fact for *everybody*, it is a fact for a specific agent too. Without forthrightly clarifying this point, CFS 2007 does not qualify as QBist. Indeed, CFS 2007 rather undermines the point, with loose talk about "two agents starting from the same facts, but different priors" and the like. A QBist description would instead involve a single agent, considering the same set of quantitative data points with either of two background meshes of belief (compare Fuchs & Schack 2009b, Stacey 2016b).

Before moving on, we note that CFS 2007 briefly discusses the Lewisian "objective chance" philosophy. Fuchs would shortly thereafter find this discussion weak enough to be irrelevant, or potentially even counterproductive (Fuchs 2014, p. 1287). Overall, CFS 2007 is the product of too many compromises among its three authors to fairly reflect the view of any of them.

2.4 The First Samizdat

Notes on a Paulian Idea (2003), later reissued as *Coming of Age with Quantum Information* (2011), is an edited collection of e-mail correspondence that Fuchs made public to "back up the hard drive" after the Cerro Grande fire destroyed his family's home in Los Alamos (Fuchs 2003). My colleagues and I have elsewhere (Appleby et al. 2017) quoted passages from this document to show how later research has fulfilled their aspirations almost to the letter. Here, I take the opposite tack.

One theme that is quite surprising to a reader familiar with QBism proper is how frequently Fuchs insists upon multiple agents being necessary to reveal the hidden ontological lesson of quantum theory. This stands in stark contrast to the "I-I-me-me-mine!" declaration of Fuchs' later manifesto (Fuchs 2010). In addition, the stories of multiple agents intermingle with the theme that the deep physical principle of quantum theory is captured by information-disturbance tradeoffs. This idea had vanished by the time QBism proper was articulated. We can point to multiple reasons why it could easily have fallen by the wayside: The quantitative expressions of it started out dense and not too illuminating (Fuchs & Peres 1996, Fuchs 1998a), and despite some promising indications (Fuchs 1998b), they never really simplified.[5] Moreover, the basic phenomenon of "no information gain without disturbance" ended up being too easy to reproduce in theories with underlying local-hidden-variable models.

In an 18 September 1996 letter to David Mermin, we find an early occurrence of the slogan that Fuchs and Schack would later spurn:

> One always assigns probabilities based on incomplete information; it is just that in quantum physics "maximal information is not complete."

We find this again in another letter to Mermin, this one dated 23 July 2000:

> The theory prescribes that no matter how much we know about a quantum system—even when we have *maximal* information about it—there will always be a statistical residue. There will always be questions that we can ask of a system for which we cannot predict the outcomes. *In quantum theory, maximal information is not complete and cannot be completed.*

Even sentences that were blessed with italics can turn out to be quite wrong-headed.

We can uncover early intimations of the QBist desire to find in quantum theory an ontological lesson, without naïvely identifying the elements in the mathematical formalism with an ontology. However, the place and manner of the search is not yet QBist. From a 4 January 1998 letter to Greg Comer:

> The "fact" that *my* information-gathering yields a disturbance to *your* predictions is the only "physical" (or ontological) statement that the theory makes; all the rest of the structure is "law of thought" subject to that consideration. To put it another way, quantum theory is a theory of "what we have the right to say" in a world where the

> observer cannot be detached from what he observes. It is that and nothing more.

And again on 22 April 1999:

> Our experimentation on the world is not without consequence. When *I* learn something about an object, *you* are forced to revise (toward the direction of more ignorance) what you could have said of it.
>
> More concisely, in a 2 September 1998 letter to Adrian Kent:
>
> Disturbance to what? To each other's descriptions, nothing more.

Likewise, we find the following in an 11 December 2000 letter to Joseph Renes:

> Why am I so obsessed with always having two players in the game? Because I want to connect all the concerns in quantum mechanics with Bayesianism as much as I can.

At this point in time, Fuchs pretty explicitly takes the convergence among agents as the meat of Bayesianism, rather than making the fundamental point the normative principle of consistency within a single agent's mesh of beliefs, with inter-agent agreement a secondary notion (when it can meaningfully be defined at all).

The most dramatic development of the theme might be in a letter to Mermin, on 20 July 2000:

> Somehow I feel that I had an epiphany in Mykonos. Do you remember the parable of "Genesis and the Quantum" from my Montréal problem set? And do you remember my slide of an empty black box with two overlays. The first overlay was of a big $|\psi\rangle$ (hand drawn in blue ink of course). I put the slide of the box up first, and said, "This is a quantum system; it's what's there in the world independent of us." Then I put the first overlay on it and said, "And this symbol stands for nothing more than [what] we know of it. Take us away and the symbol goes away too." I then removed the $|\psi\rangle$. "But that doesn't mean that the system, this black box, goes away." Finally I put back up the $|\psi\rangle$ over the box, and the final overlay. This one says: "Information/knowledge about what? The consequences of our experimental interventions into the course of Nature."
>
> Well, now I've made another overlay for my black box slide. At the top it asks, "So what is real about a quantum system?" In the center,

> so that it ends up actually in the box, is a very stylistic version of the word "Zing!" And at the bottom it answers, "The locus of all information-disturbance tradeoff curves for the system." In words, I (plan to) say, "It is that zing of the system, that sensitivity to the touch, that keeps us from ever saying more than $|\psi\rangle$ of it. This is the thing that is real about the system. It is our task to give better expression to that idea, and start to appreciate the doors it opens for us to shape and manipulate the world." What is it that makes quantum cryptography go? Very explicitly, the zing in the system. What is that makes quantum computing go? The zing in its components!
>
> Anyway, I'm quite taken by this idea that's getting so close to being a technical one—i.e., well formed enough that one might check whether there is something to it. What is real of the system is the locus of information-disturbance (perhaps it would be better to say "information-information") tradeoff curves. The thing to do now is to show that Hilbert space comes about as a compact description of that collection, and that it's not the other way around. As I've preached to you for over two years now, this idea (though it was in less refined form before now) strikes me as a purely ontological one ... even though it takes inserting an Alice, Bob, and Eve into the picture to give it adequate expression. That is, it takes a little epistemology before we can get to an ontological statement.

The number of agents is getting out of hand — not just an Alice, with whom a QBist narrative would content itself, but a Bob and now an Eve. This is quite the dramatic excess in the light of QBism's single-user focus (a later development codified in response to drilling down on the issue of Wigner's Friend (Fuchs 2014, p. xli)). It's a little like a story told by Jack Falstaff, where the number of assailants that Falstaff has fought off keeps growing: "What, four? thou saidst but two even now". "Four, Hal; I told thee four".

2.5 Fuchs (2002)

Having grounded ourselves in Fuchs' less formal solo-author writings from the late 1990s, we are now in a better position to examine "Quantum Mechanics as Quantum Information (and only a little more)" (Fuchs 2002). I have deferred discussion of this essay until now, because much of it is technical development, and the conceptual passages where it comes off most strongly non-QBist are best appreciated after becoming familiar with the *Notes on a Paulian Idea* era. Of the three distinguishing features of QBism, we can discern the "reliance on the mathematical tools of quantum information theory to reshape the look and feel of quantum theory's formal structure", at least in a preliminary way. And, not bound by the length and style constraints of a journal article's "Conclusions" section, Fuchs takes the opportunity to press the "start of a great journey" theme. But there is still only one level of personalism: Probabilities are personal, but *experiences* are not. A brief excerpt suffices to show that Fuchs attempts to launch the "great journey" just as he did at Mykonos:

> The wedge that drives a distinction between Bayesian probability theory in general and quantum mechanics in particular is perhaps nothing more than this "Zing!" of a quantum system that is manifested when an agent interacts with it. It is this wild sensitivity to the touch that keeps our information and beliefs from ever coming into too great of an alignment. The most our beliefs about the potential consequences of our interventions on a system can come into alignment is captured by the mathematical structure of a pure quantum state $|\psi\rangle$. Take all possible information-disturbance curves for a quantum system, tie them into a bundle, and *that* is the long-awaited property, the input we have been looking for from nature. Or, at least, that is the speculation.

The *first* sentence would read fine coming from a QBist, but the rest goes barreling down a blind alley.

The technical discussions hold up rather better, and the paper is noteworthy as an early example of the MIC-as-reference-measurement idea. The particular class of MIC it discusses, later designated the *orthocross MICs*, still has some open conjectures about it (DeBrota, Fuchs & Stacey 2021).

Fuchs gives an argument for why the tensor product rule for composing state spaces follows from Einstein locality and a Gleason-type context-independence condition. This proof may be of significance to a category theorist (Riehl 2016, §§2.3), as it deduces the tensor product from the requirement that the functions of interest be linear on both halves of the composite system. However, it does go somewhat against the grain of later reconstruction work with Schack and others. Those efforts focused on deriving the state and measurement spaces of a single system, which can then be resolved into components if desired. In other words, the emphasis shifted from *composition* to *decomposition*.

The 2002 paper leaves open the question of why the joint states for a bipartite system should be specifically *positive semidefinite* operators on the tensor-product space. The later literature provides at least one answer to this question (Barnum et al. 2010, de la Torre et al. 2012), but at the cost of assumptions that may feel unsatisfying on account of being physically under-motivated or mathematically over-powered. (For example, why in the grand scheme of things should the set of entangled pure states form a continuum?) There may yet be a theorem or two worth proving here.

Examining the motivations interleaved between the equations, we find another conceptual issue that marks the 2002 paper as not yet QBist. It is the distinction between doxastic *consistency conditions* and *update rules,* a point that Fuchs and Schack did not fully resolve until the better part of a decade later (Fuchs & Schack 2012). Quoting DeBrota and Stacey (2018),

> Adopting a personalist Bayesian interpretation of probability does *not* mean treating all changes of belief as applications of the Bayes rule. This is shocking to some people! And distancing ourselves from the dogmatists who claim to follow that creed is one reason why we prefer *QBism* over "Quantum Bayesianism".
>
> In the tradition of Ramsey, Savage and de Finetti, there are consistency conditions that an agent's probability assignments should meet at any given time, *and then* there are guidelines for *updating* probability assignments in response to new experiences. Going from the former to the latter requires making extra assumptions — the two are not as strongly coupled as many people think. The Bayes rule is not a condition on how an agent *must* change her probabilities, but rather a condition for how she should *expect* that she will modify her beliefs in the light of possible new experiences. For this observation, we credit Hacking, Jeffrey and van Fraassen.

Fuchs' writing in 2002 had not yet distinguished the crucial gap between a rule for how Alice *must* change her beliefs and a criterion for how Alice should *expect* today that she will act tomorrow.

This is a slip-up we encounter now and then in conversations with people who have only heard a little about QBism, usually secondhand. (Other confusions — "But how does QBism explain X?" — typically occur when the interlocutor has unwittingly switched from subjective probability to objective, or from a first-person perspective to third-person, midway through a thought process. These are habits which take discipline to avoid, at least at first.) Evading this mental trap is another good reason not to take Fuchs' 2002 salvo as a definitive, genuinely QBist position statement.

2.6 Fuchs–Peres (2000)

On occasion, we have seen CFS 2002 cited on its own to define QBism (for example, in Vaidman 2019 and Laudisa and Rovelli 2019). A similar yet more egregious misattribution occurs in an article by Jaeger (2019), which equates QBism with the 2000 *Physics Today* piece coauthored by Fuchs and Peres, "Quantum Theory Needs No 'Interpretation'" (Fuchs & Peres 2000). I must regretfully report that Asher Peres was no QBist.

The specific point at which Jaeger elides the difference between QBism and *fin-de-siècle* Fuchs–Peres is the following, which attributes an opinion of the latter to the former:

> One QBist claim is that "quantum theory does *not* describe physical reality. What it does is provide an algorithm for computing *probabilities* for the macroscopic events ('detector clicks') that are the consequence of our experimental interventions. This strict definition of the scope of quantum theory is the only interpretation ever needed, whether by experimenters or theorists."

But this is not a QBist claim. Of course, it predates QBism chronologically, but also, it contradicts what QBism actually stands for, and in a series of rather blunt and obvious ways. It is helpful here to quote a letter from Fuchs' second samizdat (Fuchs 2014, p. 1011), sent on 19 June 2005 to Greg Comer:

> First off, I wish I had never said, "quantum theory does not describe physical reality"—I really only meant "the wave function does not describe reality" and should have stuck with that formulation. But more importantly, what precisely are these "consequences of our interventions"? From the wording we used, one surely gets the impression that, whatever they are—we said "detector clicks," but what a glib phrase!—they somehow live outside of the agent performing the experiment. And I guess that's what I thought at the time.

So, "detector clicks" is misleading. Moreover, "macroscopic" is a red herring, a relic of earlier generations' shifty grasp on what might differentiate quantum from classical. For example, an agent whose species has evolved eyes just a bit better than human ones might have seen individual photons flashing on a cold and lonely night. Such an agent might regard the direct personal experience of a single photon as a microscopic event, but they can employ the "user's manual" that is quantum theory just as well as humans do. In brief, the micro/macro distinction is not, to a QBist, fundamental.

Would a QBist agree that a "strict definition of the scope of quantum theory is the only interpretation ever needed, whether by experimenters or theorists"? No, *ever needed* is all wrong there. Nothing that is all that's *ever needed* can be the start of a great adventure.

The Fuchs–Peres collaboration has a very un-QBist reliance upon the first-person plural. As David Mermin has noted (Mermin 2018),

> There is a little remarked upon but important ambiguity in the first person plural. When Heisenberg says that quantum states are about *our* knowledge, "our" can mean all of us collectively or it can mean each of us individually. [...] To avoid ambiguity it is better to say "My (your, Alice's) quantum state assignments encapsulate my (your, her) belief" to avoid misreadings based on implicit assumptions of a unique state assignment or of common knowledge.

The Fuchs–Peres essay has some affinities with the Rudolf Peierls opinion piece from a decade earlier (Peierls 1991) that they discussed during the writing process (Fuchs 2003). Mermin observes that Peierls would sound more QBist than perhaps any other figure from the early generations of quantum physicists, if he had not used the first-person plural collectively. Instead, he propagated the old confusions. This applies with equal force to the Fuchs–Peres collaboration.

Having dismissed "our", "macroscopic", "detector clicks", "ever needed" and "does not describe physical reality", what about "algorithm"? This, too, reflects an understanding that had not yet matured. An algorithm is a step-by-step procedure that can be executed mechanically.[6] For example, taking the trace of the product of two matrices is a task for which an algorithm might be written. In that sense, computing a quantum-mechanical probability is algorithmic — but there is a deeper level of meaning, too, that the word *algorithm* misses. In QBism, a state vector $|\psi\rangle$ is not more ontologically fundamental than, say, the probability of getting a "+" outcome in a spin-z experiment. True, quantum theory provides a rule for calculating $p_z(+)$ given a $|\psi\rangle$, but when in life is one ever given a $|\psi\rangle$, other than the first line of a textbook problem demanding that the student "assume the state vector is $|\psi\rangle$"?

The deeper truth is that quantum theory provides a *normative lesson*: When Alice contemplates two or more von Neumann measurements upon a system, she *should strive* to make her expectations for those different, mutually exclusive scenarios all consistent with the Born Rule. But if she detects an inconsistency within her mesh of beliefs — if she finds that there is no density operator ρ with which her varied probabilities are all in accord — the quantum formalism itself *provides no algorithm* to resolve that awkwardness (Fuchs 2017b).

The novice at any art often begins by following a procedure — say, the exact volume measurements and timings given by *The Joy of Cooking*, or the rubric for cranking through questions on the AP Physics exam. With further experience, one learns how to season for taste, what can be substituted for chicken or for eggs, how to linearize around the fixed points and so on. The procedures are always there to be relied upon when required — chopping a root or solving by radicals — but they are not the soul of the matter. So, too, for quantum theory: Algorithms are what we use, not the sum total of what we need.

Fuchs and Peres present a version of the Wigner's Friend thought-experiment. In their portrayal, Erwin applies quantum mechanics to his colleague Cathy, who in turn applies it to a piece of cake. The Fuchs–Peres discussion is, from a QBist standpoint, rather unforgivably sloppy about the distinctions between ontic degrees of freedom, epistemic statements about ontic quantities and doxastic statements regarding future personal experiences.

2.7 Three Jamesian Pragmatists in a Trenchcoat

According to Dillon, Merleau-Ponty thought Husserl retained too many vestiges of Cartesian dualism, so Merleau-Ponty's "lived world" is not the same as Husserl's "life-world" (Dillon 1998). To those for whom that is bread and butter, the contrasts between QBism and its predecessors should be easy to appreciate.

Over the preceding sections, we have unearthed fossil traces of how "Quantum Bayesianism" became QBism. We can identify important themes in this development. First, the interpretation of probability left behind the "objective Bayesianism" of Jaynes in favor of true personalism. Second, measurement outcomes became personal: "a measurement outcome is an experience, the *agent's side* of a joint agent-object event" (Fuchs & Stacey 2020). Lightly paraphrasing a repeating theme from Fuchs' second samizdat, a quantum system is like a philosopher's stone, catalyzing a transmutation within the agent. Quantum theory *hints* at the transmutations that may happen on the system's side, but its focus is upon the pole that the agent holds; agent and system are distinguished, yet participate synergistically in moments of fact creation (Fuchs 2014).

Concomitantly with that second layer of personalism being introduced, the additional agents dropped out of the basic story. The older "Quantum Bayesianism" troubled itself with *what you see when you do something different than I do*. In contrast, QBism concerns *what changes for me if I do something else instead.*

QBism has been influenced by William James more than any other single philosopher. Much of the development we have traced here could be summarized by saying that the physics became more fully Jamesian. Fuchs writes,

> Far from thinking the world is an empty place, a place only with me in it. I think it is full of things, overflowing with things. *All* distinct things, from head to toe. And literally so. It is not a world made of

> six flavors of quarks glued together in various combinations. It is not a world that maps to a single algorithm running on Rob Spekkens's favorite version of Daniel Dennett's mechanistic cellular automaton. It is a world of heads and toes and doorknobs and dreams and ambitions and every kind of particular. (And that is not a typo: It is a world in which even dreams and ambition have substance.)

This is a Jamesian ethos, brought into a modern science library. James himself quoted the poet Benjamin Paul Blood:

> Not unfortunately the universe is wild,—game-flavored as a hawk's wing. Nature is miracle all; the same returns not save to bring the different. The slow round of the engraver's lathe gains but the breadth of a hair, but the difference is distributed back over the whole curve, never an instant true,—ever not quite.

There is a direct line from the poetic exhortation "ever not quite!" to discarding von Neumann measurements in favor of POVMs. But, oh, the training that an ear requires to follow that verse! It is as yet a lonely kind of poetry to appreciate.

Recall our earlier discussion of how the Born rule serves as a replacement for the classical Law of Total Probability. As we saw, a reference measurement for the simplest possible quantum system must have at least four possible outcomes. This is a fact about the quantum formalism that is more primitive than any choice of quantum state: One fixes the size of a sample space before writing a probability distribution across it. QBism takes this as a *lesson about nature*, at least indirectly. Were the character of the natural world different, the number of outcomes required for a minimal reference measurement would also differ.

The QBist essays at reconstructing quantum theory have presumed that, once we fix the size of a reference measurement's outcome set, only a single reference measurement is necessary. That is, in order to characterize the preparation of a system, an agent is not required to contemplate multiple, mutually exclusive actions and assign probabilities to the potential outcomes of each one. This is a point that, I think, warrants a deeper conceptual probe. For starters, it distinguishes the QBist work from much that has been done in the "convex-operational" or "generalized probabilistic theory" tradition (Wilce 2016). In a more historical vein, one might also ask whether it is contrary to the spirit of Bohr's complementarity. Would it be so bad if it were? Or, should the reference measurement be thought of as "complementary" to all others (Fuchs 2017b)? Explorations in this area might reinvigorate discussions of the founders, while also suggesting new ways in which quantum theory could be generalized.

Another area which might benefit from a conceptual or philosophical inquiry is the meaning of a "classical limit" in QBism. We might provisionally define a "quasiclassical regime" as a type of circumstance in which an agent can act as if their participation were inconsequential. To illustrate this idea, suppose that I receive a box from a friend who promises me that it contains delicious cake. I place the box on my table, stand over it and prepare to open it. I might have any edge of the box facing me; when I reveal the cake, I could see it from any one of many different angles. Prior to opening the box, my expectations for what I might find given different choices of angle all meld together. If I expect that conditional upon one choice of angle a fondant rose will be on my left, then I also expect that another choice of angle will find it on my right. Once I do see the decorations, I can say confidently what I would have seen had I opened the box at a different angle instead.

This is quite unlike how choices of angle work when sending photons through polarized filters or atoms through Stern–Gerlach magnets! How, then, do we interpolate between the two regimes?

A big part of the answer appears to be *imprecision*. This can manifest in many different ways: Fuzziness of a quantum-state assignment, of measurement operators, in the specification of a Hamiltonian, in how we draw the boundary around a system, even in which part of a system we sample first (Peres 1993, Rau 2021). All of these varied kinds of vagueness tend to make a situation "more classical": An agent can act as though unperformed measurements still had outcomes, and fear less risk of contradiction when doing so. Assumptions that uncertainty can be explained as ignorance of "hidden variables" become more practically tenable. Using the Law of Total Probability instead of the Born rule to intermediate between different choices of action starts to be an acceptable dodge. As the situation becomes more classical, the extent of what is effectively co-given expands.

There seem to be many axes along which this transition might occur, with only hints as yet about a unifying picture. What, I wonder, might spark the development of that picture?

As Pienaar points out elsewhere in this volume, there has so far been rather little investigation by phenomenologists into the details of probability theory. To wax optimistic, this means the frontier is wide open. We might play off Pienaar's discussion in the following way:

Part of the atmosphere of an event can be a sense of rarity. Shock, after all, is as undeniable an emotion as any other, and melancholy can derive from the recognition that an event would once have been surprising but is now commonplace. Probability theory furnishes a means of attaching numerical degrees of *surprisingness* to events. By adopting this technique, we say that arithmetic manipulations of those gradations make sense.

To take Pienaar's example, I might regard a set of three billiard balls as being a concrete instance of the integer 3, provided that my interest is in counting them; approaching them in the context of a counting process carries the implication that I could have imagined the count coming out differently. If I am willing to attach the integer 3 to the situation, then I am saying that at least some of the mathematical manipulations of that integer would have a physical meaning. Part of the context of the experience of "finding three billiard balls" is recognizing that "finding four billiard balls" would be similar.

Here, we are edging toward the idea that part of the context of assigning one probability value to an event is recognizing that we could have assigned another. ("The inevitable stales", wrote Benjamin Paul Blood.) This, in turn, suggests that *gambling on one's own future probability assignments* is fair game. Probability theory does address this contingency, with van Fraassen's *reflection principle* being a key idea (Goldstein 1983, Shafer 1983, van Fraassen 1984). We have employed this principle in discussing how the strangeness of quantum theory might wash out into quasi-classicality (Fuchs & Schack 2012), and we have speculated that it may be useful in the reconstruction project (Stacey 2019b). The full story of it, I suspect, remains to be told.

2.8 Ever Not Quite

Basically nothing posted on the arXiv before 2009 should be cited as an example of QBism, no matter who the authors are. All the older writings fail in one or another readily apparent way to recognize at least one point that later investigation found to be necessary for a self-consistent interpretation of quantum mechanics. That said, various technical matters raised in those pre-QBist papers continue to be interesting even though the metaphysical frontier has left their original motivations far behind.

Why philosophize about physics, if there is not at least the occasional opportunity to do better physics? Perhaps most of the job is merely checking the plumbing, but now and then, a plumber ought to help build a home. Every physicist faces the challenges of setting priorities and communicating results. Every teacher of physics must begin somewhere. Surely the history and philosophy of our subject can redound to our benefit when these questions inconveniently arise. I am the wrong person to ask if we can practice physics more kindly, but surely we can do it more justly, and with a greater awareness of our foibles.

For conversations and correspondence, I thank Gabriela Barreto Lemos, Carlton Caves, John B. DeBrota, Christopher Fuchs, Jacques Pienaar and Rüdiger Schack.

Notes

1 Busch discovered this result while salvaging von Neumann's attempt at a no-hidden-variables proof. The structure of POVMs allows one to make an additivity condition that replaces the unwarranted assumption in von Neumann's argument, the postulate that had been criticized by Hermann and probably by Einstein (Hermann 1935a, 1935b, Wick 1995, Mermin & Schack 2018, Stacey 2019c).

2 For a pedagogical introduction to the "probabilities for future probabilities" thinking, see Stacey (2015, §§ 5.1). For a technical result motivated by "expected changes in expectation" concerns, see (DeBrota & Stacey 2019).

3 Caves has confirmed to me (e-mail, 24 November 2019) that he does not subscribe to QBism.

4 Indeed, such a list of predecessors occurs in Fuchs and Schack's *Reviews of Modern Physics* article (Fuchs & Schack 2009a, 2009b). Cursory inspection reveals that the views tallied there should not be identified with QBism or with each other. For example, the list includes Appleby's "Facts, Values and Quanta" (Appleby 2005), which takes pains to remain distinct from the other Quantum Bayesian writings of that period: "I should say that I do not entirely agree with them about that. [...] My feeling is that a completely satisfactory theoretical account has yet to be formulated." And so forth.

5 Asks Fuchs in 1998b, "Why is the world so constituted that binary preparations can be put together in a way that the whole is more than a sum of the parts, but never more so than by $Q \approx 0.202$ bits?" Note that the bound of $Q \approx 0.202$ bits is attained when the two alphabet states are drawn from two different MUB; while by another measure of "quantumness" in that paper, the average global fidelity, two qubit states are "most quantum with respect to each other" when they are drawn from a SIC.

6 At least, so it was in the year 2000. Nowadays, rather than the Knuthian sense of a procedure for a machine of known architecture, published so its performance can be analyzed, an algorithm is a trade secret that runs "in the cloud" and whose goal is to disguise injustice and inequality as objective logic (O'Neil 2016). O tempora, o mores.

References

Appleby, M. (2005). Facts, values and quanta, *Foundations of Physics 35*, 627, arXiv:quant-ph/0402015.

Appleby, M., C. A. Fuchs, B. C. Stacey and H. Zhu (2017). Introducing the Qplex: A novel arena for quantum theory, *European Physical Journal D 71*, 197, arXiv:1612.03234.

Baez, J. C. (2003). Bayesian Probability Theory and Quantum Mechanics, http://math.ucr.edu/home/baez/bayes.html

Barnum, H., S. Beigi, S. Boixo, M. B. Elliott and S. Wehner (2010). Local quantum measurement and no-signaling imply quantum correlations, *Physical Review Letters 104*, 140401, arXiv:0910.3952.

Brandão, F. G. S. L. and A. Harrow (2017). Quantum de Finetti theorems under local measurements with applications, *Communications in Mathematical Physics 353*, 469–506, arXiv:1210.6367.

Bub, J. (2019). "Two dogmas" redux, arXiv:1907.06240.

Bub, J. and I. Pitowsky (2010). Two Dogmas About Quantum Mechanics. In *Many Worlds?: Everett, Quantum Theory, & Reality*. Oxford: Oxford University Press. arXiv:0712.4258.

Busch, P. (2003). Quantum states and generalized observables: A simple proof of Gleason's theorem, *Physical Review Letters 91*, 120403, arXiv:quant-ph/9909073.

Cabello, A. (2017). Interpretations of Quantum Theory: A Map of Madness. In *What Is Quantum Information?* Cambridge: Cambridge University Press. arXiv:1509.04711.

Caticha, A. (2007). From objective amplitudes to Bayesian probabilities, *AIP Conference Proceedings 889*, 62, arXiv:quant-ph/0610076.

Caves, C. M., C. A. Fuchs, K. K. Manne and J. M. Renes (2004). Gleason-type derivations of the quantum probability rule for generalized measurements, *Foundations of Physics 34*, 193–209, arXiv:quant-ph/0306179.

Caves, C. M., C. A. Fuchs and R. Schack (2002a). Quantum probabilities as Bayesian probabilities, *Physical Review A 65*, 022305, arXiv:quant-ph/0106133.

Caves, C. M., C. A. Fuchs and R. Schack (2002b). Unknown quantum states: The quantum de Finetti representation, *Journal of Mathematical Physics 43*, 4537, arXiv:quant-ph/0104088.

Caves, C. M., C. A. Fuchs and R. Schack (2007). Subjective probability and quantum certainty, *Studies in the History and Philosophy of Modern Physics 38*, 255–74, arXiv:quant-ph/0608190.

Conover, E. (2018). It's official: We're redefining the kilogram, *Science News*.

de la Torre, G., L. Masanes, A. J. Short and M. P. Müller (2012). Deriving quantum theory from its local structure and reversibility, *Physical Review Letters 109*, 90403, arXiv:1110.5482.

DeBrota, J. B., C. A. Fuchs and B. C. Stacey (2020). Symmetric informationally complete measurements identify the irreducible difference between classical and quantum systems, *Physical Review Research 2*, 013074, arXiv:1805.08721.

DeBrota, J. B., C. A. Fuchs and B. C. Stacey (2021). The varieties of minimal tomographically complete measurements, *International Journal of Quantum Information 19*, 204005, arXiv:1812.08762.

DeBrota, J. B. and B. C. Stacey (2018). FAQBism, arXiv:1810.13401.

DeBrota, J. B. and B. C. Stacey (2019). Lüders channels and the existence of symmetric-informationally-complete measurements, *Physical Review A 100*, 062327, arXiv:1907.10999.

Dillon, M. C. (1998). *Merleau-Ponty's Ontology*. Chicago, IL: Northwestern University Press.

Frauchiger, D. and R. Renner (2018). Quantum theory cannot consistently describe the use of itself, *Nature Communications 9*, 3711, arXiv:1604.07422.

Fuchs, C. A. (1997). Nonorthogonal quantum states maximize classical information capacity, *Physical Review Letters 79*, 1162, arXiv:quant-ph/9703043.

Fuchs, C. A. (1998a). Information gain vs. state disturbance in quantum theory, *Fortschritte der Physik 46*, 535–65, arXiv:quant-ph/9611010.

Fuchs, C. A. (1998b). Just *two* nonorthogonal quantum states, arXiv:quant-ph/9810032.

Fuchs, C. A. (2002). Quantum mechanics as quantum information (and only a little more), arXiv:quant-ph/0205039.

Fuchs, C. A. (2003). *Notes on a Paulian Idea.* Växjö: Växjö University Press.

Fuchs, C. A. (2010). QBism, the perimeter of Quantum Bayesianism, arXiv:1003.5209.

Fuchs, C. A. (2014). *My Struggles with the Block Universe.* Edited by B. C. Stacey, with a foreword by M. Schlosshauer, arXiv:1405.2390.

Fuchs, C. A. (2017). On Participatory Realism. In *Information and Interaction: Eddington, Wheeler, and the Limits of Knowledge.* Berlin: Springer, arXiv:1601.04360.

Fuchs, C. A. (2017b). Notwithstanding Bohr, the reasons for QBism, *Mind and Matter 15*, 245–300, arXiv:1705.03483.

Fuchs, C. A., N. D. Mermin and R. Schack (2014). An introduction to QBism with an application to the locality of quantum mechanics, *American Journal of Physics 82*, 749–54, arXiv:1311.5253.

Fuchs, C. A. and A. Peres (1996). Quantum state disturbance vs. information gain: Uncertainty relations for quantum information, *Physical Review A 53*, 2038–45, arXiv:quant-ph/9512023.

Fuchs, C. A. and A. Peres (2000). Quantum theory needs no "interpretation", *Physics Today 53*, 70–71.

Fuchs, C. A. and R. Schack (2009a). Quantum-Bayesian coherence, arXiv:0906.2187. A condensed version was later printed as *Reviews of Modern Physics* 85 (2013). 1693.

Fuchs, C. A. and R. Schack (2009b). Priors in quantum Bayesian inference, *AIP Conference Proceedings 1101*, 255–59, arXiv:0906.1714.

Fuchs, C. A. and R. Schack (2012). Bayesian conditioning, the reflection principle, and quantum decoherence, *Probability in Physics*, 233–47, arXiv:1103.5950.

Fuchs, C. A., R. Schack and P. F. Scudo (2004). De Finetti representation theorem for quantum-process tomography, *Physical Review A 69*, 062305, arXiv:quant-ph/0307198.

Fuchs, C. A. and B. C. Stacey (2019). QBism: Quantum Theory as a Hero's Handbook. In *Proceedings of the International School of Physics Enrico Fermi, Course 197 – Foundations of Quantum Physics*, edited by E. M. Rasel, W. P. Schleich and S. Wölk. Amsterdam: Italian Physical Society. arXiv:1612.07308.

Fuchs, C. A. and B. C. Stacey (2020). QBians Do Not Exist, arXiv:2012.14375.

Garrett, A. J. M. (1993). Making Sense of Quantum Mechanics: Why You Should Believe in Hidden Variables. In *Maximum Entropy and Bayesian Methods (Paris, France, 1992)*, edited by A. Mohammed-Djafari and G. Demoment. Alphen aan den Rijn: Kluwer.

Gleason, A. M. (1957). Measures on the closed subspaces of a Hilbert space, *Indiana University Mathematics Journal* 6, 885–93.

Goldstein, M. (1983). The prevision of a prevision, *Journal of the American Statistical Association 78*, 817–19.

Good, I. J. (1983). 46,656 Varieties of Bayesianism. In *Good Thinking: The Foundations of Probability and Its Applications.* Minnesota: University of Minnesota Press.

Granström, H. (2006). *Gleason's Theorem.* Master's thesis, Stockholm University.

Hermann, G. (1935a). Die Naturphilosophischen Grundlagen der Quantenmechanik, *Die Naturwissenschaften 42*, 718–21.

Hermann, G. (1935b). Die Naturphilosophischen Grundlagen der Quantenmechanik, *Abhandlungen der Fries'schen Schule* 6, 69–152.

Hudson, R. L. and G. R. Moody (1976). Locally normal symmetric states and an analogue of de Finetti's theorem, *Zeitschrift für Wahrscheinlichkeitstheorie und Verwandte Gebiete 33*, 343–51.

Jaeger, G. (2019). Information and the reconstruction of quantum physics, *Annalen der Physik 531*, 1800097.

Jaynes, E. T. (2003). *Probability Theory: The Logic of Science*. Cambridge: Cambridge University Press. Cited in CFS 2002 as unpublished but available at the Jaynes memorial website.

Jeffrey, R. (1989). Reading "Probabilismo", *Erkenntnis 31*, 225–37.

Koberinski, A. and M. P. Müller (2018). Quantum Theory as a Principle Theory: Insights from an Information-Theoretic Reconstruction. In *Physical Perspectives on Computation, Computational Perspectives on Physics*. Cambridge: Cambridge University Press. arXiv:1707.05602.

Laudisa, F. and C. Rovelli (2019). Relational Quantum Mechanics, *Stanford Encyclopedia of Philosophy*. Winter 2021 Edition, Edward N. Zalta (ed.), https://plato.stanford.edu/archives/win2021/entries/qm-relational/.

Leifer, M. S. and R. W. Spekkens (2013). Towards a formulation of quantum theory as a causally neutral theory of Bayesian inference, *Physical Review A 88*, 052130, arXiv:1107.5849.

Mermin, N. D. (2018). Making better sense of quantum mechanics, *Reports on Progress in Physics 82*, 012002, arXiv:1809.01639.

Mermin, N. D. and R. Schack (2018). Homer nodded: von Neumann's surprising oversight, *Foundations of Physics 48*, 1007–20, arXiv:1805.10311.

O'Neil, C. (2016). *Weapons of Math Destruction*. New York, NY: Crown Publishing.

Peierls, R. (1991). In defence of "measurement", *Physics World 4*, 19–20.

Peres, A. (1993). *Quantum Theory: Concepts and Methods* Alphen aan den Rijn: Kluwer.

Pitowsky, I. (2002). Betting on the outcomes of measurements: A Bayesian theory of quantum probability, arXiv:quant-ph/0208121.

Ramsey, F. P. (1931). Truth and Probability. In *The Foundations of Mathematics and other Logical Essays*, edited by R. B. Braithwaite. London: Routledge & Kegan Paul Ltd.

Rau, J. (2021). *Quantum Theory: An Information Processing Approach* Oxford: Oxford University Press.

Riehl, E. (2016). *Category Theory in Context* New York, NY: Dover.

Schack, R., T. A. Brun and C. M. Caves (2001). Quantum Bayes rule, *Physical Review A 64*, 014305, arXiv:quant-ph/0008113.

Schaffer, K. and G. Barreto Lemos (2021). Obliterating thinginess: An introduction to the "what" and "so what" of quantum physics, *Foundations of Science* 26, 7–26.

Shafer, G. (1983). A subjective interpretation of conditional probability, *Journal of Philosophical Logic 12*, 453–66.

Spekkens, R. W. (2007). Evidence for the epistemic view of quantum states: A toy theory, *Physical Review A 75*, 032110, arXiv:quant-ph/0401052.

Stacey, B. C. (2015). *Multiscale Structure in Eco-Evolutionary Ecology*. PhD thesis, Brandeis University. arXiv:1509.02958.
Stacey, B. C. (2016a). Von Neumann was not a quantum Bayesian, *Philosophical Transactions of the Royal Society A 374*, 20150235, arXiv:1412.2409.
Stacey, B. C. (2016b). SIC-POVMs and compatibility among quantum States, *Mathematics* 4, 36, arXiv:1404.3774.
Stacey, B. C. (2019). On QBism and Assumption (Q). arXiv:1907.03805.
Stacey, B. C. (2019b). Quantum theory as symmetry broken by vitality, arXiv:1907.02432.
Stacey, B. C. (2019c). From Gender to Gleason: The Case of Adam Becker's *What Is Real?*, https://www.sunclipse.org/?p=2658.
Stacey, B. C. (2019d). Book review: *What Is Quantum Information?*, *Theoria 34*, 153–55.
Vaidman, L. (2019). Derivations of the Born Rule, http://philsci-archive.pitt.edu/15943/.
van Fraassen, B. C. (1984). Belief and the will, *Journal of Philosophy 81*, 235–56.
von Baeyer, H. C. (2016). *QBism: The Future of Quantum Physics*. Cambridge, MA: Harvard University Press.
Wick, D. (1995). *The Infamous Boundary: Seven Decades of Heresy in Quantum Physics*. Boston, MA: Birkhauser.
Wilce, A. (2016). A royal road to quantum theory (or thereabouts). arXiv:1606.09306.
Wright, V. J. and S. Weigert (2019a). A Gleason-type theorem for qubits based on mixtures of projective measurements, *Journal of Physics A 52*, 055301, arXiv:1808.08091.
Wright, V. J. and S. Weigert (2019b). Gleason-type theorems from Cauchy's functional equation, *Foundations of Physics 49*, 594–606.
Youssef, S. (1994). Quantum mechanics as Bayesian complex probability theory, *Modern Physics Letters A 9*, 2571–86, arXiv:hep-th/9307019.

3 QBism, Where Next?

Christopher A. Fuchs

3.1 Introduction

QBism [1–4] comes from humble beginnings. It was not born fully formed as Botticelli's Venus was, nor as the Bohmian and Everettian interpretations of quantum mechanics purport to be. Most strictly, QBism has always been a *research program*. Its long goal—to say something deep about the character of reality—was always at the top of the mind, but after 96 years of the quantum debate,[1] a slow and careful methodology seemed called for. Less cheap, guesswork ontology,[2] more surgical dissection of the theory and an honest reckoning with what its structure has been trying to tell us all along. QBism's tack was to ask over and over, what is it about the world that makes us well-advised to use the calculus of quantum mechanics for structuring our probabilities? Our *Bayesian* probabilities [5]. Said this way, it became clear that the pertinent way to move forward was to get the "epistemics" of the theory right before anything else: Getting reality right would follow for those who had patience enough to pass the marshmallow test [6].

In fact the first phase of QBism might be likened to a grand exercise in apophatic method: We won't yet tell you what reality is, but what it is not. In particular, on the supposition that probabilities (even instances of probability-1) are not part of reality, after careful analysis, we'll tell you lots of other things that cannot be part of it either. In this way, first the quantum state fell as a potential element of reality, then more surprisingly the operators used to describe quantum measurements, and then perhaps even shockingly Hamiltonians and unitary operators. So it went with nearly every *individual* term of the theory.

In July 2002, after a frenzied year of applying the method more thoroughly than ever, I compiled a 229-page samizdat [7] documenting how I was forced to my current position on these things by fighting tooth and nail with my colleagues Carlton Caves, David Mermin, and Rüdiger Schack. (They were all initially quite reluctant to go so far, and eventually Caves

DOI: 10.4324/9781003259008-4

even jumped ship.) For the abstract of it, I used a passage written already in October 2001:

> Collecting it up, it's hard to believe I've written this much in the little time since Växjö. I guess it's been an active time for me. I think there's no doubt that I've gone through a phase transition. For all my Bayesian rhetoric in the last few years, I simply had not realized the immense implications of holding fast to the view that "probabilities are subjective degrees of belief." Of course, one way to look at this revelation is that it is a *reductio ad absurdum* for the whole point of view, and this will certainly be the first thing the critics pick up on. But, you wouldn't have guessed less, I'm starting to view it as a godsend. For with this simple train of logic, one can immediately stamp out the potential reality/objectivity of any of a number of terms that might have clouded our vision. With so much dead weight removed, the little part left behind may finally have the strength to support an ontology.

How prescient that passage was: For after 21 years, I don't remember even a single standard-style philosopher of physics agreeing with my perceived "godsend." To their eyes the method of QBism left nothing whatsoever behind. I like the way my friend and QBism enthusiast Amanda Gefter [8–10] once put it, "Philosophers of physics will never accept the kind of reality QBism points them to because they can't see it as a form of reality of any variety."

So, what is that form of reality? To this day, even QBists have the thinnest glimmer of it—but that is why QBism is a project, a research program. The hints are strong that we will end up with something along the lines of what Will Durant once expressed so profoundly [11, p. 673]:

> The value of a [QBist pluriverse], as compared with a universe, lies in this, that where there are cross-currents and warring forces our own strength and will may count and help decide the issue; it is a world where nothing is irrevocably settled, and all action matters. A monistic world is for us a dead world; in such a universe we carry out, willy-nilly, the parts assigned to us by an omnipotent deity or a primeval nebula; and not all our tears can wipe out one word of the eternal script. In a finished universe individuality is a delusion; "in reality," the monist assures us, we are all bits of one mosaic substance. But in an unfinished world we can write some lines of the parts we play, and our choices mould in some measure the future in which we have to live. In such a world we can be free; it is a world of chance, and not of fate; everything is "not quite"; and what we are or do may alter everything.

But how to get there genuinely? That's where the real struggle still awaits us. Pluriverse? Unfinished world? All action matters? Everything is not quite? What do all these things really mean? How does one see them play out in the particular mathematical structure of quantum theory, and how can that structure *teach* philosophy something new to consider?[3]

QBism as it stands so far is only a scaffolding for building to those heights. Key to everything though, is that it has done so with a scientific precision which general philosophizing cannot. As I put it in [4], "One volunteers a philosophy, but one does not volunteer a physics. A physics either flies in the world, or it falters and is eliminated by Darwinian selection. That our most encompassing physical theory yet might lead to a philosophy once volunteered by temperament is a very powerful development."

Yet, reciprocally, QBism was not born in a vacuum. It has always needed inspiration from philosophy even in quite technical matters. Perhaps the most direct example of this can be found in the quantum de Finetti theorem [12], one of the first successes of the still-tentative QBist research program. This is because if one could prove a quantum de Finetti theorem, it would give a subjectivist Bayesian account of the phrase "unknown quantum state" in quantum state tomography [13]—a procedure that has often been flaunted as demonstrating that quantum states are objective after all.

Well, where did the idea of a quantum de Finetti theorem come from? It came from our having recently learned in 1998 of the classic representation theorem of de Finetti himself from the 1930s. In that case, the task at hand was to give a subjectivist account of the phrase "unknown probability distribution." De Finetti understood that the existence of such a theorem would be essential to shore up his doctrine of "probabilism" [14]. What few people realize, however, is that probabilism was not formulated in the service of statistical analysis, but was philosophy all the way down. As Richard Jeffrey wrote [15], "For de Finetti the years before publication of *Probabilismo* (1931) were a time of explosive mathematical activity fired by his philosophical vision." In fact during its formation [16], de Finetti was deeply under the influence of the Italian pragmatist movement of Giovanni Papini, Adriano Tilgher, Antonio Aliotta, and others of whom William James wrote so approvingly [17]. De Finetti himself confirmed in his autobiographical conclusion to *Probabilismo*,

> I found many things [in Burali-Forti and Mach] conforming to my ideas [but] there has recently been added a third and definitive base for my point of view: probabilism. It corrects and integrates the other two in the points that I could not accept: *those in which anything seemed to be considered as having an absolute value, transcending the psychological value it has for me, and independent of it.*
>
> [my emphasis]

Thus the development of QBism was under the influence of pragmatism before anyone was even conscious of it![4]

So without doubt, QBism has always needed help from philosophy and has on occasion gotten it. It is just that little to none of the help ever came from standard philosophy-of-physics circles. Philosophers of physics either wanted to make a fool of QBism[5] or at best temper its innovations until they could fit a semblance of them into a block-universe conception of nature seemingly so crucial to the milieu [18–21]. The years have taught us that, with a few exceptions,[6] there is probably not much reason to engage with that community further.

However, thankfully, standard philosophers of science do not exhaust the philosophical landscape relevant to physics. As Berghofer and Wiltsche write in the introduction to this volume, "The question is not whether QBists and phenomenologists should attempt to join forces, but what has taken us so long?" To be sure, it is unlikely that QBism has exhausted the inspiration it might gain from a deeper plunge into the pragmatist writings of William James, John Dewey, F. C. S. Schiller,[7] and Richard Rorty, or other non-phenomenological philosophers who also took "experience first" such as Shadworth Hodgson [22] or Richard Avenarius [23], but an unexpected workforce of eager phenomenologists is a most welcome development. Who knows what might arise from this synergy?

The big thought on my own mind is to use our new collaboration to much deepen our understanding of the "said form" of reality that current QBism seems to indicate. Will we land on Merleau-Ponty's "flesh" as the basic ontological element, instead of James's "pure experience," Dewey and Bentley's "transaction," or Whitehead's "actual occasions?" Or will it be still something else since none of those philosophies are responses to the particular mathematical details of quantum theory? Whatever the outcome, I would love our final story to be *so blue pure perfect* that it may well lead us to the next stage of physics. The volume collected here is our first step in that journey.

To that end, I will spend the better part of my contribution trying to give the most comprehensive exposition I can of where QBism stands *today*. This way, all of us will be on the same page before proceeding to deeper contributions. What exactly means QBism? Indeed, I had already emphasized that QBism is an evolving project, but a newcomer may not be aware of how drastic the evolution has been—one should read Blake Stacey's contribution to this volume [24] to see just how much so. As he writes in his conclusion, "Basically nothing posted on the arXiv before 2009 should be cited as an example of QBism, no matter who the authors are." That is, the mathematics is to be trusted, but maybe not the philosophy. Nonetheless, even some of the most authoritative expositions since 2009 [1–4,25,26]—all important readings in their own right and necessary for giving texture to the project—often have different emphases than we commonly use now in our research groups at UMass Boston and Royal Holloway University of London. Particularly, none of our writings to date have so emphasized the *normative aspects* of the mathematical structure of quantum theory. This will hopefully be corrected here.

After Section 3.2 (just described), which is the core of the chapter, Section 3.3 will reiterate how, though QBism may now have a firm hold on its "eightfold path" toward relieving the suffering at the "quantum interpretations conventions,"[8] it is still in search of its "noble truth": What is the precise ontology that compels the eightfold path? In this regard, it currently strikes me that the best way forward will consist of deeper analyses of the Wigner's friend scenario [27] and its extensions [28–30] than it has hitherto received. Thus Section 3.4 takes on Wigner's original argument from a QBist perspective to lay some groundwork for that future discussion. What becomes clear is that in QBism, Wigner and his friend must be treated symmetrically, much like Merleau-Ponty did with his hand argument. The chapter concludes in Section 3.5 with some hopeful remarks on what we might squeeze out of such further analyses.

3.2 QBism, Where Currently?

In Section 3.2.1 I first describe what is meant by the terms "agent" and "user of quantum mechanics" from a QBist perspective. (Boy, the phenomenologists might really help us here! The shortcomings will be quite apparent.) I then devote the remaining Sections 3.2.2–3.2.9 to the much more developed parts of QBism, explicating eight of its key tenets in significant detail.

3.2.1 *Agents and Users of Quantum Mechanics*

For QBism, the quantum formalism is a tool decision-making agents are advised to adopt in light of the peculiar uncertainties we find in our world. Namely, the theory guides its users in how to better gamble on the personal consequences ("experiences" or "lived experiences") of their actions on physical systems. Particularly in QBism, the quantum formalism plays a *normative* role for its users; it does not play a descriptive role concerning exactly how the world is. Its focus is on how a user *should* gamble.

But then what is a "user of the theory?" In the following, we will make a distinction between agents broadly speaking and the users of quantum mechanics:

- An *agent* is an entity that can freely take actions on parts of the world external to itself and for which the consequences of its actions matter to it.

- A *user of quantum mechanics* is an agent who applies the quantum formalism normatively for better decision making.

While our definition of a user is narrow, our definition of an agent is broad. An agent is any part of the world that can act autonomously on other parts of the world and can be analyzed fruitfully in teleological terms. Hence the phrase *matter to it*. Our definition does not rule out dogs, euglena, or even artificial life (if there can be such a thing) as agents. However, it does exclude a computer hard wired from the outside to deterministically "choose" its actions from a look-up table (essentially a Turing machine). It also excludes electrons: For though an electron may act autonomously on its external world, it is hard to think that the consequences of its actions matter to it.

On the other hand with regard to the notion of a user of quantum mechanics, as Khrennikov emphasized of QBism in Ref. [31], "The idea is that QM is something used only by a privileged class of people. Those educated in the methods of QM are able to make better decisions (because of certain basic features of nature) than those not educated in the methods of QM." By this light, Werner Heisenberg was an agent in 1924, and likely even a user of probability theory, but he was not yet a user of quantum mechanics. He himself did not enter that privileged class until 1925–1926. Will IBM ever be able to construct an entity without DNA, one made of silicon and copper, that would count as a user of quantum mechanics? Maybe. Maybe not. But that is irrelevant to the view of quantum theory as an addition to decision theory. To put it in a slogan, "Quantum mechanics is a user's manual ready and waiting for anything that can make use of it."

Hereafter, we will assume most of the agents we are speaking of are in fact users of quantum mechanics, but not always. So please stay aware of the context.

There certainly exists a range of definitions of agency in the literature, some overlapping with our definition to different degrees, some not remotely recognizable from our point of view (see any of the works of Daniel Dennett for instance). For comparison or contrast on how another physicist influenced by the phenomenological tradition—this time in the form of Martin Heidegger—has tackled the issue, see Refs. [32, 33].

One very interesting issue still in need of greater fleshing out is this. Suppose a team of scientists sharing notebooks, calculations, observations, etc., can be considered an "entity freely taking actions on its external world." Then according to the above definitions it can count as a single agent and even a user of quantum mechanics [34, 35]. Why shouldn't Napoleon's Grand Armée count as a single agent for some purposes? Why shouldn't the experimentalists and theorists of the first-ever continuous-variable quantum teleportation experiment [36] together count as a single user of quantum mechanics?

Or even backing off from quantum mechanics per se, consider this elegant example recently put forth by Jacques Pienaar [37]:

> Think of a highly trained volleyball team. The ball appears: who will take it? Where will they aim to hit it? For a team that is sufficiently cohesive, these decisions will be made without explicit speech. Subtle cues like bodily stances and small movements may contribute to indicating who should act, and how. Such movements may only be perceived subconsciously, and decisions made intuitively. The feeling would be that each member of the group knew that "that was the right move", yet nobody could say in which one of their minds the idea originated, and nor did it have to be explicitly communicated from one mind to all the rest. The same thought arose in all of them, all experienced it, but it is experience as a thought that belongs to none of them in particular, and the thought could only arise when they are all together in that situation.

Nonetheless, there may be unforeseen nuances in this way of thinking that *may or may not* contradict other bits and pieces of QBism. Surely more needs to be said! Perhaps the issue is in part captured by what Cavalcanti [30] calls a "Wigner bubble," but we leave it as a tantalizing possibility for one of the many technical directions still to be explored.

3.2.2 A Quantum State Is an Agent's Personal Judgment

In QBism, the *exclusive* purpose of the quantum formalism is to help an agent make better decisions.[9] That may sound instrumentalist, but it is not. It only means that if something is to be inferred about reality from the formalism, then it has to be done in the way of the archeologist: Why did this civilization design this tool for the terrain it lived in? What are the extant conditions that made *quantum theory as a tool* the natural choice for agents to gamble best with reality? With regard to this way of putting it, QBism excels above any of its sci-fi competitors [38], which really don't care where the formalism came from. History has shown that the rigorous use of the quantum formalism enables an agent to make more successful gambles in navigating the world than he would have otherwise. Why? It's QBism's take that when an answer is found we will finally understand John Wheeler's "How come the quantum?" [39]. But the first step toward the goal is to get straight what quantum theory is actually about. This is the reason for QBism's unflinching stance that the exclusive purpose of the quantum formalism is to help an agent make better decisions.

It is unfortunate that the term gamble evokes images of games of luck, but we use it in a sense that is meant to encompass any action an agent can take where the consequences matter to the agent. Any physics experiment is thus a gamble. As we will explain in more detail in Section 3.2.5, the quantum formalism can be viewed as an addition to classical decision theory [2, 4]. Particularly, following the approach to decision theory pioneered by Bruno de Finetti, Frank P. Ramsey, and L. J. Savage [40, 41], QBism takes all probabilities to be specific to the agent using them—they are personal, quantified degrees of belief. And when the subject matter is quantum mechanics, the probabilities involved are an agent's personal degrees of belief concerning their future measurement outcomes.

Personalist probabilities [42, 43] acquire an operational meaning by their use in decision making. A simple case of these considerations gives the so-called Dutch-book argument for the probability calculus. There, an agent's "probability" $P(D)$ for an event D is identified with her *valuation* of a lottery ticket which pays \$1.00 if D occurs and \$0.00 otherwise. A ticket like this is said to have a valuation of x if the agent (privately, perhaps secretly) commits herself to the following: Whenever offered a ticket, she will buy it for any amount less than \$$x$, and whenever asked to sell, she will do so for any amount offered larger than \$$x$. Similarly, one can speak of a "conditional probability" $P(D \mid H)$ in terms of the valuation of a lottery ticket for a compound event. In this case, \$$P(D \mid H)$ corresponds to the threshold price for a ticket that returns \$1.00 if both H and D occur, but if H does not occur, all transactions are returned to both buyer and seller.

What are the "correct" valuations for these tickets? Clearly they can only depend upon what the agent believes about the events H and D—it is about the agent's money after all and the risk she is willing to take. Most importantly, the valuations are not something determined by the agent's external world, but are genuinely personal, in some measure corresponding to her own autonomy.[10] The last thing one would want to say is that the valuations are somehow properties of the events H and D themselves or even the lottery tickets.

Further, note that the word "probability" as associated with these lotteries is *so far* a mere placeholder: It is a notion mildly evocative of how probability is used, but we might have called it by any other name since there is so little structure. As it stands, there is no mathematical specification to the merely symbolic $P(D)$.

Remarkably, however, the full structure of probability theory can in fact be derived by adding one simple normative requirement to an agent's valuations: That whatever assignments she makes for $P(H)$, $P(D)$, $P(D \mid H)$, etc., they should never be such that there exists a strategy of buying and selling which leads to a *sure loss* for the agent—i.e., a net loss for the agent no matter which outcomes occur. If such a strategy exists, then one says that a "Dutch book" can be made against the agent. If a Dutch book cannot be made, then one says that the agent is *Dutch-book coherent*, or simply *coherent*. Thus, when Dutch-book coherence is satisfied, one has every right to call these lottery valuations probabilities in the proper sense of mathematical probability theory.

More specifically, from the requirement of coherence, it follows that valuations must be nonnegative and bounded, $0 \leq P(H) \leq 1$ and $0 \leq P(D \mid H) \leq 1$, etc. When H and D are mutually exclusive, valuations must be additive

$$P(H, D) = P(H) + P(D). \tag{3.1}$$

Bayes' rule must be satisfied,

$$P(H)P(D|H) = P(D)P(H|D), \tag{3.2}$$

and similarly for all the more elaborate statements of probability theory.[11]

One such more-elaborate statement which arises directly from Dutch-book coherence is the Law of Total Probability (or LTP). Since the LTP will play a significant role in our later discussions, it is worth expressing it in the present context. Consider an agent who contemplates two sets of events $\mathcal{R} = \{R_1, R_2, \ldots, R_n\}$ and $\varepsilon = \{E_1, E_2, \ldots, E_m\}$, each of which is mutually exclusive within itself. (We use distinct indices n and m because there is no need for the sets to have the same cardinality.) Taking into account all the lotteries and compound lotteries which can be formed for these events, the agent derives that in order to be coherent she must satisfy

$$P(E_j) = \sum_{i=1}^{n} P(R_i)P(E_j \mid R_i) \qquad \forall j \in \{1, 2, \ldots, m\}. \tag{3.3}$$

Note one thing about this statement: In its very set-up, it is assumed that both the $\mathcal{R}$ event and ε event will come to be recognized—so that all lottery tickets can either be returned or paid off appropriately. This will be an important point when our discussion turns to the role of the Born rule in quantum theory in Section 3.2.5.

So much for characteristics of personalist probability theory in the most general setting. The way this makes a connection to quantum theory is through the fact that any quantum state can be identified with an "expectation catalog" for the outcomes of all possible quantum measurements—a point first emphasized by Schrödinger in 1935 [44]. QBism's strategy has thus been to understand probability first and quantum mechanics next, as directed by E. T. Jaynes [45]. After trials with other potential meanings for probability in the 1990s (as frequencies, propensities, objective chances, etc.), the originators of QBism ultimately settled on the idea that the only consistent and operationally meaningful interpretation of probability is the personalist Bayesian one. Consequently a quantum state, from the point of view of QBism, must be understood as a catalog of personal expectations if it is to be anything meaningful at all.

To put it succinctly, where Bruno de Finetti declared "probability does not exist" to convey the idea that probabilities cannot be considered properties of the agent's external world [42, 43], QBism declares "quantum states do not exist" in the same sense. Just as all probabilities are personal judgments, within QBism all quantum states are personal judgments. Of course, each probability assignment in the catalog must be Dutch-book coherent, but what is interesting in quantum theory is the way in which all these Dutch-book coherent judgments are implicitly tied to each other. We will explain this in more detail in Sections 3.2.3 and 3.2.5. The lesson for the moment though is that a quantum state is a personal judgment and not something mandated by the world external to the agent.

3.2.3 A Quantum Measurement Is an Agent's Action upon Its External World

A measurement is an action of an agent on its external world, where the consequences of the action (its *outcomes* in more usual terminology) matter to the agent. They should matter enough that the agent would be willing to gamble upon them.

Like our definition of agent before, this definition of measurement is very broad. Basically anything an agent can do to its external world—from opening a box of cookies, to crossing a street, to performing a sophisticated quantum optics experiment—counts as a measurement in our sense. The only thing that sets a quantum measurement (as normally construed) apart from a more pedestrian example is whether it is fruitful or worth one's while to apply the quantum formalism in analyzing it. In many situations, there is little to be gained by analyzing the consequences of one's actions through the aid of raw probability theory, much less the full-blown apparatus of the quantum formalism. Think of crossing a street: Already one has an intuitive-enough feel for when and when not to cross that pausing to make a calculation would be self-defeating. However, there are some situations where it is absolutely crucial to invoke the quantum formalism if one wants to try to be maximally prepared for nature's consequences. One such case is obviously the quantum optics lab, but another, more extreme case concerns an agent's actions on another living system when treated with the full generality of quantum theory. This, for instance, is the case for the Wigner's friend thought experiments, where the usage of quantum theory is pushed to its most extreme, and perhaps most revealing, analysis.

Let us work toward developing what we mean by applying the quantum formalism to analyze one's gambles. First, consider two general actions ε and ε' an agent might take on some object in his external world. For instance, the agent might be a boxer in training who is very carefully watching and analyzing his practice games before going out for a real match-up. The two actions contemplated might be aiming a right cross or a left hook at his opponent. Throughout, we will implicitly identify the actions an agent can take with the sets of potential experiences or consequences they will lead to for him.[12] Thus, ε and ε' can be expanded as $\varepsilon = \{E_1, E_2, \ldots, E_n\}$ and $\varepsilon' = \{E'_1, E'_2, \ldots, E'_m\}$, where the E_i and E'_j represent, for instance, the *mutually exclusive* potential subsequent jabs the boxer could receive from his opponent in return for his initial choice of ε or ε'. (Again, there is no need that the cardinality of the two sets be the same.) To say that an agent will gamble on the consequences of these actions means that he must settle his mind on probability distributions for the consequences of each: Sets of nonnegative numbers $P(E_i) \geq 0$ and $P(E'_j) \geq 0$ such that

$$\sum_{i=1}^{n} P(E_i) = 1 \quad \text{and} \quad \sum_{j=1}^{m} P(E'_j) = 1. \tag{3.4}$$

Through these numbers, the agent may then use decision theory to help make his choices of when to take which action, and this is where a normative theory for guidance plays its role. If the boxer incorporates the lessons learned from practice into his split-second thinking he will be much better prepared for his game.

The same is true of any two actions that might be considered in any setting. However, note that in the case of the boxer it is essentially impossible for him to perform both a left hook and a right cross at the same time. This is reminiscent of the complementarity one sees in some quantum measurements, like position and momentum. With this in hand, we are ready to be more precise about the meaning of "applying the quantum formalism."

To apply the quantum formalism means to ask for further guidance than unadorned probability theory, as represented by Eq. (3.4), can supply. This is because probability and decision theory take into account no details of the worlds in which they are used: Their edicts are independent of the characteristics of the worlds their agents inhabit. Quantum theory on the QBist view, however, is an *addition* to probability theory which very much takes into account the unique characteristics of our given world. If our world were a different world, agents would not be well advised to use the quantum formalism.

This is to say, to apply the quantum formalism an agent decides it is to his benefit to associate a Hilbert space with the object he intends to act upon and, by one means or another, establish an association

$$E_i \leftrightarrow \hat{E}_i \quad \text{and} \quad E'_j \leftrightarrow \hat{E}'_j \quad \forall\, i, j \tag{3.5}$$

between his expected experiences and the elements of some positive-operator valued measure (POVM) on the Hilbert space. (Note the hats on the right-hand symbols to denote that they are operators, rather than direct expressions of the experiences.) This means sets of operators $\{\hat{E}_i\}$ and $\{\hat{E}'_j\}$ such that

$$\langle \psi \,|\, \hat{E}_i \,|\, \psi \rangle \geq 0 \quad \text{and} \quad \langle \psi \,|\, \hat{E}'_j \,|\, \psi \rangle \geq 0 \quad \forall\, i, j, \text{ and } |\psi\rangle, \tag{3.6}$$

and

$$\sum_i \hat{E}_i = \hat{I} \quad \text{and} \quad \sum_j \hat{E}'_j = \hat{I}, \tag{3.7}$$

where $\hat{I}$ is the identity operator on the Hilbert space.[13]

POVMs are well-known and essential to quantum information science, but appear to be not so familiar to philosophers of physics. The key reason QBism adopts this notion of "applying the quantum formalism" is that POVMs represent the most general kind of measurement one can perform in quantum mechanics [46]. They therefore can model any action an agent can take upon its external world. Moreover, without POVMs (i.e., restricting to von Neumann measurements alone[14]), QBism's mathematical project of rewriting quantum theory to make its normative content manifest would be stymied. There will be much more to say on this in Section 3.2.5.

Once this much is done, the way quantum theory completes its normative guidance is by suggesting the agent strive to find a *single* quantum-state assignment $\hat{\rho}$ (pure or mixed) such that his declared probabilities $P(E_i)$ and $P(E'_j)$ may be calculated according to the Born rule:

$$P(E_i) = \text{tr}\, \hat{\rho}\hat{E}_i \quad \text{and} \quad P(E'_j) = \text{tr}\, \hat{\rho}\hat{E}'_j\,. \tag{3.8}$$

What this expresses is that the agent's *probability assignments* for the outcomes of various hypothetical measurements (or even, say, the performed and unperformed experiments Asher Peres contrasted in [47]) should not be loose and unhinged from each other. Similarly so with three hypothetical measurements, four hypothetical measurements, or any number. Indeed, when an agent contemplates how he will gamble upon the outcomes of one imagined measurement, he ought to take into account how he will gamble upon the outcomes of all others he might imagine.

Notice how this differs from a more common presentation of probabilities within quantum mechanics. There, the quantum state $\hat{\rho}$ is almost always treated as something sacrosanct, and probabilities are *derived from it*. Where the specification of the quantum state comes from in the first place is usually unquestioned—for instance, it is often simply supposed as in so many textbook exercises—but here the tables are turned. A quantum state, rather than having a status logically prior to the probabilities in Eq. (3.8), is something practically *born* with them, as the agent tries his best to take into account the peculiarities of the quantum world.

The normative thrust of quantum theory is that it is a kind of glue for probability assignments over and above the requirements of raw probability theory. From the QBist perspective, it is this glue which indicates the physical content of quantum theory. The world is such that agents should adopt this extra requirement on their gambles.

But what if given an agent's assignments $P(E_i)$ and $P(E'_j)$, there is no $\hat{\rho}$ such that Eq. (3.8) is satisfied? Then it means the agent should rethink why he believes what he believes. Perhaps his mapping from experiences to POVM elements in Eq. (3.5) should be rethought. Perhaps he should rethink the dimensionality of his chosen Hilbert space for the system. Perhaps he should simply rethink whether he really wants to assign $P(E_i)$ and $P(E'_j)$ in the first place. Do these valuations genuinely fit with his larger mesh of beliefs beyond those for the outcomes of the two contemplated measurements? Maybe further thinking or detailed calculation should be made before the bold leap of writing down $P(E_i)$ and $P(E'_j)$. However, on exactly what needs to be adjusted, quantum theory gives no guidance. It merely indicates that *something* needs to be adjusted, and this is what it means that the quantum formalism is a normative suggestion rather than a direct description of reality. Again, we will expand on this in Section 3.2.5.

One final thing for this section: Note that we were careful to apply the term measurement only to actions an agent takes on his external world. We thus require a strict separation between the agent performing the measurement and the system this measurement is an action upon. Why would this be? Why make such a restriction? In fact, it is no restriction at all, but a logical requirement for our setting. What could it mean for the agent to take actions upon himself without conceptually backing off from the very distinction between the agent (the autonomous seat of action) and its external world we have taken as our starting point? The two would have to be identified after all, contrary to the very spirit of QBism which aims to replace the notion of a block universe with a Jamesian pluriverse [4, 48–51].

Furthermore this separation means that there is no sense in QBism to an agent assigning a quantum state $\hat{\rho}$ to himself. This follows from the fact that there are no normative conditions of the form Eq. (3.8) to try to satisfy. The quantum foundations literature is rife with discussions of agents assigning quantum states to themselves, but in QBism it is a *contradictio in terminis*.

3.2.4 Measurement Outcomes Are Personal to the Agent

QBism owes much of its development to the influence of Wolfgang Pauli [52], but this is one idea of Pauli's that QBism ultimately had to disabuse itself of:

> [T]he objectivity of physics is fully preserved in quantum mechanics in the following sense. Although according to the theory it is in principle only the statistics of test series that are determined by law, the observer cannot influence the result of his observation—such as the response of a counter at a certain moment—even in the unpredictable individual case. The personal characteristics of the observer are in no way included in the theory; rather, the observation can be carried out using objective recording devices, the results of which are objectively available for everyone's inspection.
>
> [53, p. 122, but improved with Google Translate]

The first part of this relies on a frequentist understanding of probability, which is disappointing, but it is the last sentence that is particularly troublesome for QBism, as it was a block to our clear thinking for quite a number of years. With hindsight, that QBism would ultimately reject the notion of "results of which are objectively available for anyone's inspection," was an inevitable consequence of QBism's long progression from the personalist Bayesian notion of probability to the subjectivity of quantum states to the personal nature of measurement operators and the realization that unitary evolution itself is a personal judgment [54]. (Also see [49, Introduction].) The final straw came from contemplating how measurement operators could be subjective judgments at the same time as having results that obtain the same meaning for anyone who sees the outcomes. The tension was insurmountable, and QBism ultimately opted to take more seriously a different utterance of Pauli's: "[I]t is allowed to consider the instruments of observation as a kind of prolongation of the sense organs of the observer ..." [55]. QBism takes this phrase to its logical conclusion. Whereas Pauli, Bohr, and the other Copenhageners always had their classically describable measuring devices mediating between the registration of a measurement's outcome and the observer, for QBism the measuring instrument was taken to be literally part of the agent and the measurement outcome his or her direct experience.[15]

When an agent performs a measurement—that is, takes an action on its external world—the "outcome" of the measurement is the consequence of this action for his or herself. A measurement outcome is personal to the agent doing the measurement. Thus two agents cannot strictly speaking experience the same outcome. Different agents may inform each other of their outcomes and thus agree upon the consequences of a measurement, but a measurement outcome should not be viewed as an agent-independent fact which is available for anyone to see [56].

This tenet has led some commentators to claim mistakenly that QBism is a form of solipsism. This claim has been thoroughly refuted, not only by the QBists themselves [1, 34, 50, 57] but even by some professional philosophers [18, 20, 58, 59]. Among other reasons, QBism is not solipsism simply and immediately because of the premise that a measurement is an action on the *world external to the agent*. A QBist assumes the existence of an external world from the outset. Furthermore, the consequences of measurement actions are beyond an agent's control—the world can surprise the agent—as Pauli himself noted in his discussion of objectivity. The external world is thus capable of exhibiting genuine novelty in response to an agent's actions—i.e., the world and the agent cannot be identified with one another. (See [50, pp. 6–10] and [4, pp. 19–20], arXiv.org versions.) The entire basis for calling QBism solipsism is just short-circuited by the concepts QBism relies upon for its very starting point.

3.2.5 *The Quantum Formalism Is Normative Rather Than Descriptive*

We have already said much about how in QBism the quantum formalism is understood normatively, rather than descriptively as it is in nearly every other quantum interpretation.[16] Yet, this can be argued still more convincingly through a detailed analysis of the Born rule. In our discussion surrounding Eq. (3.8), the topic was the gluing together of probabilities for the outcomes of distinct quantum measurements. However, even for a single measurement the Born rule should be viewed as placing additional constraints on an agent's probability assignments—extra constraints upon which probability theory is simply silent. Indeed QBism sees the Born rule as the most fundamental addition to probability theory that the quantum world hands its agents. To understand the quantum formalism, one must first understand the Born rule.

As in Section 3.2.3, we consider the expression of the Born rule as it applies to the most general notion of quantum measurement: A positive operator-valued measure, or POVM, $\{\hat{E}_1, \hat{E}_2, \ldots, \hat{E}_n\}$. The agent's probability $Q(E_j)$ for the experience E_j, formally expressed by a positive semi-definite operator $\hat{E}_j$, is given by

$$Q(E_j) = \mathrm{tr}\, \hat{\rho}\hat{E}_j\,, \tag{3.9}$$

where $\hat{\rho}$ is some density operator. This time we use the notation $Q(E_j)$ to call attention to the fact that our probability assignment comes from a quantum mechanical calculation. However just as before, the operator $\hat{\rho}$, the distribution $Q(E_j)$, and the association

$$\{E_1, E_2, \ldots, E_n\} \leftrightarrow \left\{\hat{E}_1, \hat{E}_2, \ldots, \hat{E}_n\right\} \tag{3.10}$$

are all to be understood as subjective judgments. If for whatever reason the chosen $\hat{\rho}$, the $Q(E_j)$, and the $\hat{E}_j$ do not satisfy Eq. (3.9), then the guidance of quantum theory is that the agent should modify at least one of the terms. This guidance however does not prescribe which of the terms to modify or how to modify them; none of the terms should be considered as having logical priority over any of the others. Indeed, all the terms live at the same conceptual level. In this way quantum theory's role is analogous to the kind of guidance Dutch-book coherence gives: It speaks of the recommended relations between the terms, but nothing of their exact values, the latter being the provenance of the agent using the theory.

This is QBism's stance, but it is one thing to declare it, and another to make it compelling to one's readers. Why, unless one has already drunk the Kool-Aid, should $\hat{\rho}$, the $Q(E_j)$, and the $\hat{E}_j$ all live at the same conceptual level? Moreover, our whole description begs a question: If the POVM $\{\hat{E}_1, \hat{E}_2, \ldots, \hat{E}_n\}$ is a personal judgment, what exactly is it a judgment of? These questions can be most easily answered by introducing the notion of a "reference apparatus" for a given quantum system.[17]

A *reference apparatus* for a d-dimensional quantum system is *any* POVM $\{\hat{R}_i\}$ with d^2 linearly independent measurement operators, along with an associated set of linearly independent post-measurement quantum states $\{\hat{\sigma}_i\}$. One can prove that there are infinitely many such structures for each d. However, the word "reference" refers to the requirement that once a choice of one of these is made, it must stay fixed for all the agent's calculations—this is like choosing a reference frame in relativity theory. What is significant about such an apparatus is that any operator can then be written as a *unique* linear combination of either of the two sets. This follows because each set of operators form a complete (though nonorthogonal) basis for the d^2-dimensional vector space in which $\hat{\rho}$ lives. Moreover the expansion coefficients for $\hat{\rho}$ in terms of the $\hat{R}_i$ are determined by none other than the Born rule probabilities $P(R_i) = \mathrm{tr}\hat{\rho}\hat{R}_i$ themselves. This follows because $(\hat{A}, \hat{B}) = \mathrm{tr}\hat{A}^\dagger B$ satisfies the properties of an inner product for the space.

This sets the stage for thinking of a quantum state as not only an expectation catalog for the outcomes of all possible measurements (Schrödinger again), but as a *single* probability distribution $P(R_i)$ full stop. With such an identification

$$\hat{\rho} \leftrightarrow P(R_i), \tag{3.11}$$

one sees that it is a matter of course for a QBist to treat a quantum state as subjective as any personalist Bayesian probability distribution: For then a quantum state *just is* a probability distribution—it has no further content above and beyond that (contra [60] to some of QBism's critics [61]).

Indeed, this mapping gives a means to rewrite the entire Born rule in purely probabilistic terms. To see this, consider the following scenario. An agent has a physical system for which she plans to carry out either one of two mutually exclusive actions or protocols on it. In the first protocol, she imagines measuring the system directly according to the operators $\{\hat{E}_j\}$ of any general POVM and thereby obtaining some outcome j. However in a second, alternative protocol, she imagines cascading two measurements—a kind of one-two punch—first performing the reference apparatus and only subsequently performing the POVM $\{\hat{E}_j\}$ of the first protocol. In this case, she would obtain two outcomes i and j, not one (Figure 3.1).

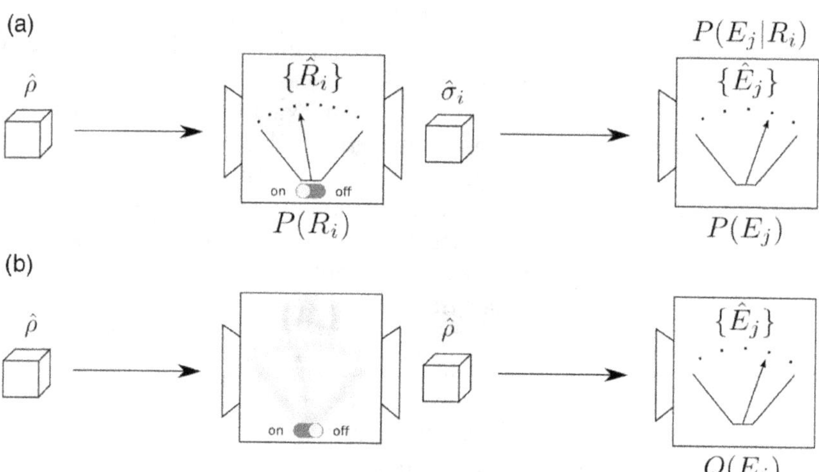

Figure 3.1 Two distinct experiments. In QBism, the Born rule is not about either one individually, but rather about the connections between their probabilities. In the top experiment, the reference device is turned on so that there are three probabilities in its telling: $P(R_i)$, $P(E_j \mid R_i)$, and $P(E_j)$. They must satisfy the Law of Total Probability, Eq. (3.12). However, in the bottom experiment the reference device is turned off—there is only one probability $Q(E_j)$ in its story. The Born rule is the narrative glue that ties the two stories together.

It must be noted, probability theory in the abstract *provides no consistency conditions* between the agent's expectations for the outcomes of these two protocols. Let P denote all of the agent's probability assignments for the consequences of following the two-step protocol and Q those for the single-step protocol. Then there is no doubt that the $P(E_j)$, $P(R_i)$, and $P(E_j \mid R_i)$ must satisfy the LTP,

$$P(E_j) = \sum_{i=1}^{d^2} P(R_i)P(E_j \mid R_i), \tag{3.12}$$

else the agent could be Dutch-booked as we explained in Section 3.2.2. This follows because all the lotteries and compound lotteries that define these valuations can in principle be settled. It is a statement purely about coherence and has nothing to do with physics.

Yet, there is no a priori reason to believe that the agent should assign

$$Q(E_j) = P(E_j). \tag{3.13}$$

In the second protocol, there are no compound lotteries by which to even define the numbers $P(E_j \mid R_i)$. Thus, if such an identification were to be made, it would require input going beyond probability theory: It would require *physics* [62, 63]. In fact, if one considers classical physics in Liouville form and takes the phase-space points of a system to be an analog of our reference apparatus, one arrives at just such an identification.

Eq. (3.13) then is not a statement of probability theory, but a statement of physics! Classical physics.[18] One thus suspects that quantum theory, if it gives anything at all beyond the negative statement

$$Q(E_j) \neq P(E_j), \tag{3.14}$$

it will give something to replace Eq. (3.13).

In fact, quantum theory is not at all silent on relating the $Q(E_j)$ to the $P(R_i)$ and the $P(E_j \mid R_i)$. Let us exhibit one such relation explicitly. There are many possibilities depending upon the choice of reference apparatus, but a unique property of them all is that Eq. (3.13) is *never* satisfied [64]. First consider the second protocol. Suppose that upon learning the outcome R_i of the first measurement, the agent ascribes a quantum state $\hat{\sigma}_i$ to the quantum system going forward to the $\hat{E}_j$ measuring device. Then, the reference apparatus will be uniquely characterized by an invertible matrix

$$[\Phi^{-1}]_{ij} := \operatorname{tr} \hat{R}_i \hat{\sigma}_j, \tag{3.15}$$

and an elementary calculation reveals that the Born rule, Eq. (3.9), becomes

$$Q(E_j) = \sum_{i=1}^{d^2} \left[\sum_{k=1}^{d^2} [\Phi]_{ik} P(R_k) \right] P(E_j \mid R_i), \tag{3.16}$$

where

$$P(E_j \mid R_i) = \mathrm{tr}\, \hat{\sigma}_i \hat{E}_j . \tag{3.17}$$

Moreover, we now know what the judgments $\hat{E}_j$ are judgments of. They are just symbolic ways to express the $P(E_j \mid R_i)$, $i = 1,\ldots,d^2$, the agent's conditional lottery valuations defined in the second protocol:

$$\hat{E}_j \leftrightarrow \{P(E_j \mid R_i)\}_{i=1}^{d^2} . \tag{3.18}$$

In all, Eq. (3.16) tells us that except for a specification of the details of the reference apparatus, the Born rule is purely a relation between probabilities and conditional probabilities. Particularly, it means that the content of the Born rule has everything to do with Asher Peres's dictum, "unperformed experiments have no results" [47]. Unperformed experiments may not have results, but that does not mean an agent shouldn't be cognizant of how she would gamble if they were to be performed [2]. Moreover, see Figure 3.2 for the immense conceptual change Eq. (3.18) brings to the table of quantum interpretation: It is perhaps the most compelling reason for the QBist contention that a measuring device must be understood as a part of the agent herself.

The form of Eq. (3.16) can be made all the more striking by introducing a completely "vectorized" notation, in which omitted subscripts signify an entire vector or matrix. That is, take $[P(E \mid R)]_{ji} = P(E_j \mid R_i)$, $[P(R)]_i = P(R_i)$, and $[Q(E)]_j = Q(E_j)$. Then the Born rule takes the compact form

$$Q(E) = P(E \mid R)\Phi P(R). \tag{3.19}$$

Note how this differs from the classical rule in Eq. (3.13), where the $Q(E)$ in the first protocol is equated with the expression derived from the LTP $P(E)$,

$$Q(E) = P(E \mid R)P(R). \tag{3.20}$$

In this language the only difference between the quantum and classical assumptions—they are both normative rules above and beyond Dutch-book coherence—is concentrated in the fact that $\Phi \neq I$ in quantum theory regardless of the reference apparatus. In fact, there is a minimal finite

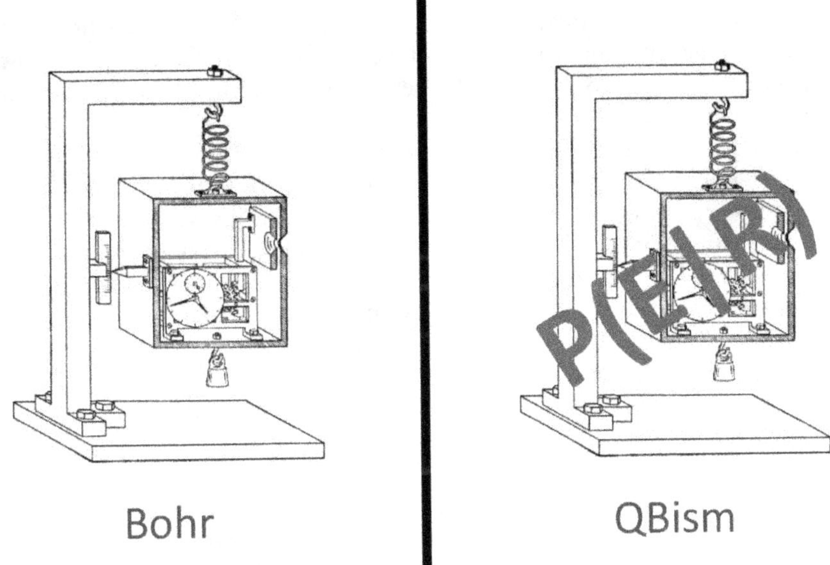

Figure 3.2 **Skeptic:** "QBists say that quantum states and measurement operators both represent personal judgments living at the same conceptual level. But that can't be right! Perhaps it is true that no one can see a quantum state, but anyone can walk into a laboratory and see a measuring device for exactly what it is. This is why Bohr [65] went to great lengths to depict measuring devices as heavy, bulky instruments, firmly bolted to their laboratory benches." **QBist:** "Bohr was wrong. Do you know how much implicit and explicit statistical analysis and calibration go into specifying the devices of even a small quantum optics experiment? Who can walk into the lab and see the personal, prior probabilities in a Bayesian experimentalist's head? The very identification of the device boils down to a very complex set of interlocking probability assignments *for him*.[19] Fortunately there is a formalism that makes the end result explicit: A set of conditional probability assignments $P(E_j \mid R_i)$, as in Eq. (3.18), no more or less personal than a quantum state itself."

separation between the two matrices [64].[20] Yet, therein lies a profound difference between these two possible worlds an agent might inhabit. The distinctive flavor of the quantum world will come into sight when we finally arrive at the Wigner's friend thought experiment in Section 3.4.

This tenet has led some commentators to claim that QBism is a form of instrumentalism. This, as with the claim of solipsism, is also easily refuted; see, e.g., Refs. [50, 59]. Indeed as emphasized in the Introduction, from its earliest days the very goal of QBist research has been to distill a statement about the character of the world from the fact that the gambling agents within it should use the quantum formalism [54]. Why would it be so? Whatever the answer turns out to be, it will be a statement about the particulars of reality. If the world were different in character, then the agents within that world would be better advised to use something other than quantum theory for their gambles.

The idea is a simple one for physicists: What QBism aims for is to reverse engineer from the formalism to a characterization of an ontology, while never straying from the progress it has made by viewing quantum theory as an addition to decision theory. This reverse engineering remains an active research program—a sign that QBism is a living subject.

Philosophers of physics seem to have more trouble than physicists with the stark admission that something is an ongoing project, as they seem to desire conclusive answers no matter how ill-considered. "Tell us your ontology *now*, or there is nothing to discuss!," as someone like Tim Maudlin would exclaim [66]. We like to quote Schrödinger [67] as a response:

> In an honest search for knowledge you quite often have to abide by ignorance for an indefinite period. Instead of filling a gap by guesswork, genuine science prefers to put up with it; and this, not so much from conscientious scruples about telling lies, as from the consideration that, however irksome the gap may be, its obliteration by a fake removes the urge to seek after a tenable answer. So efficiently may attention be diverted that the answer is missed even when, by good luck, it comes close at hand. The steadfastness in standing up to a *non liquet*, nay in appreciating it as a stimulus and a signpost to further quest, is a natural and indispensable disposition in the mind of a scientist.

Nonetheless, the methodology has already led to a number of strong ontological claims on the part of QBism—from the world being capable of genuine novelty and being in constant creation [4], to the Born rule expressing a novel form of structural realism. To put a term on the books and contrast with it the ontic and epistemic varieties of structural realism discussed by the philosophers [68], we might call this part of what QBism aims for a *normative structural realism* [69]. As the philosopher Craig Callender once paraphrased the idea, the Born rule would then represent nature's whisper to its agents [70].

3.2.5.1 *Symmetric Informationally Complete (SIC) Reference Apparatuses*

In the previous part of Section 3.2.5 we have been completely catholic in our choice of *reference apparatus*. Eq. (3.19) holds for of any of them. We did the general case without further detail so that the bare bones of the extra normativity implied by the Born rule would be on stark display. That extra normativity comes about in the relevance of the Born rule to Figure 3.1, rather than the LTP. The root cause of this is that we live in a world where *all action matters* (Durant)[21]—the reference apparatus as an action is conceptually ineliminable.

However, in much of the QBist literature going back as far as [71], extra emphasis has been given to a uniquely interesting class of reference apparatuses *should they exist*. These are the so-called symmetric informationally complete quantum measurements or SICs [72–74] in combination with assuming Lüders' rule for the post-measurement quantum-state assignments. Since the exploration of these structures is at the cutting edge of the technical side of QBism and some of the papers in this volume may reference this representation (as for instance Boge's does [75]), it seems worth giving an explanation of them here. In fact, it is a great laboratory for exhibiting the interplay emphasized in the Introduction between QBism's conceptual development and the mathematical demands it makes upon the theory. The philosophy really can't be divorced from the mathematics. Finally, this small excursion from the main goals of Section 3.2 will give an opportunity to show how the SICs play a conceptual role in some of the latest QBist thinking, where they give rise to a kind of "QBist Planck's constant" for finite dimensional quantum systems.

Comparing Eqs. (3.19) and (3.20) raises an interesting mathematical question for QBism. Depending upon which reference device the agent chooses for Eq. (3.19), it can be made to look less like or more like the classic LTP. If one could find a reference device so that $\Phi = I$, then one would have the LTP identically, and quantum theory's "extra" normative rule would not be extra at all. As already explained, though, there is no such reference apparatus [54].

So, how close Φ can be made to look like the identity matrix I? The answer would establish an important fact about quantum mechanics, namely the "essential difference" between the Born rule and the classical intuition that would seek to set $Q(E) = P(E)$ if it could.

In [64], this question was quantified by introducing a class of distance functions based on unitarily invariant norms [76],

$$d(I,\Phi) = \| I - \Phi \| . \tag{3.21}$$

A unitarily invariant norm is a matrix norm for square matrices such that $\| UXV \| = \| X \|$ for any unitary matrices U and V. Such norms form the most significant class of norms in matrix analysis [76]. The class includes the trace norm, the Frobenius norm, the operator norm, all the other Schatten p-norms, and the Ky Fan k-norms. The class of Φ matrices that achieve the minimal distance d_{Q} from the identity I define the *essential quantumness* of the Born rule: It establishes the essential gap between the classical and quantum normative rules.

To set ourselves up for expressing the essential quantumness, let us first define the notion of a SIC. A SIC is a POVM with d^2 elements for which all the $\hat{R}_i$ are rank-1, i.e., of the form

$$\hat{R}_i = c_i |\psi_i\rangle\langle\psi_i|, \tag{3.22}$$

and for which this stringent symmetry condition holds:

$$\text{tr}\, \hat{R}_i \hat{R}_j = c \quad \forall i \neq j. \tag{3.23}$$

When this is so, one can prove that the $\hat{R}_i$ must be linearly independent, i.e., they form a basis for the space of operators. One can further prove that the value of the c_i are fixed to $c_i = 1/d$ and

$$c = \frac{1}{d^2(d+1)}. \tag{3.24}$$

Thus a SIC with post-measurement states $\hat{\sigma}_i = d\hat{R}_i$ (which is what would come about if the measurement induces Lüders' rule in the wake of its action) will make for a perfectly good reference apparatus.

SICs have yet to be proven to exist in all finite dimensions d, but they are widely believed to exist without exception [74],[22] and have even been experimentally demonstrated in some low dimensions [77–79]. Proving their existence has been a consternating problem in that the effort has been going on for 23 years now, enmeshing more and more researchers [74]. However, the longer it goes, the more tantalizing the hope becomes that the payoff will be big in terms of previously undreamt of physics. For instance in the past six years a connection between SIC existence and Hilbert's (still open) 12th problem—a problem to do with algebraic number theory—has been uncovered and become quite the research rage [80, 81]. This strikes of some very deep mathematics going on here—maybe not Fermat's Last Theorem, but something in that direction—and leaves one with the question of why on earth basic quantum theory should care about exotic algebraic number fields?[23] What gem for physical understanding is hidden in this?

Thus we return to the question of the essential quantumness. Let Φ_{SIC} denote a Φ for the special case of a SIC reference apparatus. As you might guess by now, the result of [64] is that for all the distance measures considered in Eq. (3.21) and for all reference apparatuses,

$$d(I,\Phi) \geq d(I,\Phi_{\text{SIC}}) \tag{3.25}$$

with equality if and only if the reference device measures a SIC and outputs post-measurement states that are also elements of a SIC. Hence, if SICs do in fact exist in all dimensions,

$$d_Q = d(I, \Phi_{\rm SIC}). \tag{3.26}$$

In the past, QBism has indeed given special attention to reference apparatuses based on SICs, but in all cases previous to the result of Eq. (3.25), it was essentially for aesthetic reasons. For instance, note the particularly simple form Eq. (3.16) takes in this case:

$$Q(E_j) = \sum_{i=1}^{d^2} \left[(d+1)P(R_i) - \frac{1}{d} \right] P(E_j \mid R_i). \tag{3.27}$$

Long before the optimization problem was formulated, it was apparent that no representation of the Born rule could be simpler than this or more aesthetically similar to the LTP. To a physicist's nose, this was already a worthy-enough lead to follow wherever it might lead [82–86]. Now however, to see that this form is not only the solution of a conceptually important optimization problem at the root of QBism, but also appears to be deeply connected to Hilbert's 12th problem, seems astounding.[24] Where it will go, no one knows.

In the meantime, we can already get a glimpse of the meaning of d_Q by hijacking and modifying a parable due to Amanda Gefter about Niels Bohr's and John Wheeler's understanding of the "quantum principle" [87]. See Figure 3.3 for Gefter's original story. When I first saw this at a talk in South Africa, my jaw dropped because it so strikingly captured not what Niels Bohr was *on about*, but rather what QBism has always *been about*. After all, Bohr always had his "agencies of observation" (whose design can be expressed in common language suitably refined by the terminology of classical physics, etc., etc.) mediating between the subject and object, but Gefter's diagrams really went for the jugular—they got to the essential point, true subject-object ambiguity, nothing in between. The metaphor was on the mark.

There remains a question of detail, though, that is worthy of further exploration: Should the size of the circle in Figure 3.3 be symbolized by Planck's constant h or perhaps something else? A clue comes from John Wheeler[25] [90]:

> How come a value for the quantum so small as $\hbar = 1.05 \times 10^{-34}\, J \cdot s$? As well as ask why the speed of light is so great as $c = 3.00 \times 10^{6}\, m / s$! No such constant as the speed of light ever makes an appearance in a truly fundamental account of special relativity or Einstein geometrodynamics, and for a simple reason: Time and space are both tools

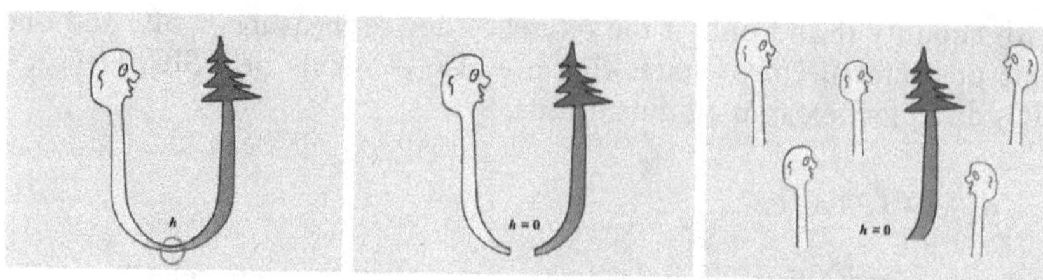

Figure 3.3 Amanda Gefter's parable [87] building on John Wheeler's famous U diagram [88, 89].

Left Frame: "You have a subject, and an object, and mostly they seem like separate things, but when you look closely enough you find that they are conjoined, that there's this piece of connective tissue where they can't be pulled apart because within that region there's no way to say which is which. ... The discreteness [we find in QM] is not a property of the object, it's a property of the subject-object relation. I think a useful way of thinking about this is to see Bohr's quantum of action as a kind of coupling constant between subject and object, between observer and observed."

Middle and Right Frames: "As you turn the strength of that coupling down, the area of overlap gets smaller and smaller, and if you turn it all the way down to zero, observer and observed can be neatly separated, it recovers classical physics, and it frees the object from its subject so that it can be shared by other subjects."

Moral: "But of course *h* is not zero and so we have to contend with the irreducible coupling between subject and object that Bohr would say is the very essence of quantum physics, and I would say ought to inform our philosophical discussions, in that realism is about an object decoupled from any subject, which you can't do, and idealism is about a subject decoupled from any object, which you also can't do. But what does that leave you?"

Analysis: "[A]s Wheeler delved deeper into the physics of gravitational collapse, he realized that you can't avoid singularities, and that at the singularity not only would spacetime itself disappear, but all the laws of physics as we know them—all conservation laws—would disappear, too. If spacetime is destroyed in gravitational collapse, it wasn't fundamental enough to be the basement-level ingredient of ultimate reality. And so Wheeler asked himself, when spacetime and all the conservation laws are destroyed in gravitational collapse, what survives? And the only answer he could come up with was: the quantum principle. What is the quantum principle? It seemed to go back to the fundamental coupling between subject and object."

> to measure interval. We only then properly conceive them when we measure them in the same units. The numerical value of the ratio between the second and the meter totally lacks teaching power. It is an historical accident. Its occurrence in equations obscured for decades one of nature's great simplicities. Likewise with $\hbar$! One day we will revalue $\hbar = 1.05\times10^{-34}\,J\cdot s$—as we downgrade $c = 3.00\times10^{6}\,m/s$ today—from constant of nature to artifact of history.

From this point of view, Planck's constant is just not something that has a meaningful size.[26] Indeed one will notice that neither *h* nor $\hbar$ made any appearance in this chapter until this very discussion.

Yet, Planck's constant surely serves a purpose in *applications* of quantum mechanics. What's the catch? For QBism, recall that quantum theory should be thought of as additive to classical theory. Here's the way we put it in [1]:

> The expectation of the quantum-to-classical transitionists is that quantum theory is at the bottom of things, and "the classical world of our experience" is something to be derived out of it. QBism says "No. Experience is neither classical nor quantum. Experience is experience with a richness that classical physics of any variety could not remotely grasp." Quantum mechanics is something put on top of raw, unreflected experience. It is additive to it, suggesting wholly new types of experience, while never invalidating the old. To the question, "Why has no one ever *seen* superposition or entanglement in diamond before?," the QBist replies: It is simply because before recent technologies and very controlled conditions, as well as lots of refined analysis and thinking, no one had ever mustered a mesh of beliefs relevant to such a range of interactions (factual and counterfactual) with diamonds. No one had ever been in a position to adopt the extra normative constraints required by the Born Rule. For QBism, it is not the emergence of classicality that needs to be explained, but the emergence of our new ways of manipulating, controlling, and interacting with matter that do.

It's in this that Planck's constant comes into the story for QBism.

Here is an example. Suppose an agent starts with a system she has been modelling as a classical harmonic oscillator when its energies are not too high (say $< E$). If she has gotten fine-tuned enough with her manipulations of it, she might decide it would be to her benefit to start treating it as a quantum system. But how should she perform such an "upgrade?" First and foremost she will need to associate a Hilbert space with it. But which one, what dimension? QBism takes the following as guidance. Since she's restricting her energies to be less than E, the potential trajectories of the system in a phase-space description of it will be bounded by an ellipse appropriate to E. To upgrade to Hilbert space, the agent simply chooses a dimension roughly equal to the area circumscribed by the ellipse divided by h, the quantum of action. It's rough guidance, but it's a start. One thing it emphasizes though is that h is discarded as soon as one starts to consistently invoke quantum theory.

So, in a way, it is Hilbert space itself that represents Gefter's circle, signifying the ineliminable connective tissue that is neither subject nor object. But we can do still better with the tools developed in this section at the same time as remaining faithful to the original metaphor. Even with a finite-dimensional Hilbert space established, the Born rule has a range of expression for how much the agent *appears to be* "coupled" (Gefter's term) to the system during the process of measurement. But one might say those are more "coordinate effects" than anything else. The role of expression (3.27) is to establish the genuine "strength" of the coupling. It is not a single number like $\hbar$ that we are dealing with now, but it still expresses some way to quantify a contingent feature of our world: Namely, just how much subject and object can be individuated from their substrata. This is the kind of contingent feature of nature calling out from quantum theory that the various philosophical ontologies courted in our concluding Section 3.5 could not have known about. The question now is how we might incorporate these details into the insights of the various "experience first" philosophers and end up with something even better?

3.2.6 *Unitary Evolution Expresses an Agent's Degrees of Belief*

In the years of slow progress that led to QBism, a number of arguments were harnessed to compel the idea that quantum states should be understood epistemically—it wasn't just a bald assertion made without any history. For instance, the quantum no-cloning theorem [91] and the existence of quantum teleportation [92] were two favorites for making the case in the early days. But if so, what was to be made of an analogous theorem in quantum information theory that unknown unitary operators could not be cloned [93, 94], just as quantum states could not? Or of the protocols that showed that unknown unitary operators could themselves be teleported [95], again just as quantum states? QBism's answer was, "Go it, and go it stronger!"[27] There was no choice [54] but to accept that unitaries too—like the quantum measurement operators before them—were cut from the same cloth as quantum states: All three entities represented personal judgments, not properties of the agent's external world. Moreover, they all must reside at the same conceptual level in the agent's attempt to conform to the normative edict of the Born rule.

As before, this is most easily seen through the formalism developed in Section 3.2.5. Suppose an agent assigns a quantum state $\hat{\rho}$ as the glue for her probability assignments to the outcomes of measurements she might perform on a system at time t_0. If the agent believes that at a later time t_1 she won't have gained or lost any predictability on balance for those same measurements, then the time lapsed must correspond to a transformation drawn from the symmetry group of the convex set of quantum states [84]—i.e., the projective unitary group. That is, at time t_1 she should assign a quantum state $\hat{U}\hat{\rho}\hat{U}^\dagger$ instead, for some unitary operator $\hat{U}$.

To put this statement into normative terms, it means assigning a probability distribution

$$P_{t_0}(R_i) = \text{tr}\,\hat{\rho}\hat{R}_i \tag{3.28}$$

to the outcomes of the reference measurement at t_0 and another other probability distribution—let us call it $P_{t_1}(R_i)$—at time t_1,

$$P_{t_1}(R_j) = \text{tr}\,(\hat{U}\hat{\rho}\hat{U}^\dagger)\hat{R}_j = \text{tr}\,\hat{\rho}(\hat{U}^\dagger\hat{R}_j\hat{U}). \tag{3.29}$$

This suggests that we focus on the measurement $\hat{R}_j' = U^\dagger\hat{R}_j\hat{U}$ and think of it just as any of the measurements we could put into the right side of Figure 3.1 to get:

$$P_{t_1}(R_j) = Q(R_j'). \tag{3.30}$$

With this, it becomes obvious what the relation between the $P_{t_0}(R_i)$ and $P_{t_1}(R_j)$ must be. It is just the Born rule as expressed in the language of Eq. (3.16), but with slightly modified variables:

$$P_{t_1}(R_j) = \sum_{i=1}^{d^2}\left[\sum_{k=1}^{d^2}[\Phi]_{ik}P_{t_0}(R_k)\right]P(R_j' \mid R_i), \tag{3.31}$$

where

$$P(R_j' \mid R_i) = \text{tr}\,\hat{\sigma}_i\hat{R}_j', \tag{3.32}$$

and Φ^{-1} is defined just as before in Eq. (3.15). In the special case when a SIC is used as the reference apparatus, this becomes

$$P_{t_1}(R_j) = \sum_{i=1}^{d^2}\left[(d+1)P_{t_0}(R_i) - \frac{1}{d}\right]P(R_j' \mid R_i). \tag{3.33}$$

With this, *the very meaning of unitary evolution* is at hand: It is captured all and only by the conditional probability assignments in Eq. (3.32). That is, it is transparently as much a personal judgment as the quantum states assignments $P_{t_0}(R_i)$ and $P_{t_1}(R_j)$ are. Moreover, the integrated Schrödinger equation $\hat{\rho} \to \hat{U}\hat{\rho}\hat{U}^\dagger$ in this notation is nothing other than a special case of the Born rule itself:

$$Q(R') = P(R' \mid R)\Phi P(R). \tag{3.34}$$

Via similar considerations, one can also show that an equation of exactly the same form holds for the full class of all possible quantum time evolutions, not just unitary evolutions. In the quantum information literature, these are known as completely positive trace-preserving maps or *quantum operations*. As it must follow because of Eq. (3.34), all such maps are to be understood as subjective judgments within QBism. In fact, just as proving a de Finetti representation theorem for "unknown quantum states" was a crucial test for QBism in its earliest days, one can prove an analogous result for "unknown quantum operations" [96, 97] thereby ensuring the consistency of the point of view.

As a final point, let us note a major distinction between the QBist way of thinking about unitarity and the workaday quantum information scientist's way. In the latter venue a unitary operation is something that is engineered, for instance via the design of a quantum circuit. One then builds a physical implementation of the circuit in the laboratory to achieve some task or to "control" an unwieldy system. Speaking of it in that manner, it may be hard to stop employing (even implicitly) an ontic understanding of unitary operations: They are expressions of the *solid stuff* that makes things happen.

However, the word "control" makes no appearance in the QBist account of unitarity. In QBism, the assignment of a unitary operation is just that: An *assignment*, a judgment that an agent makes to reasonably represent her beliefs. It is not an *action* she takes upon a system, forcing it to do one thing or another, as is the case with a quantum measurement (with its generally unpredictable outcome). This stance also sets QBism apart in the spectrum of quantum interpretations. To anticipate one aspect of the Wigner's friend thought experiment in Section 3.4, a QBist for instance would never say, as Baumann and Brukner do in [29],

> We note that the specific relative phase between the two amplitudes ... is determined by the interaction Hamiltonian between the friend and the system, which models the measurement and is assumed to be known to and *in [the] control of* Wigner.
>
> [our emphasis]

Wigner is in as much control of his external world with a unitary-operation assignment as he is with any other personal probability assignment—namely, none. A unitary instead represents an agent's hard won degrees of belief. Those beliefs may make the agent very certain of the outcome of an appropriate measurement, but a belief can always be shattered by a surprise.

3.2.7 *Even Probability-One Assignments Are Judgments*

No Bayesian of any of the 46,656 varieties classified by I. J. Good [98] would disagree that a probability assignment $P(D) = p$ for some p in the range $0 < p < 1$ represents a personal judgment—i.e., that it is not a fact of nature. Things only become testy when one starts to discuss the limit points 0 and 1, and that is where QBism makes one of its most distinctive stands. QBism regards even probability 0 and 1 assignments as personal judgments.

Assigning probability-1 to an outcome expresses the agent's supreme confidence that the outcome will occur, but it does not imply that there is something in nature to guarantee that the outcome *actually* must be the case. QBism makes a strict category distinction between the truth (or facticity) of measurement outcomes and probability-1 (as a belief) that other Bayesian interpretations, say that of Jaynes [99], ignore.

It was first recognized that such an extreme interpretation for probability-1 was crucial for QBism's consistency in 2001 [7], and later it turned into an indispensable tool for analyzing the locality of quantum theory [3], but QBists have only recently learned that this notion already had a respectable pedigree in 1991 in the thinking of the philosopher and logician Richard Jeffrey [100]:

> The ("Bayesian") framework explored [here] replaces the two Cartesian options, affirmation and denial, by a continuum of judgmental probabilities in the interval from 0 to 1, endpoints included—or what comes to the same thing—a continuum of judgmental odds in the interval from 0 to ∞, endpoints included. Zero and 1 are probabilities no less than 1/2 and 99/100 are. Probability 1 corresponds to infinite odds, 1:0. That's a reason for thinking in terms of odds: to remember how momentous it may be to assign probability 1 to a hypothesis. It means you'd stake your all on its truth, if it's the sort of hypothesis you can stake things on. To assign 100% probability to success of an undertaking is to think it advantageous to stake your life upon it in exchange for any petty benefit. We forget that when we imagine that we'd assign probability 1 to whatever we'd simply state as true.

Mundane examples of the distinction between probability-1 and truth abound so long as one is not wedded to an objective, agent-independent notion of probability. Here is an example from the world of prenuptial agreements. Anyone entering a marriage knows that their partner, being another free agent, is free to be faithful or cheat in the relationship. Marriage is a decision one does not make lightly. Thus, consider an agent who through an extensive exploration of the largest set of intertwined beliefs she can muster—all the many things she believes of her partner, the things she believes of her partner's family, their religious views, known financial matters, the society in which they live, ... any number of things, even weighing the advice of her lawyer—aims to make a probability assignment for her partner's being faithful. It would be an enormous computational task, but never mind that. What is clear from personalist Bayesian probability theory is that when a mesh of beliefs is wide enough, it can be enormously restrictive to one's probability assignments [101]. In this case, imagine our agent ends up with $p = 1$ that her partner will stay faithful. Yet, the agent's lawyer advises a prenuptial agreement. Is the agent wrong?

What does this have to do with physics? Because of the intimate connection between quantum states and the probabilities derived from them, QBism regards the assignment of all quantum states (mixed or pure) as an agent's personal judgments. This implies in particular that even a statement such as "this outcome is certain to occur for this quantum measurement" reflects an agent's judgment rather than a fact of nature. In other words, nothing in nature metaphysically guarantees that an outcome to which an agent has assigned probability-1 will in fact occur. As the world is genuinely indeterministic according to quantum theory, an agent's judgments are genuinely fallible.

This point cannot be emphasized enough, as it seems a number of commentators on QBism simply do not recognize its importance for the QBist response to a number of conundrums. For instance, consider an agent who takes the action of placing a Stern-Gerlach device in front of an electron and has just registered spin-up for it in the z-direction as her consequent experience. She will thus assign a quantum state $|z = +1\rangle$ for any subsequent measurements on the electron. However, in QBism this does not amount to a statement of fact but a statement of belief. The assignment of this state amounts to, among other things, a belief—a monumentally strong belief—that taking the same action with the Stern-Gerlach device will give rise to exactly the same consequence, namely the experience of spin-up in the z-direction. But what means "same?" Even "same" is a judgment if one is going to have a consistent subjective view of quantum states [96]. Thus QBism must say that the notion of "same measurement" is itself a belief, not a fact of nature [54]. It might be a supremely strong belief because a long measure-remeasure sequence has given the same result an inordinate number of times previously, but from the QBist conception this does not negate that it is a belief. Think, for instance, of Hume's argument that induction can never be guaranteed by any principle other than itself [57].

Thus, the world from a QBist conception can always surprise, no matter the certainty of the agent using the quantum formalism—and this even if the formalism is applied to something vastly more complicated than an electron. Writing down $|z = +1\rangle$ does not ensure that an electron will deliver at the agent's command if he performs the right measurement, but only with what the agent believes with his heart of hearts. Why should it be different for an atom, for a molecule, for a long piece of DNA, for a euglena, or even Wigner's friend? If we grant autonomy to a single electron so that it might genuinely surprise an agent, why would we not for a monstrously complicated system like Wigner's friend? One of the lessons of this chapter will be that whatever QBism says of an electron, it must say the same of Wigner's friend, and reciprocally whatever we learn from Wigner's friend we should take to our understanding of the world at large.

3.2.8 *Subjective Certainty of What an Outcome Will Be Does Not Negate That Unperformed Measurements Have No Outcomes*

Asher Peres's slogan, "unperformed experiments have no results" [47], had a powerful influence on the development of QBism. Maybe in the language of this chapter we should say "unperformed measurements have no outcomes" instead, but the sentiment is obvious. Here is the nice way David Mermin put it in our joint paper [3]:

> "*This experiment has no outcome until I experience one.*" QBism personalizes the famous dictum of Asher Peres. The outcome of an experiment is the experience it elicits in an agent. If an agent experiences no outcome, then for that agent there is no outcome. Experiments are not floating in the void, independent of human agency. They are actions taken by an agent to elicit an outcome. And an outcome does not become an outcome until it is experienced by the agent. That experience *is* the outcome.

This has all been said in previous sections, but it does not hurt to explore it from every angle. What has not been discussed so far is how this conception interacts with the notion of probability-1. Is there any special difficulty introduced when one considers such extreme probability assignments? It is common enough to find such a sentiment in the literature. An example contemporaneous with when QBism first came to its realization is this quote by Brukner and Zeilinger [102], whose positions in other aspects sometimes come somewhat close to QBism:

> Only in the exceptional case of the qubit in an eigenstate of the measurement apparatus the bit value observed reveals a property already carried by the qubit. Yet in general the value obtained by the measurement has an element of irreducible randomness and therefore

> cannot be assumed to reveal the bit value or even a hidden property of the system existing before the measurement is performed.

Maybe they have developed since then, but historical examples can be found in the writings of Heisenberg, Dirac, von Neumann, Messiah, and many others [103]. It is what the philosophers of physics call the "eigenstate-eigenvalue link."

QBism predictably takes the stand that even when an agent assigns probability-1 to one of the possible outcomes of a measurement, there is nothing in the agent's external world that metaphysically ensures it to come about. For "unperformed measurements have no outcomes" is a statement about the character of the world—that it is not a block universe—whereas a probability-1 assignment is only a *belief* (supremely strong, but nonetheless a belief) someone happens to have in the moment. To say it differently, one is an expression about the world's creative character, while the other is about a user of quantum theory's momentary state of mind.

Of course, the latter expression comes from our analysis in the previous subsection of probability-1 in general. But the former goes much deeper: QBism in fact takes it to be the great lesson of all the multitude Bell-inequality and Kochen-Specker analyses and experiments of the past half century.[28] More contemporaneously, QBism sees this lesson further reinforced by the recent "no-go theorems" of Pusey, Barrett, and Rudolph (PBR) [104] and Colbeck and Renner (CR) [105], perhaps to the dismay of the authors' original intentions.[29] Yet, QBism traces these results back to the more primordial idea that the normative advice of quantum theory is given by Eq. (3.19), instead of the classical advice Eq. (3.20) co-opted from the LTP.

Eq. (3.19) used the tools of quantum information theory to express the Born rule as a relation between probabilities that works for any quantum state and any possible measurement, but the idea that the inequivalence of the Born rule to the LTP is the root cause of *all the quantum mysteries* is an idea that goes back at least to Richard Feynman in the 1940s. In a 1951 paper titled "The Concept of Probability in Quantum Mechanics," Feynman writes [106],

> The new theory asserts that there are experiments for which the exact outcome is fundamentally unpredictable, and that in these cases one has to be satisfied with computing probabilities of various outcomes. But far more fundamental was the discovery that in nature the laws of combining probabilities were not those of the classical probability theory of Laplace.
>
> I should say, that in spite of the implication of the title of this talk the concept of probability is not altered in quantum mechanics. When

> I say the probability of a certain outcome of an experiment is p, I mean the conventional thing .... I will not be at all concerned with analyzing or defining this concept in more detail, for no departure from the concept used in classical statistics is required.
>
> What is changed, and changed radically, is the method of calculating probabilities.

Of course, Feynman is expressing the transition to the amplitude calculus here. In his original 1948 paper on path integrals [107], he describes it as "essentially a third formulation of non-relativistic quantum theory" (after matrix and wave mechanics). What is essential to us is the way he directly contrasts the new combination laws to the LTP. He does this by considering three successive experiments A, B, C, with outcomes a, b, c, and denotes the conditional probability for finding b given a as P_{ab}, etc. In a prescient passage, Feynman [107] writes,

> Now, the essential difference between classical and quantum physics lies in [the LTP]. In classical mechanics it is always true. In quantum mechanics it is often false. We shall denote the quantum-mechanical probability that a measurement of C results in c when it follows a measurement of A giving a by P^q_{ac}. [The LTP] is replaced by this remarkable law: There exist complex numbers φ_{ab}, φ_{bc}, φ_{ac} such that
>
> $$P_{ab} = |\phi_{ab}|^2, \quad P_{bc} = |\phi_{bc}|^2, \quad \text{and} \quad P^q_{ac} = |\phi_{ac}|^2. \tag{3.35}$$
>
> The classical law ...
>
> $$P_{ac} = \sum_b P_{ab} P_{bc} \tag{3.36}$$
>
> is replaced by
>
> $$\varphi_{ac} = \sum_b \varphi_{ab} \varphi_{bc}. \tag{3.37}$$
>
> If Eq. (3.37) is correct, ordinarily Eq. (3.36) is incorrect. The logical error made in deducing Eq. (3.36) consisted, of course, in assuming that to get from a to c the system had to go through a condition such that B had to have some definite value, b. ...

> Looking at probability from a frequency point of view Eq. (3.36) simply results from the statement that each experiment giving a and c, B had some value.[30] The only way Eq. (3.36) could be wrong is the statement, "B had some value," must sometimes be meaningless. Noting that Eq. (3.37) replaces Eq. (3.36) only under the circumstance that we make no attempt to measure B, we are led to say that the statement, "B had some value," may be meaningless whenever we make no attempt to measure B.

In Feynman's case, as in ours, the lesson is the same: If one adopts any method of gluing one's hypothetical gambles together which does not boil down to an application of the Born rule, then one does it at one's own peril. Any probability assignment for a potential experience, including $Q(E_j) = 1$, had better come about by some association $R_i \leftrightarrow (\hat{R}_i, \hat{\sigma}_i)$ and $E_j \leftrightarrow \hat{E}_j$ along with an application of Eq. (3.16), or one is violating the Born rule and consequently ignoring quantum theory's normative advice.

Even when an agent is certain what the outcome will be, unperformed measurements have no outcomes.

3.2.9 *Quantum Theory Is a Single-User Theory for Each of Us*

It hardly needs to be said by now, but every formulation we have made in the previous tenets has always referred to the concerns of a single agent. Even when we briefly considered a "team of scientists sharing notebooks" in Section 3.2.1, we treated the collective as a single agent. This is an inheritance from QBism's insistence that quantum theory should be understood as a normative addition to personalist Bayesian probability theory. D. V. Lindley, a prominent Bayesian statistician, put the point succinctly [108],

> The Bayesian, subjectivist, or coherent, paradigm is egocentric. It is a tale of one person contemplating the world and not wishing to be stupid (technically, incoherent). He realizes that to do this his statements of uncertainty must be probabilistic. This is important on its own for it rules out a large class of behavior patterns, like sampling-theory statistics, but is it enough? Once I coined the aphorism "Coherence is all." Was I right?

QBism clearly leans in that direction, modulo the addition of the Born rule. But, Lindley goes on to say,

> It is when we consider two coherent Bayesians that new features arise. Does their egocentric behavior allow them to talk to one another? In one respect it does. Suppose that you and I are both

coherent and you tell me that your probability for A, were B to be true, is α, say. How does this knowledge affect my probability for A given B? It is easy for me to do the coherent calculations in terms of my assessments [i.e., my probabilities] that were A and B both true, you would say α, and that were $\neg A$ and B both true, you would say α. For to me α is just data and the likelihood ratio updates my probability in the usual way. A weather forecaster who announces rain whenever it is subsequently dry, and vice versa, is badly calibrated but very useful. So I can respect your egocentricity; and you, mine.

With this, QBism agrees wholeheartedly. Indeed, one may view the recent QBist analyses [35] of the Baumann-Brukner [29] and Frauchiger-Renner [28] variations on Wigner's friend as extended exercises in seeing to it that Wigner and his friend respect each other's egocentricities. The friend may ask Wigner his quantum-state assignment for this or that, but if she does so it can only come about by her taking a physical action upon him (as with any measurement) and using the subsequently induced outcome as a datum that *may* or *may not* make much impact on her future quantum-state and probability assignments. *The extent to which it will* depend upon the details of her larger mesh of beliefs, as it does in Lindley's example. It follows that there can be no overarching principle or prescription for how two agents' quantum states for the same system must be related to each other, and QBism takes this fully onboard. Quantum theory is a single-user theory for each of us.

Contrast this attitude of QBism's to some earlier—what one might call "halfhearted"—approaches to understanding quantum states in epistemic terms. This line of thinking was initiated by Rudolf Peierls, who wrote, "there are limitations to the extent to which [two agents'] knowledge may differ" [109]:

> It is possible for two observers to have some knowledge of the same system, and the knowledge possessed by one may differ from the other. ... In this situation the two observers will use different density matrices. ... However, the information possessed by the two observers must not contradict the uncertainty principle. ... This limitation can be expressed concisely by saying that the density matrices appropriate to the two observers must commute with each other. ... At the same time, the two observers should not contradict each other. This means the product of the two density matrices should not be zero. [110]

By simple examples from quantum cryptography [111], one can see that the first part of the Peierls "compatibility criterion"—i.e., the part about necessary commutivity—cannot be maintained. But that didn't stop a number of authors from thinking that even if Peierls wasn't quite right, there should still be some criterion from physics alone to limit how much quantum-state assignments could differ. For instance Brun, Finkelstein, and Mermin (BFM) in [112] write, "We derive necessary and sufficient conditions for a group of density matrices to characterize what different people may know about one and the same physical system," and "[A] special case of our condition is that two pure-state density matrices are compatible if and only if they are identical." In that work, there was an implicit accommodation to an Everettian-style world-view (anathema to QBism to begin with), but a further analysis [113, 114] pointed out that the BFM-compatibility criterion could in fact be given a non-Everettian, purely Bayesian interpretation, so long as one recognized that it was only one among a number of inequivalent criteria for two agents to judge when they are in sufficient agreement. It followed that there was nothing necessary and sufficient about any of the criteria as expressions of "physics alone."

Ultimately, by choosing to completely sever any connection between quantum-state assignments and external, agent-independent facts, QBism had to reject the idea of compatibility requirements for distinct agents' state assignments. The two had to go hand in hand. In QBism, an agent is judged as making a proper use of quantum theory not by which quantum states she assigns, but by whether she consistently makes use of the formalism when making those assignments. What her neighbor does in making his assignments is immaterial to her.

To add a sidelight to this, let us give an example [2] for anyone who still holds out hope that what we have argued is just so much sophistry: One can see that the various compatibility criteria were always unstable notions to begin with. Consider two agents sitting in a common laboratory, and in agreement on everything to do with a qubit measurement apparatus in front of them—they each make the same POVM assignment to it. It is thus within each agents' rights to consider the measuring apparatus an extension of him or herself. Suppose it to be a measurement of the $\{|0\rangle, |1\rangle\}$ basis.

Nonetheless, imagine the agents do differ on the quantum-state assignments they make for a two-qubit system, one qubit of which they are going to measure, while keeping the other safely in the distance. Say one agent professes a quantum state ρ_+ while the other ρ_-, where

$$\rho_\pm = \frac{1}{2}\left(|0\rangle\langle 0| \otimes |0\rangle\langle 0| + |\pm\rangle\langle \pm| \otimes |\pm\rangle\langle \pm|\right) \tag{3.38}$$

and

$$|\pm\rangle = \sqrt{\frac{1}{2}}\left(|0\rangle \pm |1\rangle\right). \tag{3.39}$$

It is easy to check that these two states are compatible by all the criteria considered in [113]—the BFM criterion, the post-Peierls criterion,[31] and others.

Let us consider the case where the first qubit is measured and outcome 1 is found—both agents deemed 0 and 1 possible, but 1 was in fact found. Then the two agents' post-measurement states for the second qubit (the one not touched yet) will be $|+\rangle\langle+|$ and $|-\rangle\langle-|$, respectively. With the outcome of a single measurement, two agents go from claiming to having compatible information by every criterion, to being incompatible by all of them![32] Of course, this is simply because $|+\rangle\langle+|$ and $|-\rangle\langle-|$ are orthogonal to each other, and for a subsequent measurement of the $\{|+\rangle,|-\rangle\}$ each agent will assign probability-1 for an outcome that the other agent assigns probability-0.

What does a QBist make of this? Nothing more than has already been said: We have two agents who have made "momentous" probability assignments (in the words of Richard Jeffrey from Section 3.2.7). This means that the agents are willing in principle to stake their lives on seeing what they expect to see, but it does not make one of the agents "right" and the other "wrong" before any measurement is performed. Nor will it make one "right" and the other "wrong" after the measurement, unless one equates right and wrong with living and dying. What else can one expect in a genuinely indeterministic world? So long as each agent had a coherent mesh of beliefs leading up to their original quantum-state assignment, then they have done all that they could do, and there a QBist must leave it. In a number of recent "no-go" theorems, part of the drama as it is presented is that one agent will predict something with probability-1, whereas another agent will predict it with probability-0, and this is taken to be an affront to quantum theory. But such things are a matter of course in QBism.

To recap: All that each individual agent has to go on in QBism is whether they have been as coherent with all applications of probability theory and the Born rule as they can be. For QBism, quantum theory is always a single-user theory, but it is so for each of us. Wigner should strive to be coherent in all his own beliefs; his friend should strive to be coherent in all hers. But there is no principle of physics that requires Wigner and the friend to have compatible quantum-state assignments: Physics suffers none by agents' incompatibilities.

3.3 QBism's Eightfold Path and the Missing Noble Truth

To summarize Section 3.2, we have identified eight key tenets of QBism (in a different order this time around):

1 A quantum measurement is an agent's action upon its external world. A quantum measurement is cut from the same cloth as any other action she might take upon her world, as for instance by crossing a street.

What makes an action specifically "quantum" is when it is worth the agent's while to analyze her expectations for its outcomes in terms of the quantum formalism.

2 A measurement's outcome *just is* the consequent personal experience of the agent taking the action. In QBism there is no notion of a mediating device between the agent and the quantum system ("decohering" because of an interaction with an environment [115]). Measuring devices are considered conceptually as parts or extensions of the agent.
3 A quantum state is an agent's personal judgment. It serves to tie together all her probability assignments for the outcomes of all measurement actions she might take upon a system.
4 Even unitary evolution operators (or more generally completely positive trace-preserving maps) are agents' personal judgments. They are not set by nature; they are not ontic. In any individual case, they are set by the agent's expectations across time.
5 The quantum formalism is normative rather than descriptive. It guides the agent toward making sure all her judgments are consistent with one another and with the "quantum" nature of the world. This is most easily seen by rewriting the quantum formalism in a form that is purely probabilistic in nature, without operators on complex vector spaces, etc. Quantum states thus become probability distributions for the outcomes of a reference apparatus, measurement operators become conditional probabilities with respect to the same, and the Born rule becomes something clearly seen as a variation on the classic LTP. Furthermore, rewriting the formalism this way nods to future physical enquiry by giving a way of quantifying the essential "quantumness" conveyed by the Born rule.
6 Even probability-1 assignments are judgments and thus metaphysically fallible. Probabilities are so disconnected from the world, they can never tell nature what to do. Probability-1 refers all and only to a momentous judgment for the agent who makes it; it means she will gamble her all on it.
7 Even when an agent is subjectively certain what the outcome of her measurement action will be, she still may not assume that the yet-to-be-performed measurement has a pre-existent determined outcome. This is part of the content of using the Born rule in place of the classic LTP.
8 Quantum theory is a single-user theory for each of us. My use of quantum theory concerns my expectations for my personal experiences. Your use of quantum theory concerns your expectations for your personal experiences. There is no requirement that those two uses must somehow dovetail together.

It was quite accidental that the number of tenets turned out to be eight, but so be it. Let us use this as a symbol. Maybe it's a poor joke, but the first thing that comes to mind is the Buddha's eightfold path! It is not that QBism is already "blue pure perfect" by any means, but this much is established: The historical reason for the eight tenets as they accumulated was in every case to "relieve the suffering" caused by one or another quantum conundrum.

However, the eightfold path in Buddhist doctrine was always subsidiary to the four noble truths. The eightfold path was a methodology for living in the world, but the noble truths were a statement about how the world is. It is to this project that we turn now. What is QBism's long-sought-for noble truth? In the past we have expressed most of our ontological thinking in the language of American pragmatism and radical empiricism. This time, however, we will start with Wigner's friend and potentially end up at Merleau-Ponty's hands. There may be overlap with the previous expressions, or there may not, but the exercise should at least give more potential paths for us to consider.

3.4 From Wigner's Friend ...

Wigner first published his famous thought experiment in a 1961 volume titled *The Scientist Speculates* [27]. Indeed, what a speculation it was! Sixty-two years later, it is still a hot topic of discussion [28–30], with even a well-funded workshop devoted to it in 2022 [116]. Most interesting from our perspective, however, is how the dissolution of the conundrum's paradoxical character involves all eight QBist tenets. At the same time it showcases a further feature of QBist thinking that has yet to be discussed.

To contrast with QBism's story, let us refer to Wigner's original description of his conundrum: We will quote it at length so that no details are missed.[33] Following his convention, we will speak of "the observer" and "the friend" in all our commentary.[34] Finally, we highlight in boldface those places where QBism has some disagreement with Wigner's account of the thought experiment.

First off, Wigner makes it clear that he intends to think in epistemic terms about quantum states. (It is another matter whether he does so consistently.) Indeed, Wigner starts off in a distinctly QBist direction. He writes,

> Given any object, all possible knowledge concerning that object[(a)] can be given as its wave function. ... If one knows [the wave function], one can foresee the behavior of the object as far as it *can* be foreseen. More precisely, the wave function permits one to foretell with what probabilities the object will make one or another impression on us if we let it interact with us either directly, or indirectly. ...

> One realises that *all* the information which the laws of physics provide consists of probability connections between subsequent impressions that a system makes on one **if one interacts with it repeatedly,**[(b)] .... The wave function is a convenient summary of that part of the past impressions which remains relevant for the probabilities of receiving the different possible impressions when interacting with the system at later times.

Note that his probabilities are not about the true states of the world, but about "impressions" or "sensations." This is on a trajectory toward QBism's "experiences,"[35] but the divergence between Wigner's understanding of quantum theory and QBism start to become apparent in the middle ellipsis above:

> The information given by the wave function is communicable. **If someone else somehow determines the wave function of a system, he can tell me about it and, according to the theory, the probabilities for the possible different impressions (or "sensations") will be equally large, no matter whether he or I interact with the system in a given fashion. In this sense, the wave function "exists."**[(c)]

Next, as a preamble to his paradox, Wigner starts with a general question about what if an observation is made "indirectly":

> It is natural to inquire about the situation if one does not make the observation oneself but lets someone else carry it out. **What is the wave function if my friend**[(d)] looked at the place where the flash might show at time t? The answer is that the information available about the *object* cannot be described by a wave function. **One could attribute a wave function**[(e)] to the joint system: friend plus object, and this joint system would have a wave function also after the interaction, that is, after my friend has looked. I can then enter into interaction with this joint system by asking my friend whether he saw a flash. If his answer gives me the impression that he did, the joint wave function of friend + object will change into one in which they even have separate wave functions (the total wave function is a product) and the wave function of the object is $|\psi_1\rangle$. If he says no, the wave function of the object is $|\psi_2\rangle$ i.e., the object behaves from then on as if I had observed it and had seen no flash. However, even in this case, in which the observation was carried out by someone else, the typical change in the wave function occurred only when some information (the *yes* or *no* of my friend) entered *my* consciousness. It follows that the quantum description of objects is influenced by impressions entering my consciousness.

With that much as an introduction, Wigner gets to what he sees as the heart of the matter:

> Does ... consciousness influence the physico-chemical conditions? In other words, does the human body deviate from the laws of physics, as gleaned from the study of inanimate nature? ... [A]t least two reasons can be given to support [a "Yes" to this]. ...
>
> In order to present [the first] argument, it is necessary to follow my description of the observation of a "friend" in somewhat more detail than was done in the example discussed before. Let us assume again that the object has only two states, $|\psi_1\rangle$ and $|\psi_2\rangle$. If the state is, originally, $|\psi_1\rangle$ the state of object plus [friend] will be, after the interaction, $|\psi_1\rangle|\chi_1\rangle$; if the state of the object is $|\psi_2\rangle$, the state of object plus [friend] will be $|\psi_2\rangle|\chi_2\rangle$ after the interaction. The wave functions $|\chi_1\rangle$ and $|\chi_2\rangle$ give the state of the [friend]; in the first case **he is in a state**[(f)] which responds to the question "Have you seen a flash?" with "Yes"; in the second state, with "No." There is nothing absurd in this so far.
>
> Let us consider now an initial state of the object which is a linear combination $\alpha|\psi_1\rangle+\beta|\psi_2\rangle$ .... It then *follows* from the linear nature of the quantum mechanical equations of motion that the state of object plus [friend] is, after the interaction,
>
> $$|\Phi\rangle = \alpha|\psi_1\rangle|\chi_1\rangle+\beta|\psi_2\rangle|\chi_2\rangle. \tag{3.40}$$
>
> **If I now ask the [friend] whether he saw a flash, he will with a probability $|\alpha|^2$ say that he did, and in this case the object will also give to me the responses as if**[(g)] it were in the state $|\psi_1\rangle$. If the [friend] answers "No"—the probability for this is $|\beta|^2$—the object's responses from then on will correspond to a wave function $|\psi_2\rangle$. The **probability is zero**[(h)] that the [friend] will say "Yes," but the object gives the response which $|\psi_2\rangle$ would give because the wave function $|\Phi\rangle$ of the joint system has no $|\psi_2\rangle|\chi_1\rangle$ component. Similarly, if the [friend] denies having seen a flash, the behavior of the object cannot correspond to $|\chi_1\rangle$ because the joint wave function has no $|\psi_1\rangle|\chi_2\rangle$ component. All this is quite satisfactory: **the theory of measurement, direct or indirect, is logically consistent so long as I maintain my privileged position as ultimate observer.**[(i)]
>
> However, if after having completed the whole experiment I ask my friend, "What did you feel about the flash before I asked you?" he

will answer, "I told you already, I did (did not) see a flash," as the case may be. In other words, the question whether he did or did not see the flash was already decided in his mind, before I asked him. If we accept this, **we are driven to the conclusion that the proper wave function**[j] immediately after the interaction of friend and object was already either $|\psi_1\rangle|\chi_1\rangle$ or $|\psi_2\rangle|\chi_2\rangle$ and not the linear combination $|\Phi\rangle$. This is a contradiction, because the state described by the wave function $|\Phi\rangle$ describes a state that has properties which neither $|\psi_1\rangle|\chi_1\rangle$ nor $|\psi_2\rangle|\chi_2\rangle$ has. If we substitute for "friend" some simple physical apparatus, such as an atom which may or may not be excited by the light-flash, this difference has observable effects and *there is no doubt that* $|\Phi\rangle$ *describes the properties of the joint system correctly, the assumption that the wave function is either* $|\psi_1\rangle|\chi_1\rangle$ *or* $|\psi_2\rangle|\chi_2\rangle$ *does not.* **If the atom is replaced by a conscious being, the wave function $|\Phi\rangle$ ... appears absurd because it implies that my friend was in a state of suspended animation before he answered my question.**[k]

It follows that the being with a consciousness must have a different role in quantum mechanics than the inanimate measuring device: the atom considered above. In particular, **the quantum mechanical equations of motion cannot be linear if the preceding argument is accepted.**[l] This argument implies that "my friend" has the same types of impressions and sensations as I—in particular, that, after interacting with the object, he is not in that state of suspended animation which corresponds to the wave function $|\Phi\rangle$.

It is not necessary to see a contradiction here from the point of view of orthodox quantum mechanics, and there is none if we believe that the alternative is meaningless, whether my friend's consciousness contains either the impression of having seen a flash or of not having seen a flash. However, **to deny the existence of the consciousness of a friend to this extent is surely an unnatural attitude, approaching solipsism, and few people, in their hearts, will go along with it.**[m]

Let us now comment on the highlighted parts.

a. **"Knowledge."** Note that nowhere in Wigner's description does he deny the epistemic character of quantum states. In fact, that they are epistemic is the bedrock of his considerations, even if he is not exactly consistent about it. However, he differs from QBism in that his epistemic states are about knowledge, not belief. This gives them a factive character that QBism denies. For if it were so, a quantum state could be right or wrong as set by the facts of the world. In the development of Wigner's argument, this distinction will play a pivotal role.

b. **"Probability connections ... if one interacts with it repeatedly?"** He may be thinking of probability in a frequentist rather than a Bayesian conception here, but it is not clear. If he is, it would help shore up why he is taking a factive notion of quantum states.
c. **"If someone determines the wave function of a system, he can tell me about it and the probabilities for the possible sensations [for each of us] will be equally large, no matter whether he or I interact with the system." In this sense, the wave function exists.** As Wigner says, *if so*, this would make a quantum state exist in a way contrary to de Finetti's dictum that "probability does not exist." QBism instead requires a much more nuanced discussion of this situation, say along the lines of that described in Section 3.2.9. If you were to tell me your quantum state for a system, I might or might not take it on myself. Or, I might or might not roll it in to my previous quantum-state assignment and end up with something completely new. What is for sure is that in QBism a quantum state assignment is not blindly transferable.
d. **"What is the wave function if my friend ...."** Notice the definite article here, *the*. The very tone of the question eschews the QBist conception of quantum states as pluralistic and always tagged to a specific agent. For QBism, there might be as many quantum states for a single quantum system as there are agents considering it.
e. **"One could attribute a wave function ...."** This is the same sin as reported in the last item. Who is the "one"? QBism never tolerates ambiguity on this point, but Wigner to some extent flicks back and forth between the two players in the story. It appears the narrative is moving toward making the observer a super-observer who in the end must be made consistent with any inside observers (e.g., the friend). But remember, Wigner had already settled on states being epistemic: Why then demand identity between the two observers' state assignments? Presumably because then one observer would otherwise get the facts wrong.
f. **"The wave functions ... give the state of the observer; in the first case he is in a state ...."** Notice how perilously close this gets to an ontic-tinged notion of quantum states. He is *in* a state? It makes states sound like properties that objects can have. This is a slippery slope to an irreversible confusion between epistemic and ontic.
g. **"If I ask the observer whether he saw a flash, he will with some probability say that he did, and in this case the object will also give to me the responses as if it were in the state ...."** As if? This tentatively privileges the state of the observer in the narrative. This will drive the idea that the observer's state assignment must ultimately be made consistent with the friend's, so to erase the "as if."

h. **"The probability is zero …."** It is hard to know exactly how Wigner is thinking of probability-zero, but I would guess he identifies it with a proposition's truth value being "false."
i. **"The theory of measurement is logically consistent so long as I maintain my privileged position as ultimate observer."** Of course, QBism argues that the theory is consistent precisely because it *does not privilege* any agents' personal point of view. But then, it does that in a very different way than Wigner is aiming for.
j. **"We are driven to the conclusion that the proper wave function …."** When quantum states are factive to begin with, then sentences like this might have a meaning. But QBism undermines that; it says that it was always a mistake to think there is a single "proper wave function" across distinct agents.
k. **"If the atom is replaced by a conscious being, the entangled wave function would imply my friend was in a state of suspended animation before he answered my question."** What might he mean by "suspended animation" here? Presumably he means that the friend has no experiences while the observer holds the state $|\Phi\rangle$. But why on earth would it imply that? $|\Phi\rangle$ was supposed to only "foretell with what probabilities the object will make one or another *impression* on us" [our emphasis]. The extra supposition of "suspended animation" for the friend violates Wigner's very starting point.
l. **"The quantum mechanical equations of motion cannot be linear if the preceding argument is accepted."** It is interesting to think about what seems to be going on here. Wigner accepts outright that quantum states are epistemic, but then he demands a dualistic intervention of matter on mind without mind gaining any information in the process. If not completely so or consistently so with quantum states, Wigner is surely thinking in ontic terms for time evolution processes.
m. **"To deny the existence of the consciousness of a friend to this extent is surely an unnatural attitude, approaching solipsism, and few people, in their hearts, will go along with it."** Again, why would the observer's assignment $|\Phi\rangle$ to the composite system amount to "denying the existence of the consciousness of the friend?" Diagnosis: Likely because, though Wigner claims $|\Phi\rangle$ to be epistemic about the observer's potential sensations, he is not so consistent about it. He also seems to hold that $|\Phi\rangle$ is directly representative of what is going on in the closed system as well. Treatment: QBism, of course. So, let's get to it.

In contrast to the way Wigner sets up the problem, QBism sees the thought experiment as being about two equal-footed agents, each of which may take actions on the system of their concern and receive deeply personal experiences as a consequence. See Figure 3.4. We know that Wigner is not treating the agents on an equal footing because everything in his story is expressed univocally from the perspective of the outside observer. It is the outside observer who specifies what quantum state the friend will hold for the object after seeing a flash or not. But does anyone ever ask the friend how he would gamble on the sensations he would receive by taking actions on the outside observer instead? Indeed, we know the outside observer is a user of quantum mechanics[36] rather than an uneducated agent, as he writes down quantum states, unitary operations, and makes probability assignments through the Born rule. Yet Wigner never bothers to tell us whether the friend too is a user of quantum mechanics. Wigner could have played the "as if" game mentioned in comment (g) all the way down, with $|\psi_1\rangle$ and $|\psi_2\rangle$ never being the friend's actual state assignments, just the ones the observer thinks the friend *ought to* make. The states $|\chi_1\rangle$ and $|\chi_2\rangle$ stand only for the answers to the dumb query, "Have you seen the flash?" The friend could answer that without knowing a bit about quantum mechanics.

Thus it is crucial for QBism's conception of quantum theory as a user's manual that it treats the two agents on an equal footing. Furthermore, QBism cannot restrict the applicability of quantum mechanics to inanimate objects, else the manual would not be universal to everything external to the agent. As already broached in Section 3.2.7, an agent can even apply the normative quantum calculus to their expectations arising from actions they might take on other agents.

Yet QBism claims that despite Wigner, all of this can be done without modifying quantum mechanics in any way. As already emphasized, it takes all the apparatus of QBism to fight off the various quantum conundrums. But one might ask really what are the most salient points for this *particular* conundrum? There are three.

First is the idea that quantum-state assignments do not "pierce into" the systems to which they are ascribed. They do not describe what is "going on in there." Their sole role is to capture the ascriber's gambling attitudes toward the consequent personal experiences that would arise from actions on the system—this is true of any quantum state $|\psi\rangle$. Consequently it is no less true when it comes to state assignments like $|\Phi\rangle$ in Eq. (3.40). The latter may look more like it is talking about the goings on inside, but if so, that is surely an artefact of the representation. Every term in it is about the outside observer's beliefs concerning his own experiences. There is no metaphysical necessity that the friend even make the state assignments $|\psi_1\rangle$ and $|\psi_2\rangle$ himself, and QBism in fact basks in the possibility that it might be so. In a slogan, "no matter what quantum state assignment is made to a system, the system is more than that." The outside observer might make the assignment $|\Phi\rangle$ and yet the friend decide to make no measurement at all ... or even do something completely absurd like *eat the object* and incorporate it into himself! If a simple spin can surprise a measuring agent by shunning its probability-1 for the outcome of some observable, how much more so might another full-blown agent? But this was already emphasized in Section 3.2.7.

The second has to do with the idea of "suspended animation." There is a tendency in many if not most discussions concerned with assigning quantum states to agents that pure-state assignments somehow contradict the very idea of agency. After all, if the assigner has made a pure-state assignment he must be in a position to apply any unitary operation he wishes to make the state anything he will. For instance, in the implied quantum evolution of the friend plus object, one has

$$|\Phi_0\rangle = (\alpha|\psi_1\rangle + \beta|\psi_2\rangle)|\chi_0\rangle \overset{\hat{U}}{\rightarrow} |\Phi\rangle = \alpha|\psi_1\rangle|\chi_1\rangle + \beta|\psi_2\rangle|\chi_2\rangle, \quad (3.41)$$

where $|\chi_0\rangle$ is a "ready" state for the friend. One says that the left state $|\Phi_0\rangle$ represents that the friend's measurement has yet to happen, whereas the right state $|\Phi\rangle$ represents that the process is complete. But then the unitary operation $\hat{U}$ that took the process forward can be reversed by $\hat{U}^{-1}$ to take it all back—so that the measurement outcome goes "poof," right back out of existence. For instance as already quoted in Section 3.2.6, Baumann and Brukner [29] write that $\hat{U}$ "is determined by the interaction Hamiltonian between the friend and the system, which models the measurement and is assumed to be known to and in [the] control of Wigner." However, one must remember that like a quantum state, within QBism $\hat{U}$ is purely a compendium of beliefs, and itself does not pierce into the metaphysical goings on the friend's closed laboratory. $\hat{U}^{-1}\hat{U} = I$ is purely a statement about the outside observer's expected experiences, not about what is happening out in the world.

In fact the disparity between QBism and the usual discussions goes deeper: If a QBist agent ever works up a state of belief that corresponds to assigning a pure quantum state to her friend, by necessity, it will be precisely because she believes him to be another free-willed agent such as herself, not in contradiction to it. Take some *system* that our QBist agent believes will pass a Turing test should she administer it, will pass Danny Greenberger's "large red button test" [117], one for which she feels a heartfelt empathy, one for which she would feel the loss of their companionship should they leave, etc., etc. One can consider a very long list of such things. If they are all things she believes, then they are all things she believes: They must be rolled into her quantum state assignment for the system. They are not thoughts to be ignored, but rather ones to be refined until they fit within a numerical framework capturing the agent's best gambles. So, what if that rolling up (along with a myriad of other details of the agent's belief system) leads to some pure-state assignment?

QBism's retort is, "Indeed, so what? There's nothing to be made of it." It only means that the pure-state assignment landed on will contain *all* those beliefs. Rather than contradict anything, all those beliefs are necessarily in there.

So, $|\Phi\rangle$ = a state of suspended animation? *Pah!* But then that leaves us with the toughest issue of all: This is the third salient point alluded to above, and it is the most profound, for it takes us into the territory of ontology. For every agent in the Wigner's friend story there is something that happens—new experience comes into existence. Yet, the quantum formalism itself is always thoroughly and exclusively first personal in any particular application. Whatever is gambled on is always "my experience," whoever the "me" happens to be. This is the "QBist Copernican principle" first proto-formulated in [1, 2, 48, 118] and perhaps most thoroughly elaborated in [4]. See also [30] for a sympathetic analysis from an independent point of view.

Yet, how can this be? Wigner's quantum state assignment cannot pierce into his friend's reality, but he can still maintain this Copernican principle. If this feels like a contradiction, it shouldn't. The state assignment lives exclusively at the level of the first personal, while the general structure of quantum theory with its normative suggestions lives at the level of the third-personal. We called it normative structural realism before. If quantum theory is a user's manual that any agent can use, then the presupposition for all agents is that they shall experience. From this point of view, the lesson of quantum theory is that *experience happens*, and the refinement of the notion that comes from Wigner's thought experiment is that there *is a sense* in which those experiences need not live in a single universe.

Here is the way we put it in [48]:

> The only glaringly mutual world there is for Wigner and his friend in a QBist analysis is the partial one that might come about if these two bodies were to later take actions upon each other ("interact")—the rest of the story is deep inside each agent's private mesh of experiences, with those having no necessary connection to anything else.
>
> But what a limited story [Wigner's friend] is: For its concern is only of agents and the systems they take actions upon. What we learn from Wigner and his friend is that we all have truly private worlds in addition to our public worlds. But QBists are not reductionists, and there are many sources of learning to take into account for a total worldview—one such comes from Nicolas Copernicus: That man should not be the center of all things (only some things). Thus QBism is compelled as well: What we have learned of agents and systems ought to be projected onto all that is external to them too. The key

> lesson is that each part of the universe has plenty that the rest of the universe can say nothing about. That which surrounds each of us is more truly a pluriverse.
>
> This always meshed well with William James's "republican banquet" vision of a pluriverse [119]:
>
> Why may not the world be a sort of republican banquet of this sort, where all the qualities of being respect one another's personal sacredness, yet sit at the common table of space and time?
>
> To me this view seems deeply probable. Things cohere, but the act of cohesion itself implies but few conditions, and leaves the rest of their qualifications indeterminate. …
>
> [I]f we stipulate only a partial community of partially independent powers, we see perfectly why no one part controls the whole view, but each detail must come and be actually given, before, in any special sense, it can be said to be determined at all. This is the moral view, the view that gives to other powers the same freedom it would have itself.

But the phenomenology of Merleau-Ponty gives us heart that the last word has still not been said on this. For all that we have learned from the Wigner's friend thought experiment, we now start to see that the standard diagramming of the story, with its spin-1/2 particle, a laboratory surrounding the friend, etc., is just so much distraction. The same holds for QBism's solution as depicted in Figure 3.4. Perhaps a more essential point can be found by stripping away these extra elements and focusing on Wigner and the friend *tout court*. What further insight might be gleaned from a simple handshake between Wigner and his friend?

3.5 … to Merleau-Ponty's Hands, Concluding Remarks

John Bell once asked, "[A]re we not obliged to admit that more or less 'measurement-like' processes are going on more or less all the time more or less everywhere?" [120] It will come as a surprise to many in the philosophy of physics community that in fact QBism agrees heartily with Bell on this point. The words will have a different intention than Bell's, but indeed, *yes we are obliged!* However, as such we genuinely do need to say more about what "measurement-like" is. This much has been settled: When we are talking about an application of the quantum formalism concerning an agent taking an action on a system, the "measurement outcome" just is the consequent experience. So "measurement-like process" =

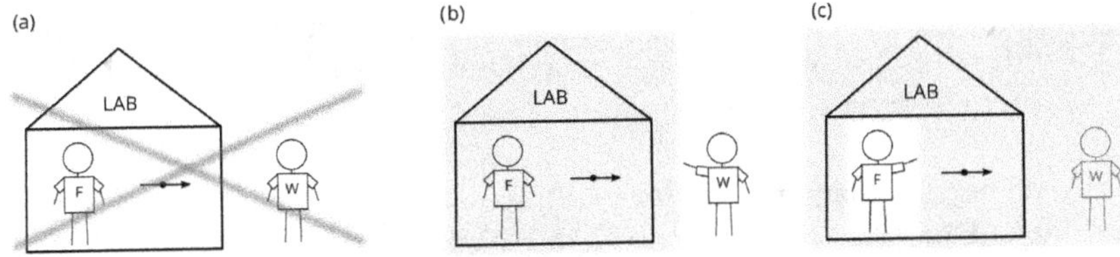

Figure 3.4 (a) In usual descriptions of various Wigner's-friend thought experiments, there is an urge to portray everything from a God's eye view. Here, we depict the thought experiment, where Wigner, his friend, and a spin-1/2 particle interact, and we symbolize QBism's disapproval of such portrayals with a big red X. In QBism, the quantum formalism is only used by agents who stand *within* the world; there is no God's-eye view. (b) Instead, in QBism, to make predictions, Wigner treats his friend, the particle, and the laboratory surrounding her (all shaded in green) as a physical system external to himself. While (c) to make her own predictions, the friend must reciprocally treat Wigner, the particle, and her surrounding laboratory (all shaded in green) as a physical system external to herself. It matters not that the laboratory spatially surrounds the friend; it, like the rest of the universe, is external to her agency, and that is what counts.

"lived experience" in this case. The suggestion then is that this applies more broadly, that experience[37] is somehow the very stuff of the world [4].

At first sight, it might appear that this move is of a spectrum with the various "perspectival" or "relational" ontologies being explored for quantum theory currently [121, 122]—for instance in Rovelli's relational quantum mechanics (RQM) [123], Brukner's relative facts [124], Healey's quantum pragmatism [125], or Glick's perspectival normative realism [20]. But if one looks closely at the apparatus expounded in this chapter, one will see that it's just not the case. The very words "perspectival" or "relational" evoke a block-universe conception of the world, a world which is in some sense already complete and unified. That for instance might be part of the psychology for the recent retreat of Adlam and Rovelli [126] from a more radical version of RQM [127, 128]. Compare that kind of thinking to this passage from QBist lore [129]:

> [I]f anything, the Bayesian account of quantum theory [now known as QBism] is essentially the opposite of solipsism. Rather than a unity to nature, it suggests a plurality. An image that might be useful (but certainly flawed) comes from Escher's various paintings of impossible objects. The viewer would initially like to think of them as 2D projections of a three-dimensional object; but he cannot. Now imagine how much worse it would get if we were to have two viewers with two slightly different paintings, each purporting to be a

> different perspective on "the" impossible object. Since neither viewer can lift from his own 2D object to a 3D one, there is no way to unify the pictures into a single whole.

So perspectivalism really does not work when it comes to QBism's ontological project. Experience is a far richer notion than a perspective on some pre-existing thing. As well it is far more than a simple relation among pre-existing things. Lived experience has an autonomy that neither of these notions capture. Each quantum measurement creates something new in the universe that is above and beyond the agent's relation to the quantum system they are acting upon. Quantum measurement in QBism is more like childbirth. Without a father and mother, there would be no child, but the child does not express merely the mother's relation to the father or vice versa. The child is something new and *sui generis*. To be "measurement-like"—in QBism's retort to John Bell—is to be like a fresh moment of creation not unakin to what is imagined of the big bang itself.[38]

Maurice Merleau-Ponty used reflections on a person touching one hand with the other to develop his ontology of chiasm and flesh. Here is a sample of the thinking [130, pp. 140–142]:[39]

> If we can show that the flesh is an ultimate notion, that it is not the union or compound of two substances, but thinkable by itself, if there is a relation of the visible with itself that traverses me and constitutes me as a seer, this circle which I do not form, which forms me, this coiling over of the visible upon the visible, can traverse, animate other bodies as well as my own. And if I was able to understand how this wave arises within me, how the visible which is yonder is simultaneously my landscape, I can understand a fortiori that elsewhere it also closes over upon itself and that there are other landscapes besides my own. If it lets itself be captivated by one of its fragments, the principle of captation is established, the field open for other Narcissus, for an "intercorporeity." If my left hand can touch my right hand while it palpates the tangibles, can touch it touching, can turn its palpation back upon it, why, when touching the hand of another, would I not touch in it the same power to espouse the things that I have touched in my own? It is true that "the things" in question are my own, that the whole operation takes place (as we say) "in me," within my landscape, whereas the problem is to institute another landscape. When one of my hands touches the other, the world of each opens upon that of the other because the operation is reversible at will, because they both belong (as we say) to one sole space of consciousness, because one sole man touches one sole thing through both hands. But for my two hands to open upon one sole world, it does not suffice that they

> be given to one sole consciousness—or if that were the case the difficulty before us would disappear: since other bodies would be known by me in the same way as would be my own, they and I would still be dealing with the same world. No, my two hands touch the same things because they are the hands of one same body. And yet each of them has its own tactile experience. If nonetheless they have to do with one sole tangible, it is because there exists a very peculiar relation from one to the other, across the corporeal space—like that holding between my two eyes—making of my hands one sole organ of experience, as it makes of my two eyes the channels of one sole Cyclopean vision. ... Now why would this generality, which constitutes the unity of my body, not open it to other bodies? The handshake too is reversible; I can feel myself touched as well and at the same time as touching ... Why would not the synergy exist among different organisms, if it is possible within each? Their landscapes interweave, their actions and their passions fit together exactly: this is possible as soon as we no longer make belongingness to one same "consciousness" the primordial definition of sensibility, and as soon as we rather understand it as the return of the visible upon itself, a carnal adherence of the sentient to the sensed and of the sensed to the sentient. For, as overlapping and fission, identity and difference, it brings to birth a ray of natural light that illuminates all flesh and not only my own.

It is pretty clear that this argument begs to be compared to the touch shared between Wigner and his friend in Figure 3.5. Could we pick up where Merleau-Ponty left off? For though M-P ends up with a metaphysics

Figure 3.5 Merleau-Ponty's hands, upgraded to the extreme. Eugene Wigner (right) and his friend Werner Heisenberg (left) passing a pen between themselves. One might as well say they are shaking hands. There is a "tingly bit" that, even if it cannot be expressed mathematically across the two agents' uses of the theory, must somehow be treated symmetrically. After all, Wigner and friend both are both agent and both system; it is only the story as related from one focus or the other that appears asymmetrical.

of "radical contingency" [131], just as QBism itself does [4, 49, 129], QBism still has something else to say. *Taken in turn*, each side of the divide in Figure 3.5 tolerates the addition of the normative quantum formalism. Why should that be? What does it tell us further about nature that the reversibility of the touch does not?

There can be no answer to these questions here, but at least now we know where to look. Moreover, there is already some good material in the literature to give us a start, some in this volume and some elsewhere [132–136]. As emphasized in the Introduction, Berghofer and Wiltsche ask, "The question is not whether QBists and phenomenologists should attempt to join forces, but what has taken us so long?" Indeed, let us forge this alliance *now*, for the battle will be hard!

Acknowledgments

First and foremost I thank those colleagues I called the "few exceptions" in the Introduction. Despite my damning statement, I have had many pleasant conversations with Guido Bacciagaluppi, Florian Boge, Harvey Brown, Jeff Bub, David Glick, Richard Healey, Jenann Ismael, Huw Price, Chris Timpson, Jos Uffink, and David Wallace. I also cherish what I learned from my friends Bill Demopoulos, Itamar Pitowsky, and Pat Suppes who are no longer with us. None of these colleagues managed to back me away from what they perceived as the dangers of QBism, but they respected the subject enough to have a conversation. Next, I thank Mario Hubert for a discussion that helped sow the seed of normative structural realism as descriptive of one aspect of QBism. Huge gratitude goes to Amanda Gefter for the generous sharing of her unpublished thoughts, which make their first public appearance here. Finally, I graciously thank the Institute for Quantum Optics and Quantum Information (IQOQI) in Vienna for its hospitality during a stage of the writing of this chapter. My academic great-great-great-grandfather Franz Exner, with his belief in indeterminism even before quantum mechanics, walked the same streets, and I feel his spirit looks in on QBism whenever I am there. This work was supported in part by National Science Foundation Grant 2210495 and in part by Grant 62424 from the John Templeton Foundation. The opinions expressed in this publication are those of the author and do not necessarily reflect the views of the John Templeton Foundation.

Notes

1 Say, starting in 1927 with Heisenberg's uncertainty principle paper and Bohr's complementarity.

2 Is it a pilot wave? Is it parallel worlds? Does it consist of flashes of wavefunction collapse? How about consistent histories? Maybe relations without

relata? The full list of speculative ontologies that have been proposed is surely longer than can fit into any reasonably sized footnote.

3 So that Bohr might revise (were he alive today) his opinion of philosophers: "I felt that philosophers were very odd people who really were lost, because they have not the instinct that it is important to learn something and that we must be prepared really to learn something of very great importance." [137]

4 Though, see my story in response to Question 14 of [48] for an amusing follow-up.

5 To be fair, some papers were at least honest efforts at it, as for instance [138], but plenty more were irredeemably misguided. Perhaps the most egregious of the latter was [139] by John Earman—a senior and quite respected man in the field—for it showed quite bluntly that he must not have read much of the QBist literature he cited [140].

6 See acknowledgements at the end of the chapter.

7 See for instance his monumental essay "Axioms as Postulates" [141].

8 See the table on page 2 of Ref. [54].

9 As an aside, one might argue that this is the best sense of "explanation" as well [142].

10 There is a wonderful dialog in the movie *The Ballad of Buster Scruggs* directed by Joel and Ethan Coen that captures this idea perfectly. The setting is that of five people squeezed into a stagecoach heading to a certain Fort Morgan. The key part of dialog is between a Frenchman (René) and a morally upright Lady (Mrs. Betjeman):

LADY: I have been living with my daughter and son-in-law these last three years.

FRENCHMAN: Ahh, the parent should not burden the household of the child. This was wrong of you, *madame*.

LADY: I was not a burden! I was welcome in my daughter's house!

FRENCHMAN: Ahh, she would say so, of course. But no doubt you could read in her facial expressions ... that your presence was not wanted. We each have a life. Each a life, only our own.

LADY: You know nothing of me or my domestic affairs!

FRENCHMAN: I know that we must each spin our wheel, play our own hand. I was once at cards with a man named Cipolski. This was very many years ago ... We were at cards. My hand was poor; I folded. But Cipolski and four others remain. Cipolski said to me, René I am in distress, you must play for me while I perform *mes nécessités*—ah, my necessaries. (*Lady giggles uncomfortably.*) I say, friend, no I cannot wager for you. He says, but of course you can, we know each other well—you wager as I would do. I say, it is quite *impossible*, no? How a man wagers, it is decided by who he is, by the entirety of his relation to poker right up until the moment of that bet. I cannot bet for you. *Pourquoi pas*? I cannot know you. Not to this degree. We must each play our own hand. No, Cipolski, I say. No, we may call each other friend, but puh we cannot know each other so ...

LADY: Poker is a gambling game. (*Frenchman nods.*) You have pursued a life of vice and dissipation—and you are no doubt expert in such pursuits. But no conclusions drawn from such an existence will apply to a life rightly lived.

FRENCHMAN: Life is life. Cards will teach you what you need to know. ...

11 Or at least probability theory over finite sets of events.
12 Note the similarity between this notion and the way R. T. Cox, the founder of E. T. Jaynes's preferred school of Bayesianism, models the notion of a *question*: A question is the set of all its possible *answers* [143].
13 In this notion of quantum measurement, there is no requirement that the operators $\hat{E}_i$ be orthogonal to each other. Nor are the cardinalities of the sets $\mathcal{E}$ and $\mathcal{E}'$ limited by the dimension of the Hilbert space. This is because $\hat{E}_i$ are formal tools for associating probabilities with the agent's mutually exclusive experiences E_i. This may be a point of confusion for those in the philosophy of physics community who are more familiar with the von Neumann notion of measurement or with propositional concepts drawn from the tradition of quantum logic.
14 As Earman's imagined QBians do [139,140].
15 See the recent paper of J. Pienaar [144] for a much fuller explication of this than anything written before.
16 Perhaps the only exception is Richard Healey's "pragmatist interpretation" of quantum theory [125], but it is so very far from QBism in what it takes to be a norm—for instance, it even posits a "correct" quantum state for a given perspective as a norm [145]—that it essentially shares nothing in common with QBism except the word "normative." Moreover, it is simply a wholly different interpretative framework: Quantum mechanics could be posed even if there were never an agent in sight, and sin of all sins, it supports a block-universe picture of nature!
17 Introducing a *reference measurement* is a conceptual technique with a long history in QBism, going back to the proof of the quantum de Finetti theorem [12] so crucial for analyzing the notion of an "unknown quantum state" from a QBist perspective. The notion of a reference apparatus is the same concept but with the addition of a set of post-measurement quantum states for the system.
18 This is to say, even classical physics can be viewed in a normative light. One is just not accustomed to thinking of it that way.
19 In this regard, it can be instructive to look at a "simple" quantum information task such as quantum teleportation. First, see it as it is presented on paper [146], where the protocol can be depicted in diagrams almost as simple and stark as Bohr's illustration above. Then compare *that* to an image of an actual laboratory implementation of the same, such as in this photograph [147] with its more than 500 mirrors and lenses.
20 For those interested, see the optional Section 3.2.5.1 for more detail.
21 See also these two early expressions in the QBist corpus [129, 148] for a more full-flavored account.
22 The current state of knowledge is that they have been *proven* to exist via exact algebraic solutions in over 150 distinct dimensions extending up to $d = 39{,}604$, and there is high precision numeral evidence for all $d \leq 193$ [149]
23 From a different direction, but also to some extent mathematically deep, see [150,151].

24 By the way, the simplicity of a form like this arising from the essential quantumness question is not at all a given. A proof of principle can be found in the well-studied foil to standard quantum theory known as real-vector-space quantum mechanics. In that theory, everything in the formalism is identical to standard quantum theory, except that it is based on real vector spaces rather than complex: Vectors always have real components with respect to any basis, symmetric operators take the place of Hermitian operators, and rotations take the place of unitaries. As it turns out, an analogue of a SIC (call it a real-SIC) generally does not exist in that theory. For instance real-SICs do exist for dimensions $d = 2,3,7$, and 23, but can be proven not to exist for any of the dimensions in between. So, what becomes of the essential quantumness in the dimensions for which a real-SIC does not exist? Recently, we explored the issue for $d = 4$ and the result was strikingly ugly [152]! In that setting, Eq. (27) is replaced by an expression requiring four complicated lines of symbols for its display, and even then the expression depends on a floating parameter according to which unitarily invariant norm in Eq. (21) one uses. Nothing appears universal about it. If one had to settle for a representation of the Born rule like that in standard complex quantum theory, one would wonder why even bother?

25 Some paraphrasing for punchiness and clarity. Also units have been adjusted to be on the same footing.

26 Perhaps this paper [153] is also relevant?

27 As William James said he almost cried aloud with "glee" and "admiration" during the 1906 San Francisco earthquake [154, p. 3].

28 Perhaps the essential point is expressed the most vividly in this paper by N. D. Mermin [155].

29 From our point of view what these two no-go theorems prove is that one cannot have the *conjunction* of two statements: A) that quantum states are epistemic, and B) that they are *epistemic about* (i.e., knowledge, information, or beliefs about) some pre-existent ontic variables. Holding fast to the notion that quantum states must be epistemic, it then follows that they cannot be epistemic about pre-existent ontic variables. Put another way, these theorems give us reason to reject the "ontic models framework" of Harrigan and Spekkens [156] where unperformed experiments *do have results*, not reject that quantum states are epistemic. It is noteworthy that the original titles for the original postings of PBR and CR were "The quantum state cannot be interpreted statistically" and "Completeness of quantum theory implies that wave functions are physical properties," respectively. Neither title could be sustained in the light of QBism.

30 Of course, QBism here relies on a coherence argument instead of a frequentist one, but the point is the same.

31 Post-Peierls compatibility is the one later popularized by Pusey, Barrett, and Rudolph [104] in their attempt to give a no-go theorem for epistemic interpretations of quantum states.

32 Except, that is, for the uniquely QBist criteria W and W' in [113], where any two quantum states whatsoever are compatible.

33 Though, in place of his original mathematical expressions we will use Dirac notation and some notational shortcuts for ease of reading.

34 As well, we will disambiguate those instances in his text where he writes "observer" when he clearly means "friend."

35 See for instance [157] for some discussion of James's "rich notion of experience." Until our recent readings of Merleau-Ponty, this has been the historical model QBism has followed.
36 Remember the definitions from Section 3.2.1.
37 I say "experience" here, but more work should be done to find the right word for this de-anthropocentrized ontological distillate of QBist considerations. Other potential terms are pure experience [158], neutral monism [159,160], phenomenon [161, pp. 178–192], elementary quantum phenomenon [162], actual occasion [163], intra-action [164], trans-action [165], chiasm [130], flesh [130], or even QBoom [49, pp. 2193–2194, 2213–2214].
38 See the discussion between John Wheeler and theologian Richard Elvee in [166]:

ELVEE: Dr. Wheeler, who was there to observe the universe when it started? Were we there? Or does it only start with our observation? Is the big bang here?

WHEELER: A lovely way to put it—"Is the big bang here?" I can imagine that we will someday have to answer your question with a "yes." ... Each elementary quantum phenomenon is an elementary act of "fact creation." That is incontestable. But is that the only mechanism needed to create all that is? Is what took place at the big bang the consequence of billions upon billions of these elementary processes, these elementary "acts of observer-participancy," these quantum phenomena? Have we had the mechanism of creation before our eyes all this time without recognizing the truth? That is the larger question implicit in your comment.

39 I apologize for introducing this lengthy quote, but it seemed important to have it displayed here and now. Nonetheless, it comes from a man who was in deep need of some paragraphing lessons.

References

1. C. A. Fuchs, "QBism, the Perimeter of Quantum Bayesianism," arXiv: 1003.5209.
2. C. A. Fuchs and R. Schack, "Quantum-Bayesian Coherence," *Rev. Mod. Phys.* **85**, 1693–1715 (2013); arXiv: 1301.3274.
3. C. A. Fuchs, N. D. Mermin and R. Schack, "An Introduction to QBism With an Application to the Locality of Quantum Mechanics," *Am. J. Phys.* **82**, 749–754 (2014); arXiv: 1311.5253.
4. C. A. Fuchs, "Notwithstanding Bohr, the Reasons for QBism," *Mind Matter.* **15**, 245–300 (2017); arXiv: 1705.03483.
5. C. M. Caves, C. A. Fuchs and R. Schack, "Quantum Probabilities as Bayesian Probabilities," *Phys. Rev. A.* **65**, 022305 (2002); arXiv: quant-ph/0106133.
6. W. Mischel, *The Marshmallow Test: Why Self-Control Is the Engine of Success* (Little, Brown and Company, New York, NY, 2014).

7. C. A. Fuchs, *Quantum States: What the Hell Are They?*, unpublished (2002).
8. A. Gefter, *Trespassing on Einstein's Lawn: A Father, a Daughter, the Meaning of Nothing, and the Beginning of Everything* (Bantam Books, New York, NY, 2014).
9. A. Gefter, "A Private View of Quantum Reality," *Quanta Magazine*, 4 June 2015.
10. A. Gefter, "Amanda Gefter on Schrödinger's QBist cat," in "The many meanings of Schrödinger's cat," IAI News, 24 February 2022.
11. W. Durant, *The Story of Philosophy: The Lives and Opinions of the Greater Philosophers* (Pocket Books, New York, NY, 2006).
12. C. M. Caves, C. A. Fuchs and R. Schack, "Unknown Quantum States: The Quantum De Finetti Representation," *J. Math. Phys.* **43**, 4537–4559 (2002); arXiv: quant-ph/0104088.
13. M. G. A. Paris and J. Řeháček eds., *Quantum State Estimation*, Lecture Notes in Physics, Vol. 649 (Springer, Berlin, 2004).
14. B. de Finetti, "Probabilism: A Critical Essay on the Theory of Probability and on the Value of Science," *Erkenntnis.* **89**, 169–223 (1989).
15. R. Jeffrey, "Reading *Probabilismo*," *Erkenntnis.* **89**, 225–237 (1989).
16. B. de Finetti, *L'invenzione della verita* (Raffaello Cortina Editore, Milan, 2006).
17. W. James, "G. Papini and the Pragmatist Movement in Italy," in *The Works of William James,* Vol. 5: Essays in Philosophy (Harvard University Press, Cambridge, MA, 1978), pp. 144–148.
18. C. G. Timpson, "Quantum Bayesianism: A Study," *Stud. Hist. Philos. Mod. Phys.* **39**, 579–609 (2008); arXiv: 0804.2047.
19. G. Bacciagaluppi, "A Critic Looks at qBism," in *New Directions in the Philosophy of Science*," edited by M. C. Galavotti, D. Dieks, W. Gonzalez, S. Hartmann, T. Uebel, and M. Weber (Springer, Cham, Switzerland, 2020), pp. 403–416; http://philsci-archive.pitt.edu/9803/.
20. D. Glick, "QBism and the Limits of Scientific Realism," *Euro. J. Philos. Sci.* **11**, 53 (2021).
21. C. A. Fuchs and B. C. Stacey, "Are Non-Boolean Event Structures the Precedence or Consequence of Quantum Probability?," arXiv: 1912.10880.
22. S. H. Hodgson, *The Metaphysic of Experience*, in four books (Longmans, Green, and Co., London, 1898); reprinted by Thoemmes Press, Bristol, England, 2001.
23. D. C. Lamberth, *William James and the Metaphysics of Pure Experience* (Cambridge University Press, Cambridge, UK, 1999).
24. B. C. Stacey, "Towards a World Game-Flavored as a Hawk's Wing," to appear in this volume; for an earlier version of the paper see, B. C. Stacey, "Ideas Abandoned en Route to QBism," arXiv: 1911.07386.
25. C. A. Fuchs and B. C. Stacey, "QBism: Quantum Theory as a Hero's Handbook," in *Proceedings of the International School of Physics "Enrico Fermi" Course 197 – Foundations of Quantum Physics*, edited by E. M. Rasel, W. P. Schleich, and S. Wölk (IOS Press, Amsterdam; Società Italiana di Fisica, Bologna, 2018), pp. 133–202; arXiv: 1612.07308.
26. N. D. Mermin, "Making Better Sense of Quantum Mechanics," *Rep. Prog. Phys.* **58**, 012002 (2019); arXiv: 1809.01639.

27. E. P. Wigner, "Remarks on the Mind-Body Question," in *The Scientist Speculates*, edited by I. J. Good (William Heinemann, Ltd., London, 1961), pp. 284–302; reprinted in E. P. Wigner, *Symmetries and Reflections: Scientific Essays of Eugene Wigner* (Ox Bow Press, Woodbridge, CT, 1979), pp. 171–184.
28. D. Frauchiger and R. Renner, "Quantum Theory Cannot Consistently Describe the Use of Itself," *Nat. Commun.* **9**, article 3711 (2018); arXiv: 1604.07422.
29. V. Baumann and Č Brukner, "Wigner's Friend as a Rational Agent," in *Quantum, Probability, Logic: The Work and Influence of Itamar Pitowsky*, edited by M. Hemmo and O. Shenker (Springer, Cham, Switzerland, 2020), pp. 91–99; arXiv: 1901.11274.
30. E. Cavalcanti, "The View from the Wigner Bubble," *Found. Phys.* **51**, 39 (2021); arXiv: 2008.05100.
31. A. Khrennikov, "Towards Better Understanding QBism," *Found. Sci.* **23**, 181–195 (2018).
32. T. Müller and H. J. Briegel, "A Stochastic Process Model for Free Agency Under Indeterminism," *Dialectica.* **72**, 219–252 (2018).
33. H. J. Briegel, "On Creative Machines and the Physical Origins of Freedom," *Sci. Rep.* 2, 522 (2012).
34. J. B. DeBrota and B. C. Stacey, "FAQBism," arXiv: 1810.13401.
35. J. B. DeBrota, C. A. Fuchs and R. Schack, "Respecting One's Fellow: QBism's Analysis of Wigner's Friend," *Found. Phys.* **50**, 1859–1874 (2020); arXiv: 2008.03572.
36. A. Furusawa, J. L. Sørensen, S. L. Braunstein, C. A. Fuchs, H. J. Kimble and E. S. Polzik, "Unconditional Quantum Teleportation," *Science.* **282**, 706–709 (1998).
37. J. Pienaar, private communication, 7 December 2022.
38. C. A. Fuchs and C. S. Powell, "Quantum Physics is No More Mysterious Than Crossing the Street: A Conversation with Chris Fuchs," Discover Magazine, 29 November 2019.
39. J. A. Wheeler, "How Come the Quantum?" in *New Techniques and Ideas in Quantum Measurement Theory*, edited by D. M. Greenberger, Ann. N. Y. Acad. Sci. **480**, 304–316 (1987).
40. L. J. Savage, *The Foundations of Statistics*, 2nd ed. (Dover, New York, NY, 1972).
41. J. M. Bernardo and A. F. M. Smith, *Bayesian Theory* (Wiley, Chichester, 1994).
42. B. de Finetti, *Theory of Probability* (Wiley, New York, NY, 1990).
43. J. Berkovitz, "On de Finetti's Instrumentalist Philosophy of Probability," Euro. *J. Philos. Sci.* **9**, 25 (2019).
44. E. Schrödinger, "The Present Situation in Quantum Mechanics: A Translation of Schrödinger's 'Cat Paradox' Paper," translated by J. D. Trimmer, *Proc. Am. Philos. Soc.* **124**, 323–338 (1980).
45. E. T. Jaynes, "Probability in Quantum Theory," in *Complexity, Entropy, and the Physics of Information*, edited by W. H. Zurek (Addison-Wesley, New York, NY, 1990), pp. 381–403.
46. M. A. Nielsen and I. L. Chuang, *Quantum Computation and Quantum Information*, 10th Anniversary Edition (Cambridge University Press, Cambridge, England, 2010).
47. A. Peres, "Unperformed Experiments Have No Results," *Am. J. Phys.* **46**, 745–747 (1978).

48. C. A. Fuchs, "Interview with a Quantum Bayesian," in *Elegance and Enigma: The Quantum Interviews*, edited by M. Schlosshauer (Springer, Berlin, Frontiers Collection, 2011). arXiv: 1207.2141.
49. C. A. Fuchs, *My Struggles with the Block Universe: Selected Correspondence, January 2001 – May 2011*, edited by Blake C. Stacey, foreword by Maximilian Schlosshauer (2014), 2,349 pages; arXiv: 1405.2390.
50. C. A. Fuchs, "On Participatory Realism," in *Information and Interaction: Eddington, Wheeler, and the Limits of Knowledge*, edited by I. T. Durham and D. Rickles (Springer, Berlin, 2016), pp. 113–134; arXiv: 1601.04360.
51. C. A. Fuchs, "Copenhagen Interpretation Delenda Est?," arXiv: quant-ph/ 1809.05147.
52. C. A. Fuchs, *Coming of Age with Quantum Information: Notes on a Paulian Idea* (Cambridge University Press, Cambridge, UK, 2010).
53. W. Pauli, *Writings on Physics and Philosophy*, edited by C. P. Enz and K. von Meyenn (Springer-Verlag, Berlin, 1994).
54. C. A. Fuchs, "Quantum Mechanics as Quantum Information (and only a little more)," arXiv: quant-ph/0205039.
55. W. Pauli, letter to Niels Bohr, dated 15 February 1955, photocopy obtained from the Niels Bohr Institute via H. J. Folse. More extensive excerpts from the letter can be found in Ref. [4].
56. C. M. Caves, C. A. Fuchs and R. Schack, "Subjective Probability and Quantum Certainty," *Stud. Hist. Philos. Mod. Phys.* **38**, 255–274 (2007); arXiv: quant-ph/0608190.
57. N. D. Mermin, "Why QBism Is Not the Copenhagen Interpretation and What John Bell Might Have Thought of It," in *Quantum [Un]Speakables II: Half a Century of Bell's Theorem*, edited by R. Bertlmann and A. Zeilinger (Springer, Berlin, 2017), pp. 83–94; arXiv: 1409.2454.
58. R. Healey, "Quantum-Bayesian and Pragmatist Views of Quantum Theory," in *The Stanford Encyclopedia of Philosophy*, edited by Edward N. Zalta (The Metaphysics Research Lab, Philosophy Department, Stanford University, Stanford, 2016). https://plato.stanford.edu/.
59. D. Wallace, "A Case for QBism," presentation at the XII International Ontology Congress, San Sebastian, Spain, 5 October 2016.
60. A. Barzegar, "QBism Is Not So Simply Dismissed," *Found. Phys.* **50**, 693–707 (2020).
61. H. R. Brown, "The Reality of the Wavefunction: Old Arguments and New," in *Philosophers Look at Quantum Mechanics*, edited by A. Cordero (Springer, Berlin, 2019), pp. 63–86.
62. J. B. DeBrota and B. C. Stacey, "Discrete Wigner Functions from Informationally Complete Quantum Measurements," *Phys. Rev.* A. **102**, 032221 (2020); arXiv: 1912.07554.
63. J. B. DeBrota, C. A. Fuchs and B. C. Stacey, "The Varieties of Minimal Tomographically Complete Measurements," *Int. J. Quant. Info.* 2040005 (2020); arXiv: 1812.08762.
64. J. B. DeBrota, C. A. Fuchs and B. C. Stacey, "Symmetric Informationally Complete Measurements Identify the Irreducible Difference between Classical and Quantum Systems," *Phys. Rev. Res.* **2**, 013074 (2020); arXiv: 1805.08721.

65. N. Bohr, "Discussion with Einstein on Epistemological Problems in Atomic Physics," in *Albert Einstein: Philosopher-Scientist*, edited by P. A. Schilpp (MJF Books, New York, NY, 1970), pp. 201–241.
66. T. Maudlin, "The Metaphysics of Quantum Theory," *Belgrade Philos. Annu.* **29**, 5–13 (2016).
67. E. Schrödinger, *Nature and the Greeks and Science and Humanism* (Cambridge University Press, Cambridge, UK, 2014).
68. J. Ladyman,"Structural Realism," in *The Stanford Encyclopedia of Philosophy*, edited by Edward N. Zalta (The Metaphysics Research Lab, Philosophy Department, Stanford University, Stanford, 2020). https://plato.stanford.edu/.
69. C. A. Fuchs, "Interview with Physicist Christopher Fuchs," with R. P. Crease and J. Sares, *Cont. Philos. Rev.* **54**, 541–561 (2021).
70. C. Callender, private communication, Grindavik, Iceland, July 2015.
71. C. A. Fuchs and R. Schack, "From Quantum Interference to Bayesian Coherence and Back Round Again," in *Foundations of Probability and Physics – 5*, edited by L. Accardi et al., AIP Conference Proceedings, Vol. 1101 (American Institute of Physics, Melville, NY, 2009), pp. 260–279.
72. G. Zauner, *Quantum Designs: Foundations of a Noncommutative Design Theory*, PhD thesis, University of Vienna (1999); translated in Int. J. Quant. Inf. 9, 445–507 (2011).
73. J. M. Renes, R. Blume-Kohout, A. J. Scott and C. M. Caves, "Symmetric Informationally Complete Quantum Measurements," *J. Math. Phys.* **45**, 2171–2180 (2004); arXiv: quant-ph/0310075.
74. C. A. Fuchs, M. C. Hoang and B. C. Stacey, "The SIC Question: History and State of Play," *Axioms.* **6**, 21 (2017); arXiv: 1703.07901.
75. F. J. Boge, "Back to Kant! QBism, Phenomenology, and Reality from Invariants," to appear in this volume; http://philsci-archive.pitt.edu/21571/.
76. R. A. Horn and C. R. Johnson, *Topics in Matrix Analysis* (Cambridge University Press, Cambridge, UK, 1994).
77. T. Durt, C. Kurtsiefer, A. Lamas-Linares and A. Ling, "Wigner Tomography of Two-Qubit States and Quantum Cryptography," *Phys. Rev. A.* **78**, 042338 (2008); arXiv: 0806.0272.
78. Z. E. D. Medendorp, F. A. Torres-Ruiz, L. K. Shalm, G. N. M. Tabia, C. A. Fuchs and A. M. Steinberg, "Experimental Characterization of Qutrits Using Symmetric Informationally Complete Positive Operator-Valued Measurements," *Phys. Rev. A.* **83**, 051801 (2011); arXiv: 1006.4905.
79. Y.-Y. Zhao, N.-K. Yu, P. Kurzyński, G.-Y. Xiang, C.-F. Li and G.-C. Guo, "Experimental Realization of Generalized Qubit Measurements Based on Quantum Walks," *Phys. Rev. A.* **91**, 042101 (2015); arXiv: 1501.05096.
80. M. Appleby, S. Flammia, G. McConnell and J. Yard, "SICs and Algebraic Number Theory," *Found. Phys.* **47**, 1042–1059 (2017); arXiv: 1701.05200.
81. M. Appleby, I. Bengtsson, M. Grassl, M. Harrison and G. McConnell, "SIC-POVMs from Stark Units: Prime Dimensions," *J. Math. Phys.* **63**, 112205 (2022); arXiv: 2112.05552.
82. C. A. Fuchs and R. Schack, "A Quantum-Bayesian Route to Quantum-State Space," *Found. Phys.* **41**, 345–356 (2011); arXiv: 0912.4252.

83. D. M. Appleby, Å Ericsson and C. A. Fuchs, "Properties of QBist State Spaces," *Found. Phys.* **41**, 564–579 (2011); arXiv: 0910.2750.
84. D. M. Appleby, C. A. Fuchs, B. C. Stacey and H. Zhu, "Introducing the Qplex: A Novel Arena for Quantum Theory," *Euro. Phys. J.* **D 71**, 197 (2017); arXiv: 1612.03234.
85. J. B. DeBrota, C. A. Fuchs, J. L. Pienaar and B. C. Stacey, "Born's Rule as a Quantum Extension of Bayesian Coherence," *Phys. Rev. A.* **104**, 022207 (2021); arXiv: 2012.14397.
86. B. C. Stacey, "SICs and Bell Inequalities," in *A First Course in the Sporadic SICs*, Springer Briefs in Mathematical Physics, Vol. 41 (Springer, Cham, 2021), pp. 39–55; for an earlier version of the paper see, B. C. Stacey, "Is the SIC Outcome There When Nobody Looks?," arXiv: 1807.07194.
87. A. Gefter, *John Archibald Wheeler & The Problem of Multiple Observership*, presentation at the Stellenbosch Institute for Advanced Study (STIAS), Stellenbosch, South Africa, 30 May 2018.
88. C. M. Patton and J. A. Wheeler, "Is Physics Legislated by Cosmogony?," in *Quantum Gravity: An Oxford Symposium*, edited by C. J. Isham, R. Penrose, and D. W. Sciama (Clarendon Press, Oxford, UK, 1975), pp. 538–605.
89. J. A. Wheeler, "World as System Self-Synthesized by Quantum Networking," *IBM J. Res. Dev.* **32**, 4–15 (1988).
90. J. A. Wheeler, "Information, Physics, Quantum: the Search for Links," in *Proceedings of the 3rd International Symposium on Foundations of Quantum Mechanics in the Light of New Technology*, edited by S. Kobayashi, H. Ezawa, Y. Murayama, and S. Nomura (Physical Society of Japan, Tokyo, 1990), pp. 354–368.
91. C. M. Caves and C. A. Fuchs, "Quantum Information: How Much Information in a State Vector?," in *The Dilemma of Einstein, Podolsky and Rosen – 60 Years Later*, edited by A. Mann and M. Revzen, Ann, Israel Phys. Soc. **12**, 226–257 (1996); arXiv: quant-ph/96010125.
92. C. A. Fuchs, "Quantum Foundations in the Light of Quantum Information," in *Decoherence and Its Implications in Quantum Computation and Information Transfer*, edited by A. Gonis and P. E. A. Turchi (IOS Press, Amsterdam, 2001), pp. 38–82; arXiv: quant-ph/0106166.
93. G. M. D'Ariano, P. Lo Presti and M. G. A. Paris, "Using Entanglement Improves the Precision of Quantum Measurements," *Phys. Rev. Lett.* **87**, 270404 (2001); arXiv: quant-ph/0109040.
94. G. Chiribella, G. M. D'Ariano and P. Perinotti, "Optimal Cloning of Unitary Transformation," *Phys. Rev. Lett.* **101**, 180504 (2008); arXiv: 0804.0129.
95. S. F. Huelga, J. A. Vaccaro, A. Chefles and M. B. Plenio, "Quantum Remote Control: Teleportation of Unitary Operations," *Phys. Rev. A.* **63**, 042303 (2001); arXiv: quant-ph/0005061.
96. C. A. Fuchs, R. Schack and P. F. Scudo, "A De Finetti Representation Theorem for Quantum Process Tomography," *Phys. Rev. A.* **69**, 062305 (2004); arXiv: quant-ph/0307198.
97. C. A. Fuchs and R. Schack, "Unknown Quantum States and Operations, a Bayesian View," in *Quantum Estimation Theory*, edited by M. G. A. Paris and J. Řeháček (Springer-Verlag, Berlin, 2004), pp. 151–190; arXiv: quant-ph/0404156.

98. I. J. Good, "46656 Varieties of Bayesians," in *Good Thinking: The Foundations of Probability and Its Applications* (Dover, Mineola, NY, 1983), pp. 20–21.
99. E. T. Jaynes, *Probability Theory: The Logic of Science* (Cambridge University Press, Cambridge, UK, 2003).
100. R. Jeffrey, "Introduction: Radical Probabilism," in *Probability and the Art of Judgment* (Cambridge University Press, Cambridge, UK, 1992), pp. 1–13.
101. D. V. Lindley, *Understanding Uncertainty* (Wiley, Hoboken, NJ, 2006).
102. Č Brukner and A. Zeilinger, "Conceptual Inadequacy of the Shannon Information in Quantum Measurements," *Phys. Rev. A.* **63**, 022113 (2001); arXiv: quant-ph/0006087.
103. M. J. R. Gilton, "Whence the Eigenstate-Eigenvalue Link," *Stud. Hist. Philos. Mod. Phys.* **55**, 92–100 (2016).
104. M. F. Pusey, J. Barrett and T. Rudolph, "On the Reality of the Quantum State," *Nat. Phys.* **8**, 475–478 (2012).
105. R. Colbeck and R. Renner, "Is a System's Wave Function in One-to-One Correspondence With Its Elements of Reality?" *Phys. Rev. Lett.* **108**, 150402 (2012); arXiv: 1111.6597.
106. R. P. Feynman, "The Concept of Probability in Quantum Mechanics," in *Proceedings of the Second Berkeley Symposium on Mathematical Statistics and Probability*, edited by J. Neyman (University of California Press, Berkeley, CA, 1951), pp. 533–541.
107. R. P. Feynman, "Space-Time Approach to Non-Relativistic Quantum Mechanics," *Rev. Mod. Phys.* **20**, 367–387 (1948).
108. D. V. Lindley, "Comment on A. P. Dawid's, 'The Well-Calibrated Bayesian,'" *J. Am. Stat. Assoc.* **77**, 611–612 (1982).
109. R. E. Peierls, "In Defense of 'measurement,'" *Phys. World.* **4**(1), 19–21 (1991).
110. R. Peierls, "Observations and the 'Collapse of the Wave Function,'" in *More Surprises in Theoretical Physics* (Princeton University Press, Princeton, NJ, 1991), pp. 6–11.
111. N. D. Mermin, "Whose Knowledge?," in *Quantum (Un)speakables: From Bell to Quantum Information*, edited by R. Bertlmann and A. Zeilinger (Springer, Berlin, 2001), pp. 271–280; arXiv: quant-ph/0107151.
112. T. A. Brun, J. Finkelstein and N. D. Mermin, "How Much State Assignments can Differ," *Phys. Rev. A.* **65**, 032315 (2002); arXiv: quant-ph/0109041.
113. C. M. Caves, C. A. Fuchs and R. Schack, "Conditions for Compatibility of Quantum State Assignments," *Phys. Rev. A.* **66**, 062111 (2002); arXiv: quant-ph/0206110.
114. B. C. Stacey, "SIC-POVMs and Compatibility Among Quantum States," *Mathematics.* **4**, 36 (2016); arXiv: 1404.3774.
115. C. A. Fuchs and R. Schack, "Bayesian Conditioning, the Reflection Principle, and Quantum Decoherence," in *Probability in Physics*, edited by Y. Ben-Menahem and M. Hemmo (Springer, Berlin, Frontiers Collection, 2012), pp. 233–247; arXiv: 1103.5950.
116. V. Baumann, Č Brukner, E. Cavalcanti, H. Wiseman and W. Zeng, organizers, Wigner's Friends: Theory Workshop, The Institute, Salesforce Tower, San Fransisco, CA, 3 Nov.–2 Dec. 2022.

117. D. M. Greenberger, "Can a Computer Ever Become Conscious?," Vienna Quantum Cafe blog, 14 April 2014.
118. C. A. Fuchs and R. Schack, "Quantum Measurement and the Paulian Idea," in *The Pauli-Jung Conjecture and Its Impact Today*, edited by H. Atmanspacher and C. A. Fuchs (Imprint Academic, Exeter, UK, 2014), pp. 93–107; arXiv: 1412.4209.
119. W. James, "On Some Hegelisms," from 1882, in W. James, *The Will to Believe and Other Essays in Popular Philosophy; Human Immortality—Both Books Bound as One* (Dover, New York, NY, 1956), pp. 263–298.
120. J. Bell, "Against 'measurement," *Phys. World.* **3**(8), 33–40 (1990).
121. A. Barzegar and D. Oriti, "Epistemic-Pragmatist Interpretations of Quantum Mechanics: A Comparative Assessment," arXiv: 2210.13620.
122. Č Brukner, "Wigner's Friend and Relational Objectivity," *Nat. Rev. Phys.* **4**, 628–630 (2022).
123. C. Rovelli, "Relational Quantum Mechanics," *Int. J. Theor. Phys.* **35**, 1637–1678 (1996).
124. Č Brukner, "A No-Go Theorem for Observer-Independent Facts," *Entropy.* **20**, 350 (2018); arXiv: 1804.00749.
125. R. Healey, *The Quantum Revolution in Philosophy* (Oxford University Press, Oxford, UK, 2017).
126. E. Adlam and C. Rovelli, "Information is Physical: Cross-Perspective Links in Relational Quantum Mechanics, arXiv: 2203.13342.
127. J. Pienaar, "QBism and Relational Quantum Mechanics Compared," *Found. Phys.* **51**, 96 (2021); arXiv: 2108.13977.
128. J. Pienaar, "A Quintet of Quandaries: Five No-Go Theorems for Relational Quantum Mechanics," *Found. Phys.* **51**, 97 (2021); arXiv: 2107.00670.
129. C. A. Fuchs, "Delirium Quantum: Or, where I will take quantum mechanics if it will let me," in *Foundations of Probability and Physics – 4*, edited by G. Adenier, C. A. Fuchs, and A. Yu. Khrennikov, AIP Conference Proceedings, Vol. 889 (American Institute of Physics, Melville, NY, 2007), pp. 438–462; arXiv: 0906.1968.
130. M. Merleau-Ponty, *The Visible and Invisible*, edited by C. Lefort, translated by A. Lingis (Northwestern University Press, Evanston, IL, 1968).
131. M. Bertram, "The Different Paradigms of Merleau-Ponty and Whitehead," *Philos. Today* **24**(2), 121–132 (1980).
132. R. Schack, "A QBist reads Merleau-Ponty," to appear in this volume; arXiv: 2212.11094.
133. M. Bitbol, "A Phenomenological Ontology for Physics: Merleau-Ponty and QBism," in *Phenomenological Approaches to Physics*, edited by H. A. Wiltsche and P. Bergofer (Springer, Cham, 2020), pp. 227–242; http://philsci-archive.pitt.edu/19512/.
134. L. de la Tremblaye and M. Bitbol, "Towards A Phenomenological Constitution Of Quantum Mechanics: A QBist Approach," *Mind Matter* **20**, 35–62 (2022).
135. M. Bitbol and L. de la Tremblaye, "QBism: An Eco-Phenomenology of Quantum Physics," to appear in this volume; http://philsci-archive.pitt.edu/20090/.

136. H. C. von Baeyer, "On the Consilience between QBism and Phenomenology," to appear in this volume; arXiv: 2201.04734.
137. Interview of Niels Bohr by T. S. Kuhn, L. Rosenfeld, A. Petersen, and E. Rudinger on 1962 November 17, Niels Bohr Library & Archives, American Institute of Physics, College Park, MD, USA, www.aip.org/history-programs/niels-bohr-library/oral-histories/4517-5.
138. W. C. Myrvold, "Subjectivists About Quantum Probabilities Should Be Realists About Quantum States," in *Quantum, Probability, Logic: The Work and Influence of Itamar Pitowsky*, edited by M. Hemmo and O. Shenker (Springer, Cham, Switzerland, 2020), pp. 449–465; http://philsci-archive.pitt.edu/16656/.
139. J. Earman, "Quantum Bayesianism Assessed," *Monist.* **102**, 403–423 (2019).
140. C. A. Fuchs and B. C. Stacey, "QBians Do Not Exist," arXiv: 2012.14375.
141. F. C. S. Schiller, "Axioms as Postulates," in *Personal Idealism: Philosophical Essays by Eight Members of the University of Oxford*, edited by H. Sturt (Macmillan and Co., New York, NY, 1902).
142. C. A. Fuchs and R. Schack, "QBism and the Greeks: Why a Quantum State Does Not Represent an Element of Physical Reality," *Phys. Scr.* **90**, 015104 (2015); arXiv: 1412.4211.
143. R. T. Cox, *The Algebra of Probable Inference* (Johns Hopkins Press, Baltimore, MD, 1961).
144. J. Pienaar, "Extending the Agent in QBism," *Found. Phys.* **50**, 1894–1920 (2020); arXiv: 2004.14847.
145. R. Healey, "Representation and the Quantum State," in *Quantum Mechanics and Fundamentality*, edited by V. Allori (Springer, Cham, Switzerland, 2022), pp. 303–316.
146. C. H. Bennett, G. Brassard, C. Crépeau, R. Jozsa, A. Peres and W. K. Wootters, "Teleporting an Unknown Quantum State via Dual Classical and Einstein-Podolsky-Rosen Channels," *Phys. Rev. Lett.* **70**, 1895–1899 (1993).
147. Unsigned photograph, "Experimental Setup of Quantum Teleportation Performed in 2013," from "Quantum teleportation on a chip," American Association for the Advancement of Science (AAAS) EurekAlert, 1 April 2015.
148. C. A. Fuchs, "The Anti-Växjö Interpretation of Quantum Mechanics," in *Quantum Theory: Reconsideration of Foundations*, edited by A. Khrennikov (Växjö University Press, Växjö, Sweden, 2002), pp. 99–116; arXiv: quant-ph/0204146.
149. M. Grassl, private communication, 26 April 2022.
150. D. M. Appleby, S. T. Flammia and C. A. Fuchs, "The Lie Algebraic Significance of Symmetric Informationally Complete Measurements," *J. Math. Phys.* **52**, 022202 (2011); arXiv: 1001.0004.
151. D. M. Appleby, C. A. Fuchs and H. Zhu, "Group Theoretic, Lie Algebraic and Jordan Algebraic Formulations of the SIC Existence Problem," *Quant. Info. Comput.* **15**, 61–94 (2015); arXiv: 1312.0555.
152. C. A. Fuchs, M. Olshanii and M. B. Weiss, "Quantum Mechanics? It's All Fun and Games Until Someone Loses an *i*," to appear in *Asian J. Phys.* (2023); arXiv: 2206.15343.

153. J. P. Ralston, "Quantum Theory without Planck's Constant," *Int. J. Quant. Found.* **6**, 48–87 (2020); arXiv: 1203.5557.
154. R. D. Richardson, *William James: In the Maelstrom of American Modernism* (Houghton Mifflin Co, Boston, MA, 2006).
155. N. D. Mermin, "Quantum Mysteries for Anyone," *J. Philos.* **78**, 397–408 (1981).
156. N. Harrigan and R. W. Spekkens, "Einstein, Incompleteness, and the Epistemic View of Quantum States," *Found. Phys.* **40**, 125–157 (2010); arXiv: 0706.2661.
157. M. Moller, "'The Many and the One' and the Problem of Two Minds Perceiving the Same Thing," *William James Stud.* **3** (2008).
158. W. James, *Essays in Radical Empiricism* (University of Nebraska Press, Lincoln, NB, 1996).
159. E. C. Banks, *The Realistic Empiricism of Mach, James, and Russell: Neutral Monism Reconceived* (Cambridge University Press, Cambridge, UK, 2014).
160. H. Atmanspacher and D. Rickles, *Dual-Aspect Monism and the Deep Structure of Meaning* (Routledge, New York, NY, 2022).
161. A. Plotnitsky, *Reality without Realism: Matter, Thought, and Technology in Quantum Physics* (Springer, Heidelberg, 2021).
162. J. A. Wheeler, "Elementary Quantum Phenomenon as Building Unit," in *Quantum Optics, Experimental Gravity, and Measurement Theory*, edited by P. Meystre and M. O. Scully (Plenum Press, New York, NY, 1983), pp. 141–143.
163. A. N. Whitehead, *Process and Reality*, corrected edition, edited by D. R. Griffin and D. W. Sherburne (The Free Press, New York, NY, 1978).
164. K. Barad, *Meeting the Universe Halfway: Quantum Physics and the Entanglement of Matter and Meaning* (Duke University Press, Durham, NC, 2007).
165. J. Dewey and A. F. Bentley, *Knowing and the Known* (1949), in *John Dewey: The Later Works, 1925–1953*, Vol. 16: 1949–1952, edited by J. A. Boydston (Southern Illinois University Press, Carbondale, IL, 1989).
166. J. A. Wheeler, "Bohr, Einstein, and the Strange Lesson of the Quantum," in *Mind in Nature: Nobel Conference XVII, Gustavus Adolphus College, St. Peter, Minnesota*, edited by R. Q. Elvee (Harper & Row, San Francisco, CA, 1982), pp. 1–23, and discussions pp. 23–30, 88–89, 112–113, and 148–149.

4 A QBist Reads Merleau-Ponty

Rüdiger Schack

In his 1814 book *A Philosophical Essay on Probabilities*, Pierre-Simon Laplace wrote the following words: "An intellect which at a certain moment would know all forces that set nature in motion, and all positions of all items of which nature is composed, if this intellect were also vast enough to submit these data to analysis, it would embrace in a single formula the movements of the greatest bodies of the universe and those of the tiniest atom; [...] the future just like the past would be present before [this intellect's] eyes" (Laplace 1902, p. 4).

This quotation summarizes the predominant scientific world view in the 19th century, which is characterized by two main ideas. Firstly, nature is governed by physical laws that constrain the "movements of the greatest bodies of the universe and those of the tiniest atom", and secondly, it is in principle possible to have an objective view of the universe from the outside, i.e., from a god's eye or third-person standpoint.

Quantum mechanics presents a challenge to these ideas. In particular, there is a strong tension between the third-person view of the universe and the fact that the observer plays a central role in a quantum measurement. Despite this, it is probably fair to say that the majority of 21st-century scientists subscribes to the two ideas above. Moreover, the field of quantum foundations is dominated by efforts to reconcile quantum theory with Laplace's 19th-century world view. For instance, the many-worlds interpretation of quantum mechanics postulates the existence of an objective wavefunction of the universe which evolves in time according to a deterministic law.

Many thinkers have been critical of a third-person account of the world. For instance, in his recent book *Looking East in Winter*, Rowan Williams wrote "Ironically, an epistemology that knows only the operation of 'from outside' knowledge applied to a material universe is in danger of occluding the reality of the body in its specific location and embeddedness in a pattern of interdependence, and of nurturing the fantasy that knowing is essentially the action of a disembodied subject working on embodied data—the default position of a good deal of post-medieval Western

DOI: 10.4324/9781003259008-5

thinking about knowledge, as if the object is always somewhere and the subject is nowhere" (Williams 2021, p. 72).

Laplace's 19th-century world view is simple and tidy, but it runs counter to one of our most basic experiences as human agents, namely that we possess free will and are able to choose our actions freely. Philosophy has come up with many different answers to the question of whether free will can exist in an objective world governed by physical law. For instance, defenders of *compatibilism* claim that free will is compatible with determinism. *Libertarians* maintain that the world is indeterministic and that free will exists. And some authors defend the view, sometimes labelled *hard incompatibilism*, that the laws of physics rule out the notion of free will, independently of the question of determinism (O'Connor & Franklin 2022).

Many scientists agree with hard incompatibilism. For instance, Stephen Hawking and Leonard Mlodinow wrote: "It is hard to imagine how free will can operate if our behaviour is determined by physical law, so it seems that we are no more than biological machines and that free will is just an illusion" (Hawking & Mlodinow 2010, p. 32).

The essential difficulty of reconciling a third-person view of a world governed by physical law with free will is vividly expressed in this quotation by Thomas Nagel in his book *The View from Nowhere*: "This naturally suggests that […] an account of freedom can be given which is compatible with the objective view, and perhaps even with determinism. But I believe this is not the case. All such accounts fail to allay the feeling that, looked at from far enough outside, agents are helpless and not responsible. Compatibilist accounts of freedom tend to be even less plausible than libertarian ones" (Nagel 1989, p. 113).

I agree with Nagel here. I thus believe that, in order to preserve freedom, one has to abandon what Nagel calls the objective view. Fortunately, there exist alternatives. In this chapter, following earlier works by Michel Bitbol (2020) and Laura de la Tremblaye (2020), which examine QBism from the perspective of phenomenology, I will exhibit points of contact between QBism and the philosophy of Maurice Merleau-Ponty, both of which explicitly reject a third-person account of the world.

QBism (Fuchs 2010, 2016, 2017; Fuchs & Schack 2013; Fuchs et al. 2014) resolves the tension between the third-person view of the world and quantum mechanics through a bold move concerning the nature of probabilities. QBism maintains that all probabilities, including those equal to 1 or 0 (which express certainty one way or another) and those that are derived from applying the rules of quantum mechanics, are valuations that an agent ascribes to his degrees of belief in possible outcomes. In this view, probabilities are always personal to the agent who assigns them. They get operational meaning from decision theory: an agent uses probabilities to decide rationally what action to take in the face of uncertainty.

The personalist, decision-theoretic approach to probability was developed in the 1920s by Frank Ramsey (1931) and Bruno de Finetti (1931) and given its modern form by Leonard J. Savage (1954) in the 1950s. This approach to probability is well established in economics and finance. When used in physics, it is often associated with the idea of ignorance: when an agent assigns a probability to the outcome of a measurement, say, of a particle position, the probability expresses the agent's ignorance of what the outcome will be, but the outcome is determined by the actual particle position, assumed to exist prior to the measurement. In this reading, a measurement uncovers a pre-existing property, and agents use probabilities only because they are ignorant of this property.

QBism's bold move was to retain the personalist nature of probability while at the same time rejecting the idea that a quantum measurement reveals a pre-existing property. According to QBism, the world is truly indeterministic. Neither a measurement outcome nor its probability are given by physical law. Measurement outcomes are personal consequences for the agent taking the measurement action and come into existence only through the measurement action itself. And their probabilities express the agent's subjective degrees of belief in the possible outcomes.

Probabilities come into the quantum formalism through the Born rule, a fundamental formula that connects three mathematical objects: a quantum state, a collection of measurement operators, and a probability distribution. In the usual reading, the quantum state is determined by the properties of the measured physical system along with the way it was prepared, while the collection of measurement operators characterizes the measurement apparatus. Both are regarded as objective quantities. In this reading, the Born rule thus functions as a physical law that specifies objective outcome probabilities given quantum state and measurement operators.

By contrast, QBism views the Born rule as a normative constraint on an (or any) agent's decision making. By calling the Born rule normative, we mean that, rather than connecting quantities that describe how a part of the world *is*, it tells the agent how he *should* act in order to make better decisions. For a QBist, not only probabilities but also quantum states and measurement operators express an agent's—typically an experimental physicist's—degrees of belief. The Born rule now functions as a consistency criterion which puts constraints on the agent's decision-theoretic beliefs.

To show what this might look like in practice, here are some steps that a physicist or agent might follow in order to apply the quantum formalism.

Step 1: Identify a part of the world (a "physical system") on which to act. This could be, for instance, an ion in a trap. This step typically involves an elaborate experimental apparatus enabling the agent to focus on the precise part of the world he is interested in. QBism considers the apparatus as an extension of the agent, allowing the agent to access the physical system directly.

Step 2: Identify a set of measurement outcomes. For a meaningful experiment, an agent needs to know what he is looking for. In the case of the ion, this could simply be an answer to the question "at a given moment, will I see the ion emit light or not?". As mentioned earlier, QBism regards the answer as a consequence of the measurement action which is personal to the agent.

Step 3: Assign probabilities to these outcomes. The probabilities could be based on, e.g., prior calibration measurements as well as the agent's prior beliefs.

Step 4: Assign each outcome a measurement operator. This and the next step requires intimate knowledge of the experimental setup and familiarity with the quantum formalism. The assignment will depend on the agent's prior beliefs, which will be shaped by his experience as an experimenter.

Step 5: Assign a quantum state.

Step 6: Check that the assignments in steps 3, 4 and 5 are consistent with the Born rule. If that is not the case, return to steps 3, 4 and 5 and try to resolve the inconsistency. There is no unique way of doing this. It could involve verifying calculations, critically examining assumptions, adjusting the experimental setup, making additional calibration measurements, etc. The normative content of the Born rule is that it requires agents to strive for consistency, without providing them with instructions as to how to get there.

According to QBism, quantum mechanics is a small theory. It does not attempt to describe the inner workings of physical systems, let alone the universe. It only provides agents with a criterion of consistency to strive for. But it is a small theory with enormous power. By applying its formalism consistently, physicists and engineers have created modern technology.

Since agents and users of quantum mechanics take centre stage in QBism, it will be useful to say explicitly what we mean by these terms: *Agents* are entities that (i) can take actions freely on parts of the world external to themselves so that (ii) the consequences of their actions matter for them. And a *user of quantum mechanics* is an agent that is capable of applying the quantum formalism normatively. These definitions make sure that it is meaningful for an agent to use decision theory without restricting the class of agents too much.

To briefly summarize our discussion of QBism so far, we have seen that QBism regards quantum mechanics as a normative tool for agents to make better decisions. Probabilities are personal to each agent and are not given by physical law. Quantum measurement outcomes are personal to the agent making the measurement and do not pre-exist the measurement. The world does not admit a third-person description and is fundamentally indeterministic. Agents possess genuine freedom.

QBism's spirit is beautifully captured in this quotation by William James: "Why may not the world be a sort of republican banquet [...], where all the qualities of being respect one another's personal sacredness, yet sit at the common table of space and time? To me this view seems deeply probable. Things cohere, but the act of cohesion itself implies but few conditions, and leaves the rest of their qualifications indeterminate. [...] [I]f we stipulate only a partial community of partially independent powers, we see perfectly why no one part controls the whole view, but each detail must come and be actually given, before, in any special sense, it can be said to be determined at all. This is the moral view, the view that gives to other powers the same freedom it would have itself [...]" (James 1956).

What remains a big challenge is to develop an explicit ontology for QBism, an ontology that agrees with the spirit of William James's "republican banquet", i.e., an ontology based on first-person experience but which accounts for a real world beyond any particular agent's experience and thus avoids the trap of idealism. It turns out that there is a significant overlap between the project of finding such a QBist ontology and the philosophy of Maurice Merleau-Ponty and other phenomenologists. Below I will sketch some points of contact between QBism and Merleau-Ponty's essay *The intertwining—the chiasm* (Merleau-Ponty 1968).

In that essay, Merleau-Ponty asks three main questions: What does it mean to experience the world through seeing and touching? What is the nature of our relationship with others? What is the status of our ideas about the world? He starts out with an implicit acknowledgement of the great difficulty of answering these questions in a philosophical framework that takes first-person experience as its starting point: "once again [philosophy] must recommence everything", he writes (Merleau-Ponty 1968, p. 130). Like QBism, he rejects both idealism and a third-person view of the world: "We have to reject the age-old assumptions that put the body in the world and the seer in body, or, conversely, the world and the body in the seer as in a box" (Merleau-Ponty 1968, p. 138).

QBism does not only rule out a third-person view of the whole world, it also rejects the notion that a quantum state can provide a complete description of any particular physical system. Firstly, an agent's quantum state for a physical system is not a description of the system at all, but an encoding of an agent's expectations regarding the consequences of his actions on the system. Secondly, before writing down a quantum state, the agent will have to decide what (necessarily) limited set of aspects to focus on, i.e., identify a set of potential measurement outcomes. For instance, this could be the range of wavelengths that a certain spectrometer is able to resolve. The outcome of the measurement will be personal to the agent and will be the answer to a very specific question. It will never capture all there is to say about the system.

Merleau-Ponty expresses the idea that an agent's experience is always much richer than any label one might put on it in a striking passage: "[...] this red under my eyes is not, as is always said, a quale, a pellicle of being without thickness, a message at the same time indecipherable and evident, which one has or has not received, but of which, if one has received it, one knows all there is to know, and of which in the end there is nothing to say". Rather, it is a boundlessly rich experience, a "punctuation in the field of red things, which includes the tiles of roof tops, the flags of gatekeepers and of the Revolution, certain terrains near Aix or in Madagascar, it is also a punctuation in the field of red garments, which includes, along with the dresses of women, robes of professors, bishops, and advocate generals, and also in the field of adornments and that of uniforms" (Merleau-Ponty 1968, p. 132).

Merleau-Ponty's answer to the question of ontology is the concept of "flesh", which he compares to the elements of ancient Greek philosophy, water, air, earth, and fire: "[flesh] is an element of being" (Merleau-Ponty 1968, p. 139). It is flesh which sustains the rich experience of colours and other "visibles". He writes that "[flesh] is the tissue that lines them, sustains them, nourishes them, and which for its part is not a thing, but a possibility, a latency [...] flesh is not matter, is not mind, is not substance [...] it is a general thing, midway between the spatio-temporal individual and the idea, a sort of incarnate principle that brings a style of being wherever there is a fragment of being" (Merleau-Ponty 1968, p. 139).

A key idea of QBism is that there is no such thing as passive observation of a quantum system. Any measurement is an active intervention. To experience a response from a system, an agent needs to touch the system, typically through a measurement apparatus which in QBism should be understood as an extension of the agent himself. This is why QBists tend to talk about measurement *actions*, rather than just measurements, and this is why QBists rarely use the term "observer", but prefer the term "agent" instead.

The touching subject plays a central role in Merleau-Ponty's philosophy. He uses the example where "my right hand touches my left hand while it is palpating the things" to show that the touching subject is part of the palpable world itself, "such that the touch is formed in the midst of the world and as it were in the things" (Merleau-Ponty 1968, p. 134). He argues that an analogous relation exists between the seeing subject and the visible, i.e., that a seeing subject is necessarily part of the visible world. I can touch and see things precisely because I am visible and tangible myself, because the visible is an archetype for the seeing and vice versa. Ultimately it is flesh—a relation of the visible with itself that "constitutes me as a seer" (Merleau-Ponty 1968, p. 140). Again quoting Rowan Williams, "In the striking phrase used by Orion Edgar in his study of the theological implications of Merleau-Ponty's philosophy, 'nature lies on both sides of perception'" (Williams 2021, p. 73).

In QBism, the idea that touching and being touched, or seeing and being seen, are two sides of the same coin is expressed through this "Copernican principle": "By one category of thought we are agents, but by another category of thought we are physical systems. And when we take actions upon each other, the category distinctions are symmetrical" (Fuchs & Schack 2013).

That there are other agents is one of the most fundamental human experiences, and both QBism and Merleau-Ponty acknowledge this explicitly. According to QBism, one agent can apply the quantum formalism to another agent in the same way as to any physical system. As for any physical system, this allows the agent to bet on the consequences of his actions on the other agent. And as for any physical system, this leaves the fundamental autonomy of the other agent intact. To repeat William James's words quoted above, this is "the moral view, the view that gives to other powers the same freedom as it would have itself [...]".

For Merleau-Ponty it is again flesh which provides a connection between the first-person experiences of different subjects. To paraphrase him, this is possible as flesh is universal, not particular to me and my body and my experience of the world. Hence it enables the synergy not only between different organs in my body, e.g., my hands and eyes, or between my body and the visible and palpable world around me, but also between myself and others. This happens "[...] by virtue of the fundamental fission or segregation of the sentient and the sensible which, laterally, makes the organs of my body communicate and founds transitivity from one body to another" (Merleau-Ponty 1968, p. 143).

Merleau-Ponty contrasts his vision of a world that has inexhaustible depth and contains first-person experiences other than our own with the "solipsistic illusion" that an agent might capture from his perspective all there is to say about the world: "But what is proper to the visible is, we said, to be the surface of an inexhaustible depth: this is what makes it able to be open to visions other than our own. In being realized, they therefore bring out the limits of our factual vision, they betray the solipsist illusion that consists in thinking that every going beyond is a surpassing accomplished by oneself" (Merleau-Ponty 1968, p. 143). Other agents are fundamental to Merleau-Ponty's philosophy: "[...] for the first time, through the other body, I see that, in its coupling with the flesh of the world, the body contributes more than it receives, adding to the world that I see the treasure necessary for what the other body sees" (Merleau-Ponty 1968, p. 144).

When an experimental physicist selects a physical system to act on, identifies a sample space of potential measurement outcomes, chooses a measurement apparatus, assigns probabilities, quantum states and measurement operators and finally takes a measurement action, this is when Merleau-Ponty's "fundamental fission or segregation of the sentient and the sensible" occurs. QBism regards the measurement apparatus as an extension of the agent. Translated into Merleau-Ponty's language, the measurement apparatus thus becomes a part of the "sentient", whereas the "sensible" (i.e., the palpable or visible) corresponds to the physical system that the agent is acting on.

For example, if I measure a qubit, the bit (0 or 1) I obtain as my measurement outcome is just a tiny aspect of the particle in front of me. What allows me to focus on this tiny aspect and ignore everything else about the particle is the measurement apparatus. Equipped with the apparatus as an extension of myself, the measurement can be seen as me touching the qubit and experiencing the outcome directly. If I understand Merleau-Ponty correctly, my touching of the qubit is an instance of the "fission of the sentient and the sensible".

Coming back to the question of other agents, QBism's Copernican principle has a compelling application to the resolution (DeBrota et al. 2020) of the famous Wigner's friend paradox (Wigner 1961). In a nutshell, the paradox arises from a thought experiment in which Wigner applies the quantum formalism to a lab containing his friend. Since Wigner is outside the lab and has assigned a quantum state to the total system including the friend, it appears as if Wigner were in a privileged position vis-a-vis his friend. This leads to the paradox. It arises from a failure to treat the friend as an autonomous agent on the same footing as Wigner himself, in violation of the symmetry required by the QBist Copernican principle.

For instance, in their recent version of the paradox, Baumann and Brukner (Baumann & Brukner 2020) argue that the friend should base her predictions on Wigner's state assignment as well as her personal experience. But if the friend is an autonomous QBist agent, in order to apply the quantum formalism consistently, she must treat Wigner and the relevant parts of the laboratory surrounding her as a physical system external to herself, just as Wigner treats her and the laboratory as a physical system external to himself. It does not matter that the laboratory spatially surrounds the friend; it, like the rest of the universe, is external to her agency, and that is what counts. Once symmetry is restored in this way, the paradox disappears.

The root of the paradox is the notion that Wigner's state assignment is able to capture everything that is relevant to the friend's experience.

One could argue that this turns the Wigner of the thought experiment into a solipsist. In Merleau-Ponty's words, he has fallen for "the solipsist illusion that consists in thinking that every going beyond is a surpassing accomplished by oneself". QBist Wigner, by contrast and quoting William James's republican banquet passage, "gives to [his friend] the same freedom [he] would have [himself]".

To conclude, there exist two radically different scientific world views. On the one hand, there is the mainstream third-person perspective of a world which evolves according to laws and in which agents are biological machines. On the other hand, there is the QBist view which rejects a third-person perspective of the world, in which agents matter fundamentally, and where laws have a normative character. It is obvious which of the two views a free agent should adopt.

References

Baumann, V. & Brukner, Č (2020). Wigner's Friend as a Rational Agent. In M. Hemmo and O. Shenker (Eds.), *Quantum, Probability, Logic: The Work and Influence of Itamar Pitowsky* (pp. 91–99). Cham: Springer.

Bitbol, M. (2020). A Phenomenological Ontology for Physics: Merleau-Ponty and QBism. In H. A. Wiltsche & P. Berghofer (Eds.), *Phenomenological Approaches to Physics*. Cham: Springer. https://doi.org/10.1007/978-3-030-46973-3_11.

de Finetti, B. (1931). Probabilismo. *Logos 14* (p. 163). Translation (1989): Probabilism. *Erkenntnis 31* (p. 169).

de La Tremblaye, L. (2020). QBism from a Phenomenological Point of View: Husserl and QBism. In H. A. Wiltsche & P. Berghofer (Eds.), *Phenomenological Approaches to Physics*. Cham: Springer. https://doi.org/10.1007/978-3-030-46973-3_12.

DeBrota, J. B., Fuchs, C. A. & Schack, R. (2020). Respecting One's Fellow: QBism's Analysis of Wigner's Friend. *Foundations of Physics 50* (pp. 1859–1874).

Fuchs, C. A. (2010). QBism, the Perimeter of Quantum Bayesianism. https://arxiv.org/abs/1003.5209.

Fuchs, C. A. (2016). On Participatory Realism. In I. T. Durham and D. Rickles (Eds.), *Information and Interaction: Eddington, Wheeler, and the Limits of Knowledge* (pp. 113–134). Berlin: Springer.

Fuchs, C. A. (2017). Notwithstanding Bohr, the Reasons for QBism. *Mind and Matter 15* (pp. 245–300).

Fuchs, C. A., Mermin, N. D. & Schack, R. (2014). An Introduction to QBism With an Application to the Locality of Quantum Mechanics. *American Journal of Physics 82* (pp. 749–754).

Fuchs, C. A. & Schack, R. (2013). Quantum-Bayesian Coherence. *Reviews of Modern Physics 85* (pp. 1693–1715).

Hawking, S. & Mlodinow, L. (2010). *The Grand Design*. New York, NY: Bantam Books.

James, W. (1956). On Some Hegelisms. In W. James, *The Will to Believe, and Other Essays in Popular Philosophy* (pp. 263–298). New York, NY: Dover.

Laplace, P.-S. (1902). *A Philosophical Essay on Probabilities*. London: John Wiley & Sons.

Merleau-Ponty, M. (1968). The Intertwining—The Chiasm. Translation by Alphonso Lingis. In M. Merleau-Ponty, *The Visible and the Invisible*. Evanston: Northwestern University Press.

Nagel, T. (1989). *The View from Nowhere*. New York: Oxford University Press.

O'Connor, T. & Franklin, C. (2022). Free Will. In E. N. Zalta (Ed.), *The Stanford Encyclopedia of Philosophy*. https://plato.stanford.edu/archives/sum2022/entries/freewill.

Ramsey, F. P. (1931). Truth and Probability. In F. P. Ramsey, *The Foundations of Mathematics and Other Logical Essays*, edited by R. B. Braithwaite (pp. 156–198). New York, NY: Harcourt, Brace and Company.

Savage, L. J. (1954). *The Foundations of Statistics*. New York, NY: John Wiley & Sons.

Wigner, E. P. (1961). Remarks on the Mind-Body Question. In I. J. Good (Ed.), *The Scientist Speculates* (pp. 284–302). London: William Heinemann, Ltd. Reprinted in E. P. Wigner (1979), *Symmetries and Reflections: Scientific Essays of Eugene Wigner* (pp. 171–184), Woodbridge: Ox Bow Press.

Williams, R. (2021). *Looking East in Winter: Contemporary Thought and the Eastern Christian Tradition*. London: Bloomsbury Continuum.

5 Unobservable Entities in QBism and Phenomenology

Jacques Pienaar

5.1 Introduction

Unlike most other interpretations of quantum mechanics, *QBism*[1] departs radically from traditional forms of scientific realism by insisting that the quantum state of a system only refers to an agent's subjective beliefs about the likelihood of outcomes of various measurements they might perform on the system. Furthermore, QBism holds that the outcome of any such measurement is an experience of the agent performing the measurement, and so is intrinsically "personal"[2] to the agent.

Despite its emphasis on the subjective aspects of quantum theory, QBism sees itself as a methodological programme that aims towards an ontology of the physically real (Fuchs, 2016, p. 118; Crease and Sares, 2021, p. 542). To this end QBism seeks to identify those elements of quantum theory which do not admit of a purely subjective interpretation. While these elements would still necessarily be grounded in agents' "experiences", their explanation would require making some definite claims about "how the world is" for *all* agents, that is, such elements would point towards the ontological content of quantum theory. QBism thereby acknowledges that an agent's "experience" does not refer to an isolated consciousness contemplating itself, but includes also some transcendent elements (i.e., elements that refer to a "world" which lies beyond the agent's consciousness) and QBism relies upon this fact to refute the charge of solipsism, i.e., the claim that only one's own mind is knowable through experience. However, in taking both the subjective and transcendent aspects of "experience" as equally fundamental, QBism contradicts the traditional view that ontology is something *a priori* objective, and consciousness is something reducible to non-conscious ontological elements (Fuchs, 2017, p. 277; Pienaar, 2021, p. 7). In order to successfully defend its unorthodox vision of "ontology" against more traditional forms of scientific realism, QBism would benefit from a metaphysical framework capable of accommodating the unique demands that it places upon "experience".

DOI: 10.4324/9781003259008-6

Framed in this way, there is good reason to hope that phenomenologists might be able to provide this guidance. Like QBists, phenomenologists seek to ground scientific claims about the world in phenomenal experience, and like QBists they reject traditional mind-world dichotomies and seek an ontology grounded in phenomena, which have an *intentional* character that makes them neither wholly immanent to consciousness nor wholly transcendent (Zahavi, 2018, p. 29). These similarities make phenomenology a natural setting for elaborating a metaphysical thesis underlying QBism, thereby clarifying and fortifying QBism's ontological position.

This chapter aims to investigate how QBism might benefit from phenomenology in articulating its response to a particularly vexing question: which of the "unobservable" entities that appear in scientific theories, such as atoms, quarks, and quantum states, have a claim to physical existence, i.e., to be regarded as worldly objects?

QBism's stance on this issue is presently unclear. QBists often tacitly assume that "systems" exist which have a mostly objective character, in the sense that the same system can be addressed by multiple agents; these systems include not only straightforwardly observable objects like tables and trees, but also scientific objects like atoms and fundamental particles whose observability may be questioned. In the latter case, the "subjective" component of these systems is relegated to the quantum state that is assigned to them, which characterizes the agent's expectations for the outcomes of measurements that could be performed on the given object.

This tacit splitting up of scientific phenomena into an objective "system" furnished with a subjective "quantum state" seems problematic on at least two counts. Firstly, particularly in quantum mechanics, there seems to be a certain arbitrariness in what constitutes a "system" as distinct from a "property of a system". For example, at low energies, one can take a single electron to be a system, insofar as it remains identifiable as distinct from other electrons. At high energies, this distinction becomes impossible due to particle creation and annihilation processes, and it is more appropriate to regard an electron as an outcome of a measurement (say, checking the "number of quanta") of a more complicated system that is a "quantum field". Secondly, both the systems and their quantum states appear to have similar roles in quantum theory, in that both are "unobservable" entities which serve to co-ordinate and constrain the theoretical tasks of prediction and explanation of observable experimental data. Given this similarity, what reasons can the QBist have for treating "systems" as being more objective than the "quantum states" assigned to them? Since some phenomenologists have explicitly discussed the metaphysical status of "unobservable" scientific entities, QBists might hope to find answers by drawing on the phenomenological literature.

This chapter is structured loosely as follows: we begin by clarifying what is meant by "unobservable scientific entities". We then discuss a selection of the phenomenological literature on this topic, which will lead us to divide the question into two key components: (i) whether scientific entities are unobservable in principle or only in practice and (ii) whether belief in the physical existence of in principle unobservable entities can nevertheless be justified. We then propose a loose connection between QBism's notion of "experience" and the phenomenologist's notion of "perception", and we use this connection to explicate QBism's answers to the above key questions. Finally, we identify two aspects of the QBist's replies that pose a challenge to phenomenologists, and we outline some avenues for reconciliation.

5.2 Unobservable Entities

What is an "unobservable scientific entity"? Many objects studied by physicists seem to be "unobservable", in the sense of not being directly perceptible in the same manner as ordinary everyday objects. We can see the legs of a table just by looking at it or touching it, but we cannot "see the atoms" in the table by any similar manner. We physicists are nevertheless disposed to claim that we are justified in believing in the atoms' existence, i.e., that although we can't see them, we know they must "be" there as physically real entities. The justification for this claim seems to require something more than what is ordinarily perceptible – but what is this "something more"?

To begin with, we must not commit the fallacy of thinking that abstract or theoretical objects are unobservable *per se*. Phenomenologists emphasize that every object given in perception is given within a certain context or "horizon", against which it takes on certain significance. This significance might include the object's being a concrete instance of some theoretical abstraction, or embodying some formal mathematical model. For example, a set of three billiard balls may take on the significance of being a concrete instance of the number "three", just in case one is interested in counting them; alternatively they may be seen as concrete instances of perfectly rigid and uniform spheres moving on a two-dimensional frictionless surface, just in case one is interested in predicting their dynamics on the billiard table with a low degree of precision. Even for less overtly analytical tasks, like recognizing that the objects are "billiard balls", makes reference to a context in which it acquires its abstract meaning, in this case the context is that of games involving cue-sticks and balls. To the extent that ideal theoretical entities like perfect spheres can be concretized in perceived objects, we can consider them "observable" despite their conceptual or abstract character;

paraphrasing Merleau-Ponty, the matter of perception is "pregnant" with its form (Merleau-Ponty, 1964, p. 12, 15; Merleau-Ponty, 2012, p. 154).

Significantly, certain scientific entities including atoms, quarks, and electromagnetic fields, do not seem to ever appear concretely as perceived objects. We know of them only indirectly, by inference from the behaviour of certain directly perceptible objects that serve to signify their presence. For example, we can use a scattering of iron filings to infer the presence and partial structure of an electromagnetic field, and the readings of a scanning-tunnelling microscope to infer the structure of atoms on a surface. In short, the phenomenal evidence for the existence of such entities typically comes in the form of graphs, numerical readings, motions of mechanical sensors, etc., which serve to indicate these entities but do not present the measured objects "bodily" in perception; thus the entities themselves are said to be "unobservable".

5.3 Subjectivity and Objectivity

The question naturally arises as to whether this unobservability is a matter of principle or of practice.

According to QBism, the answer depends on the nature of the unobservable entity. QBists implicitly make a distinction between two types of unobservable entity: those that have the character of being "system-like", versus those that have the character of being "probability-like". The former kind are entities that are regarded by the agent as external objects to which measurements can be specifically addressed; they are uniquely identifiable by some label and a set of characteristic properties that define their type (in the case of an electron, these would be: a point-like spatial extent, a mass of 9.109×10^{-31}kg, quantum spin of one half, etc.). Such entities may include planets, chairs, fields, atoms, and fundamental particles. The second kind of entity are those which are explicable as "catalogs of probability assignments". The most basic of these is just a probability distribution over a complete set of mutually exclusive experiences the agent could have. More complicated examples are quantum states and Hamiltonian operators, which the QBist interprets as being ultimately equivalent to sets of probability assignments (or in some cases, conditional probability assignments) about possible experiences that the agent could have, specifically, experiences of the outcomes of certain measurements (i.e., elaborate actions) they could perform.

For a QBist, the system-like entities have a more "objective" character, while probability-like entities have a more "subjective" character. This terminology is unfortunate, because these terms are a relic of the Cartesian split between a mind-bearing "subject" placed in categorical opposition to an independent world of "objects". A close reading of QBism suggests that they reject such an absolute division.

To see why, let us set aside ontological questions for the moment and consider QBism strictly within the context of decision theory. In this restricted context, QBism asserts that quantum theory is a "tool of reasoning", that is, a set of norms or best practices that guide the decisions of agents. Implicit in this formulation is the idea that the users of quantum theory belong to a special class of agents, i.e., not only must they be capable of reasoning explicitly using symbolic language, but they must have specific concerns for which the usage of quantum theory confers an advantage, such as building a quantum computer, or designing high-precision particle detectors. To the extent that QBism distinguishes system-like entities from probability-like entities, this distinction is relative to the particular "set of possible agents" who are implicated as the users of the specific decision theory (in this case quantum theory). Given a specification of the community of possible agents, one may call "subjective" those quantities or entities which may vary from one agent to another within the community. Conversely, those entities which preserve their form across the whole community may be called "objective", provided we understand this word as meaning "invariant between subjects" and not "independent of subjects" (which is how physicists usually understand this word).

Seen in this way, it also becomes clear that the traditional terms "subjective" and "objective" are inadequate to capture the full meaning of the distinction. Firstly, even the most "subjective" elements – the experiences of the individual agents in all their richness and idiosyncracy – are only conceivable within the broader "objective" norms that were posited at the beginning to define "the set of possible agents". The mere fact that we have taken up a specific decision theory (e.g., quantum theory) which is only relevant to a specific class of agents means that we have imposed a set of implicit "objective" features that are shared by all agents, and shape their experiences. Secondly, even the most "objective" elements – like the abstract rules of quantum theory themselves – have their origin and grounding in the "subjective" raw material that comprises the history of observations by individual agents within the community, subsequently communicated, refined, modified to accommodate new experiments, and ultimately processed by that community until invariance was achieved (at which point the rules become "laws" and are disseminated in textbooks and institutional curricula). In summary: variance can only be defined with respect to the invariant, and new invariants can only be extracted from a substrate of initial variance; there is always a seed of the objective in the subjective and vice-versa.

So long as QBism remains only a kind of decision theory, these subtle matters need not trouble it. For a decision theory does not require the class of possible agents to be explicitly named and characterized, in the same way that you do not need to be explicitly aware that you have opposable thumbs in order to use a hammer – you simply pick it up and start using it. This is why QBists are constantly frustrated by criticisms to the effect that they do not bother to carefully define "what is an agent". Their reply is the same as the decision theorists' reply: there's no need for such theorizing, just try to use the "hammer": if you can, you are an agent of the relevant sort; otherwise, you are not. Thus, QBists often use the word "subjective" to characterize various parts of quantum theory (quantum states, Hamiltonians, and so on) without really bothering to give it a precise definition, aside from remarking that these things are subjective by virtue of being able to differ markedly from one agent to the next. As such, the "subject" vs "object" distinction persists in QBist language, despite the fact that their own metaphysical commitments ultimately push them to go beyond this naïve dichotomy.

This tension becomes impossible to ignore when QBism aims beyond mere decision theory and tries to articulate an ontology. For, if QBism follows the example of phenomenology, it must be acknowledged that the "agent" which appears at the decision theoretic level must not be separable from their "external world" at the fundamental level. QBism must seek a notion of "pre-reflective" ontological being which is rich enough to give a historical account of the evolution of the agent-world split. In the process, we must lose any sharp distinction between "objective" and "subjective", these being regarded as idealized extremes on a gradient of degrees, and the classification of phenomena as being "more objective" versus "more subjective" will then be seen as a contingent process which is never completed, but is constantly evolving and being revised as the emergent "self" contends with the emergent "world" through the phenomena in which they are mutually bound. Incidentally, this is also why it is impossible for QBists to talk about the transformation or mutation of the agent without also making ontological commitments that go beyond decision theory, as in, e.g., Pienaar (2020, p. 1906).

This is relevant because in QBism some unobservable entities have a "more objective" character, and others a "more subjective" character, and this has bearing on whether they can be observable in principle or not[3]. System-like entities are "objective" in the following sense: an agent can experience that there is a system before them upon which they can do measurements; moreover it is unproblematic to assume that this system can be identified by a unique label over time and can be shared with other agents who can be said to perform measurements on the "same" system. On the other hand, the probability-like entities have a "subjective" character in the sense that they represent the probabilities (i.e., quantified degrees of belief) of the agent, and therefore neither stand before an agent like an object, nor can be simply taken as shared between agents[4]. In particular, by virtue of the personal nature of probability-like entities (i.e., their variability across agents), QBists insist that they cannot be regarded as physical entities, i.e., entities which exist as things in the world external to the agent; therefore they are not even in principle observable in the sense of being perceptible as physical things.

We pause to remark that this view about the character of probability-like entities is a direct consequence of QBism's commitment to a subjective Bayesian interpretation of probabilities (Bernardo and Smith, 1994). According to this view, there is no such thing as a "wrong" prior probability assignment. In the present context, we may think of this as providing QBism with a constructive principle that says something about how we ought to define the "class of possible agents", namely, we should define it to be as permissive as possible, by including all agents whose beliefs can represented by some valid probability function.

5.4 The Prolongation Thesis

In contrast to probability-like entities, there seems to be a precedent in QBism for regarding system-like entities, including atoms, electromagnetic fields, and the like, as being in principle observable. The crucial ingredient is the QBist principle which states that measuring instruments are to be regarded as "prolongations" of the agent's senses.

It is worth explaining in more detail what the QBists mean by this. As recalled in Fuchs (2017, p. 20), the principle originates from a remark by Wolfgang Pauli in a letter to Niels Bohr in 1955, where Pauli suggests that "it is allowed to consider the instruments of observation as a kind of prolongation of the sense organs of the observer". Pauli later retreated from this idea, but QBism adopted it as a core principle. QBists saw this as a necessary move in order to resolve the "Wigner's Friend" paradox. We need not review the details of this debate here (see Fuchs, 2010; DeBrota et al, 2020; Crease and Sares, 2021); it suffices to note that QBism criticizes the Copenhagen tradition (as variously propounded by Bohr, Heisenberg, and Pauli) as being unable to address this paradox, because of its refusal to allow measurement outcomes to be "personal" to the agent, which is implied if one takes the instruments to be prolongations of the agent's senses. As Fuchs puts it:

> QBism [...] takes [Pauli's idea] deadly seriously and runs it to its logical conclusion. This is why QBists opt to say that the outcome of a quantum measurement is a personal experience for the agent gambling upon it. Whereas Bohr always had his classically describable measuring devices mediating between the registration of a

measurement's outcome and the individual agent's experience, for QBism the outcome just is the experience.

(Fuchs, 2017, p. 266)

We will encapsulate this idea in the following statement, which we call the "Prolongation Thesis":

> Prolongation Thesis: Those measuring instruments which the agent regards as being the source of their experiences (i.e. of their measurement outcomes) are to be regarded as prolongations of the agent's body, and thus having the same metaphysical status as the bodily sense organs of the agent.

It is important to note that QBists do not intend this to be a claim about "measuring instruments" as defined by some external criterion. Rather, in asserting that measuring instruments are prolongations of the agent, QBists are proposing that this is *definitive* of what a "measuring instrument" means for the agent.

Specifically, the QBist is here implying that to call something a "measuring instrument" is to assert that its presence to oneself and its usage is comparable to that of any other of one's given bodily sense organs. In particular, the experience of perceiving an object through the instrument must be no less direct, immediate, effortless, and unmediated as the experience of perceiving an object via one's familiar sense perceptions. Thus an instrument whose outcomes are inferred by explicit theoretical work, tedious adjustments, or any other cognitive processes exceeding those normally employed in the usage of one's "natural" bodily functions, must strictly speaking not be a measuring instrument on the QBist definition.

This terminology is admittedly awkward. It implies that QBists do not accept the meaning given to "measuring instruments" in common scientific parlance, where things such as microscopes and telescopes are taken to be *de jure* measuring instruments, irrespective of the operator's skill and experience while using them. Given the fact that most scientists likely do not feel the same facility with their instruments as they do with their own hand or eyes, it appears that most things the physics community considers to be measuring instruments do not meet the QBist's definition.

One way to navigate this overloading of terminology is to follow (Pienaar, 2020, section 3) and make a principled distinction between apparatuses external to the agent that satisfy the conventional desiderata of scientific instruments, as opposed to those apparatuses with which the agent is so fluent that they may be considered equivalent to additional sense organs. Here we endorse this distinction by referring to "extracorporeal" versus "intracorporeal" measuring instruments, respectively.

Returning to the question of "unobservable entities", we see that the observability of the system-like entities depends on whether the initially *extra* corporeal instruments that signify the system's presence can become *intra* corporeal after sufficient training and practice; that is, whether the instrument used to detect the system could ever in principle be taken as being on par with the agent's own sense organs. If it could, then we would have to conclude per the Prolongation Thesis that the systems which can be made detectable to the agent through their intracorporeal instruments are in principle "observable", in the same manner that any other physical thing is observable to the agent.

We must therefore ask: is it conceivably possible for an initially extracorporeal instrument to become intracorporeal? As a possible counterexample, consider that the detection of the Higgs Boson by the Large Hadron Collider involves an apparatus that is not only incredibly huge, but involves a collaboration of thousands of scientists, engineers, and other specialists, all working in concert to collect, prepare, and analyse the data. Although it is certainly tempting to claim that such an instrument could never conceivably be "intracorporeal" for any agent, there are some reasons to hesitate in making this judgement.

Firstly, following decision theory, QBism by no means restricts "agents" to be individual people; in fact QBists have at times explicitly endorsed the idea that groups of people could be regarded as agents in the relevant sense. If one could make this idea explicit, say, by positing what constitutes an "experience" for a collective, then the human operators of the Large Hadron Collider (LHC), the world's largest particle collider, might well be considered a single agent, and (exotic as it sounds) the LHC considered as one of their "sense organs".

Secondly, even if the LHC specifically cannot be conceivably treated as intracorporeal to an agent (even a collective), we cannot easily exclude the possibility that one could build a more compact and automated device for detecting Higgs Bosons, which might itself be amenable to such fluent usage as to become intracorporeal to the user.

Since neither of these possibilities can at this stage be readily negated or affirmed – i.e., because we lack both a systematic and rigorous account of what it could mean for a collective to be an "agent", and we lack any arguments for or against the plausibility of a hand-held Higgs detector – we must leave it open as to whether Higgs Bosons are in principle observable or not, according to QBism's Prolongation Thesis.

Fortunately, for present purposes, it is enough to note that at least some unobservable system-like entities are detectable by instruments which could plausibly become intracorporeal for an agent, thereby rendering such entities observable. A case in point is the example of a scientist using a scanning tunnelling microscope (STM) to measure atoms. In *Meeting the*

Universe Halfway (Barad, 2007, p. 355), Karen Barad recounts physicist Don Eigler's dramatic live demonstration of the usage of the STM to the audience of National Public Radio's 1996 broadcast of the Morning Edition. Barad leaves little doubt of the visceral nature of the experience for both the operator of the device and the listeners:

> With a few clicks of the computer mouse, Eigler maneuvers the STM tip so close to a gadolinium atom sitting on the surface of a piece of niobium that it begins to bond with the gadolinium atom. He moves the tip sideways, pulling the gadolinium atom across the niobium surface to a new location, and then pulls the tip back, releasing the atom. The listening audience is treated to a sonic display of the single-atom manipulation, courtesy of Eigler's clever connection of the STM to a stereo that converts the strength of the "tunneling current" (used to sense the presence of an individual atom) to an audible tone [...] The proof is in the hearing. During the sideways tug of the gadolinium atom across the niobium surface, the audience hears distinct "thunks" as the atom is pulled across the unit cell structure formed by the spaces between the niobium atoms on the surface: that is, one can hear the atom being moved.
>
> (Barad, 2007, p. 355)

This account suggests that it is possible to create an interface to the STM which seamlessly connects a user's bodily motions to sensory feedback from the measured atoms, giving them an especially intimate experience of the measurement. Perhaps with enough practice, the user might come to feel as if they really are hearing or touching atoms as they move them around. However, this is merely a plausibility argument: we do not claim to have established that such an interface would be sufficient to give the atoms a "bodily presence" in perception.

In fact, this argument poses the first challenge to phenomenologists: can one make sense of the QBists Prolongation Thesis in a phenomenological context? This seems likely to be a contentious question in phenomenology. On one hand, in defending the in principle unobservability of atoms and the like, Wiltsche explicitly refutes the "argument from mutation" (Wiltsche, 2012, p. 119), which states that advanced scientific instruments may become organic parts of an observer's sensory apparatus through evolution or prosthesis – essentially the argument we have just presented in support of the Prolongation Thesis. On the other hand, phenomenologists including Merleau-Ponty and Samuel Todes take the body to be the mutable and changeable ground of perception: the difference between bodily and non-bodily objects is on their account a difference of degree but not of kind (Merleau-Ponty, 1968; Todes, 2001). This seems to leave open the door for a revival of the argument from mutation, perhaps in a more sophisticated formulation. Here is not the place to speculate; we leave the issue open to debate.

5.5 Justifications for Physical Existence

In light of the foregoing, there may yet be some entities which are not observable even in principle. This provokes a second question concerning such entities, namely: is their in principle unobservability due to their having a purely conceptual character (like probabilities), or can these entities be present as things in the world like tables and trees, in which case their unobservability is due to them being ordinarily invisible (like X-rays or viruses)? In the first case such entities would be seen as glorified "book-keeping devices" (*Hilfsmittel* as Husserl would put it) that we use for scientific reasoning and prediction, and would not be considered "scientifically real". In the second case we would have to provide independent arguments (i.e., besides the possibility of direct observation) that could establish the existence of these entities as physical things within the world, about which we could be justified in saying that they are "really present before us", even though we could not ever directly observe them. We may frame this as the question of whether we can have *epistemic justification* for believing that something exists as a thing in the world, given that it is in principle unobservable.

In the phenomenology literature, Wiltsche interpreted Husserl as endorsing a particular criterion of epistemic justification (Wiltsche, 2012), which Berghofer subsequently labelled the "Originality Thesis of Justification" (OTJ), giving it this definition:

> Epistemic justification is limited to what can be originally given in the sense that if X cannot be given in an *originary presentive intuition*[5], then one cannot be justified in believing that X exists or obtains.
>
> (Berghofer, 2017, p. 4)

Adopting this criterion implies that if an entity cannot be observed in principle, then one cannot be justified in believing that it is a really existing thing in the world[6]. However, Berghofer points out that Husserl's position is more ambiguous, and that phenomenology does have the resources to accommodate a criterion of justification that would allow for justifiably believing in the existence of unobservable entities (Berghofer, 2017). Berghofer's strategy is to appeal to the phenomenological concept of a *horizon*: when something is given originarily in perception, it is not simply given as a dead and lifeless fact, but as a carrier of associated further possible experiences which are *co-given* with it. Some of what is co-given are things that can themselves in principle be given originarily, for instance, when one perceives a house from the street one also perceives that it has an interior, and this co-given interior can become originarily given by entering the house and perceiving it.

Berghofer argues that one can also consider co-given entities which cannot be given originarily in perception even in principle, but whose existence can still be justified through their implications for other things that can be given originarily (Berghofer, 2017, p. 10). For example, although we cannot perceive muons directly, they are co-given in our perceptions of the various forms of observable evidence for them, such as tracks in cloud chambers or detector clicks, which serve to indicate their physical presence. They can thereby shape our expectations about what will happen in future observations, which can subsequently be fulfilled in originary perceptions. Thus, following Berghofer, we can still be justified in believing in the physical existence of muons even if we consider them unobservable in principle. Berghofer names this alternative the "Criterion of Justification" (CoJ):

> Justification with respect to content C or object O is only possible if C/O can either be originally given or if C/O can be reasonably associated with expectations that can be verified by originary presentive intuitions.
>
> (Berghofer, 2017, p. 11)

We note that Berghofer does not claim that this criterion is sufficient for justification: he merely opens the door for such justification to be possible. Nevertheless, this criterion is more permissive than the QBist would like, because it seems to permit some things to exist as physical things which the QBist would very much like to exclude from that category.

As we have mentioned, QBism adheres to a subjective Bayesian interpretation of probability, and considers quantum states (among other parts of the quantum formalism) to be equivalent to probability assignments, which in turn have a strictly conceptual character: they are symbols used by agents for the purposes of reasoning about their experiences. Significantly, the agent does not experience probabilities as things that stand before them in the world: though they might "exist" to the agent in a certain general sense, they do not exist *as physical things* (according to QBism). The fact that they may have *correlates* in things which are physical – such as patterns in the firing of the agent's neurons, or bets the agent subsequently places in the casino – does not refute the QBist's fundamental position that probabilities themselves are subjective in the highest degree, and thus are not "things" like rocks or trees.

The OTJ could provide QBism with the resources it needs to defend this position. One only has to argue that probabilities are not themselves directly observable in principle, that is, while they can be *inferred* from observable phenomena (neuronal firings and bets at the casino), there is not and can never be any instrument that would let one agent directly "see" (or hear, touch, etc.) another agent's beliefs and hence their probability assignments. To the extent that such an argument is defensible, the OTJ would allow us to conclude, following de Finetti, that "probability does not exist" (de Finetti, 2017, p. ix). By extension, this would justify the claim that quantum states are not physical things, as the QBist desires.

By contrast, Berghofer's CoJ is no help to the QBist; since probabilities can indeed be "reasonably associated with expectations that can be verified by originary presentive intuitions", like neuronal firings and casino bets, CoJ alone is powerless to prevent us from declaring probabilities (and quantum states) to physically exist in a manner not fundamentally different to rocks and trees, forcing the QBist to cast about for independent arguments to exclude this possibility.

Another case in a similar vein is the possible existence of so-called "hidden variables", which cannot be directly observed in principle, but may be invoked in theoretical models as supplementary properties of physical systems that serve to determine the outcomes of future measurements on them. No-go theorems such as Bell's theorem establish that these models must have extremely counter-intuitive behaviour (such as exhibiting faster-than-light transfer of information), and this is one of the many reasons why QBism rejects the existence of such hidden variables. Indeed, the presumed existence of hidden variables would render QBism irrelevant as an interpretation. Here again, OTJ would prove a useful tool for the QBist: since hidden variables are unobservable in principle, OTJ implies that we cannot be justified in believing that they exist as real properties of physical things. This conclusion vindicates QBism's rejection of hidden variables. By contrast, CoJ alone provides no grounds for such a rejection, since proponents of hidden variable models can argue that they do have observable consequences.

In conclusion, if QBism is to draw upon phenomenology to address the problem of the existence of unobservable entities, it must do so in a peculiarly opportunistic fashion: on one hand it must side with (Berghofer, 2017) against (Wiltsche, 2012) in considering OTJ to be too restrictive, because it excludes things such as atoms, electromagnetic fields, etc., from the possibility of existence; on the other hand, ironically, it is precisely OTJ's restrictiveness that would permit QBism to exclude probabilities from being physically real things. To reconcile this, we propose that QBism endorse a slightly more permissive version of OTJ that expands the scope of what is in principle observable through the Prolongation Thesis. This "prolongated OTJ" would be permissive enough to escape many of Berghofer's criticisms of OTJ, while still being restrictive enough to keep the door closed to the physical existence of probabilities and hidden variables.

5.6 Husserl and Subjective Probability

If QBism does adopt the OTJ as we suggest, then probabilities (and hence all probability-like entities, including quantum states) must be excluded from the domain of worldly existence. Probabilities must rather be thought of as having a conceptual or symbolic character, manifesting themselves not as features of perceived things, but as features of the horizons of perceived things (de La Tremblaye, 2020). When something is given in perception, a horizon of possible further perceptions is co-given with it, and we may perhaps think of probabilities (hence also quantum states) as modalities or "weights" of these possibilities within the horizon, corresponding somehow to a sense of "how likely" each of them is.

This gives rise to our second challenge to phenomenologists: can they provide an interpretation of probabilities that can accommodate QBism's subjective interpretation of probabilities, quantum states and other probability-like entities? There are two reasons why this poses a challenge, one historical and one interpretational.

Historically, after Husserl's incomplete and somewhat confusing treatment of the phenomenology of probability, whose most mature expression is found in his lectures *Logik und allgemeine Wissenschaftstheorie* (Husserl, 2019), Husserl's followers have displayed surprisingly little interest in the topic of probability. Merleau-Ponty, Sartre, Heidegger, and other famous phenomenologists barely mention probabilities, except perhaps very obliquely in their discussions of signs and symbols. An exception to this trend is Lobo (2019), who revisits Husserl's phenomenology of probability and offers some hypotheses as to why it has been neglected by later scholars. Lobo's particular interest is to revive a programme, initiated by Husserl and subsequently developed by Gian-Carlo Rota, whose ambitious aim is to reform classical logic to ground it in probability theory, which Husserl came to regard as being the more fundamental formal structure. Our present aim concerns a different programme, which is to provide a phenomenological analysis of probability, for which our most recent primary reference still appears to be Husserl's incomplete sketch from over a century ago (Husserl, 2019).

The interpretational challenge presented there is Husserl's ambiguous stance with regard to the question of whether probabilities are "objective" or "subjective". Given our earlier discussion of the meaning of these words, it may be anticipated that Husserl, like QBism, does not take these words at face value. Indeed, in what follows we shall argue that Husserl's view of the meaning of probability is compatible with that of QBism, provided that what Husserl meant by the "objective" aspects of probability are taken to correspond to what the QBists mean by the "normative" aspects of probability. As we will see, this connection has the added benefit of deepening and enriching the QBist's notion of "normative" and connecting it to contemporary philosophy of normativity along the lines of Robert Brandom's work on inferentialism (Brandom, 2000).

We begin by reviewing the essential points of Husserl's analysis of probability. In Husserl (2019) one finds three key points about how to interpret probabilities in phenomenology. We can loosely think of the first two points as addressing the "subjective" aspect of probability, while the third point addresses their "objective" aspect.

Husserl's first point is that probabilities, at their most basic level, are not themselves objects of consciousness, but are rather modes of intentionality, i.e., they are about something – an event or occurrence – which is not itself a probability. More specifically, they are qualities of judgements that one makes about things, and the quality of something being "probable" is not subservient or reducible to any mixture of absolute certainty and uncertainty, but rather stands on its own as an independent sentiment, existing separately alongside related notions like necessity and questionableness in the "sphere of judgements". Each of them is a distinct phenomenal quality that "has its own source", as Husserl explains (our emphasis):

> Subjectively speaking, we make an affirmative judgment about possibility or probability when we say, "It is possible that S is P", "It is probable, questionable, that S is P". The concepts "possible", "questionable", "probable" are thereby not specifically apophantic categories, if we understand apophansis as judgment, as supposed truth. *They just arise from their own sources*. Supposed truth in certain ways stands apart in a parallel manner: supposed possibility, probability, questionableness.
>
> (Husserl, 2019, pp. 249–250)

Husserl's second key point has to do with the mysterious "sources" of probability judgements. As Marian Evans (as George Eliot) wryly remarked in her famous novel *Middlemarch*, judgements often seem to depend upon "a sense of likelihood, situated perhaps in the pit of the stomach or in the pineal gland" (Eliot, 1965, p. 482). But Husserl cautions that there is much more to probability assignments than the accompanied feelings, whether emotional or physiological. Husserl was cognizant of the fact that any successful phenomenological definition of probability would have to be able to account for the existence of strict rules governing its structure, namely the rules of the probability calculus. These rules appear to serve as good advice to anyone, regardless of their particular feelings, and this robustness against variation from one individual to the next must also be accounted for by the phenomenal character of probabilities.

To achieve this, Husserl argues that probabilities have their source in a very special type of "possibility", which must be distinguished from other types. Specifically, probabilities are those possibilities that we may call *presumptions*, which have the character of being weighted or "loaded" by the contextual presence of other facts and things that either "speak" for or against it. In Husserl's words (our emphasis),

> Empty possibilities are possibilities in whose favor nothing argues. They are mere figments of the imagination or imaginable things. A well-founded "possibility", or something presumable with a reason for it, is something entirely different. In all presuming, something figures to the consciousness as presumable, i.e. just as something argues in favor of this or that. [...] But, of course, it is a presumption and not a belief. It is not expressed as a judgment, as an affirmative positing. In our eyes it is not yet truth, but only "probability". Probability – originally and understood outside of any relationship – is the same as the possibility occurring here, and again the same presumability. *And, for its validity, every probability in this sense requires a reason for being probable.*
>
> (Husserl, 2019, p. 252)

Husserl's third point is that probabilities can thereby be compared and placed within an ordered structure of gradations. Since reasons can "speak" more or less strongly in favour of presumptions, it is possible to compare presumptions according to their relative gradations of how strongly they are "spoken for". In discussing this comparison between probabilistic weights, the word "objective" appears (our emphasis):

> Now, however, presumptions come into relationships. You immediately understand what is meant if I point to the familiar way of speaking, "This argues in favor of it, that other thing also argues in favor of it, and still more is in its favor". I am speaking from a subjective perspective of stronger and weaker presuming. We say that the presumption grows stronger with the number of reasons for presumability. However, we also say that the more there is that argues in favor of something, namely, in favor of the fact that something exists, the greater is its probability; the more that argues against it, the more the probability diminishes. This is an objective way of speaking [...] *[A] unique originary relation of precedence and gradation belongs to the presuming-meanings, and accordingly also objectively*

> *valid laws of gradation.* In these relationships of gradation what, considered in isolation, was already called probability, is then referred to as mere possibility, and possibilities are weighed against other possibilities [...] with respect to the reasons for them or the weight that they carry.
>
> (Husserl, 2019, p. 252)

Although it might seem paradoxical that Husserl should emphasize the "objective" character of probabilities, having earlier emphasized their "subjective" character, we can unscramble Husserl's omelette of subjectivity and objectivity by appealing to our earlier strategy of translating these concepts into statements about variability and invariance with respect to a given set of possible agents.

To this end, we note that in chapter 5 of Husserl (2019), Husserl distinguishes "simple" judgements from "composite" judgements. As Lobo points out (Lobo, 2019, p. 517) we should therefore correspondingly distinguish "first-order" from "second-order" probability statements. First-order statements are those in which probabilities figure in judgements about something non-probabilistic, such as "I consider it very probable that it will rain", whereas second-order statements take probabilities assignments as their objects, such as "the probability of rain is 1 minus the probability of not-rain". While first-order statements are evidently always subjective (since they depend on the one making the judgement) second-order statements can capture relations between probability assignments which are independent of the particular values of the assignments themselves, hence are "objective" in the relevant sense. And if one carefully reads the last passage we quoted from Husserl, it appears that the kinds of statements he cites as "subjective" and those he cites as "objective" do indeed fit into the respective categories of first-order and second-order probability statements.

Having thus clarified Husserl's three points about the interpretation of probabilities, let us compare them to QBism. Husserl's first point is evidently quite compatible with QBism, since it asserts that probabilities are (in Husserl's terminology) "qualities of judgement" and therefore may vary between subjects, and so are "subjective" in the sense relevant to QBism.

Husserl's second point differs from QBism in its emphasis: whereas Husserl emphasizes that agents must supply "reasons" that "speak in favour" of their probability assignments, QBism ignores the agents' reasons for their assignments, following instead the decision-theoretic line of only demanding that agents commit to purchasing or selling hypothetical lottery tickets by which they stand to gain or lose according to their probability assignments. While not an outright contradiction, this difference in emphasis deserves to be elucidated, and we shall return to it in the next section, after dealing with Husserl's third and final point.

Husserl's third point might seem problematic because of its reference to some "objective" aspect of probability. Given our explication of what "objective" means in this context, however, it should be noted that in QBism the rules of the probability calculus are presumed to be the same for all agents; indeed, this is what QBists mean when they posit that agents strive to be "rational" and to be "coherent". Therefore despite maintaining the subjective status of probability assignments, QBists allow that these enter into abstract relations whose form is invariant among the agents, and so we may call it "objective" in the relevant sense, although QBists prefer to use the word "normative" to avoid giving the false impression that the equations describe something independent of the agents as a group.

We may therefore reconcile Husserl's third point with QBism, if we read Husserl's usage of "objective" as indicating the rules of probability theory in a "normative" capacity. We can say, along with Husserl, that the rules of the probability calculus are "second order" constraints on probabilities, and are "objective" (in Husserl's words) only in the sense that they represent "normative" rules (in QBism's words) which all rational agents adhere to.

5.7 Probability and Normativity

Finally, let us return to the interesting tension regarding Husserl's point about the sources of probability weights. At a glance, there might seem to be an inconsistency here: does a probability weight indicate the relative strength of the "reasons that speak for it", or does it represent the amount of money at which an agent would be prepared to commit to buying or selling a hypothetical lottery ticket? We now argue that the answer must be: both!

To see that the two views are compatible, note that despite QBism's emphasis on interpreting probabilities in terms of hypothetical bets (an attitude inherited from subjective Bayesian probability theory), it also provides the conceptual infrastructure for any agent to provide "reasons" for their probability assignments, if questioned. Formally, following Bayesian methods for reasoning with probabilities, an agent begins at some point in the past by making a set of "prior" probability assignments (a Bayesian prior). If the agent is surprised by subsequent experiences, they revise their probability assignments to recover "coherence" among their beliefs and their experience, as represented by certain formal constraints on their probabilities (such as Bayes' rule). Therefore, for a QBist, the "reasons" behind a probability assignment comprise the agent's prior assignment, a history of their experiences, and the normative rules they employ to keep their probabilities mutually coherent in light of those observations.

It is worth emphasizing that despite the abstractness of this formulation, it can perfectly well capture what we ordinarily consider to be reasons for probability assignments. For example, if someone were to ask me why I think there is "a 75% chance of rain this afternoon", I could tell you that from my experience I have come to believe that here in New England there is a 50% prior chance of rain on any given day, and moreover that rain is often portended by grey clouds. Using Bayes' formula I can take into account the datum that I saw grey clouds in the sky this morning, which leads me to revise my probability for afternoon rain upwards to 75%. Although this is more detail than they probably wanted to hear, it would certainly satisfy the questioner.

While QBism is therefore compatible with interpreting probability assignments in terms of the "reasons" that "justify" them, it certainly downplays this interpretation. This is because, as mentioned in Section 5.3, QBism places minimal restrictions on what agents' prior beliefs can be: any arbitrary probability assignment is assumed to be *ab initio* justifiable by some (unspecified) reasons. Put another way, the QBist tacitly assumes that *reasons can always be contrived to justify any initial prior*. Under this assumption, one is not interested in finding out why agents believe what they do; rather, one is interested in calling upon the agents to act upon their beliefs, say, by having them place bets with meaningful stakes and seeing how they fare.

Bringing QBism in contact with Husserl's conception of probabilities therefore suggests that we should adopt a more balanced view that gives equal emphasis both to the reasons behind probability assignments (as captured by agents' priors and past history) and the future actions that the agents commit themselves to in making their assignments.

Fortunately, such a balanced approach has already been articulated in detail by the contemporary pragmatist philosopher Robert Brandom. Within his framework of "inferentialism", Brandom proposes that the fundamental conceptual elements are "assertions", which are defined as "essentially performances that can both serve as and stand in need of reasons" (Brandom, 2000, p. 189). Moreover, Brandom considers an assertion as akin to a "move" made in the context of a "game of giving and asking for reasons". The "game" is comprised of a set of normative rules, to which all agents under consideration are assumed to adhere. As Brandom points out, the two-sided nature of an assertion gives rise to two kinds of normative rules, which Brandom names "commitments" and "entitlements". As Brandom explains:

> [A]ssertional games must have rules of this sort: rules of *consequential commitment.* Why? Because to be recognizable as assertional, a move must not be idle, it must make a difference, it must have consequences for what else it is appropriate to do, according to the rules of the game. [...] Understanding a claim, the significance of an assertional move, requires understanding at least some of its consequences, knowing what else (what other moves) one would be committing oneself to by making that claim. [...] For this reason we can understand making a claim as taking up a particular sort of normative stance toward an inferentially articulated content. It is endorsing it, taking responsibility for it, committing oneself to it.
>
> (Brandom, 2000, pp. 191–192)

And:

> Giving reasons for a claim is producing other assertions that license or entitle one to it, that justify it. Asking for reasons for a claim is asking for its warrant, for what entitles one to that commitment. Such a practice presupposes a distinction between assertional commitments to which one is entitled and those to which one is not entitled. Reason-giving practices make sense only if there can be an issue as to whether or not practitioners are entitled to their commitments. Indeed, I take it that liability to demands for justification – that is, demonstration of entitlement – is another major dimension of the responsibility one undertakes, the commitment one makes, in asserting something.
>
> (Brandom, 2000, p. 193)

From these excerpts one can clearly see that what counts as "good reasons" for an assertion, and what counts as "consequent" upon an assertion, are established by normative rules, which are the "entitlements" and "commitments", respectively, that govern a specific "game of giving and asking for reasons". By considering the various ways that these normative rules relate to one another, Brandom develops what he calls the "normative fine-structure of rationality" (Brandom, 2000, p. 195).

One is immediately struck by the idea that a probability assignment may be interpreted precisely as a special kind of "assertion" in Brandom's sense, whereby QBism's apparent divergence from Husserl on his second point may be understood as due to a difference in emphasis: Husserl's emphasis on an agent's "entitlements" to a probability assertion on the one hand, and QBism's emphasis on an agent's "commitments" in making a probability assertion; on Brandom's account these are two sides of the same coin. Furthermore, QBism so far only deals with the issue of how an agent's probability assignments *commit* them to certain actions (like buying lottery tickets); Brandom's concept of entitlement could enrich QBism by widening its scope to include how an agent's beliefs may or may not *entitle* them to probability assignments of a certain form[7].

The possible relevance of Brandom's work to quantum mechanics has been recognized previously in the work of Richard Healey (2012). The key difference between our present proposal and Healey's approach can be traced to two fundamentally different methodological starting points regarding the appropriate "set of possible agents". As already discussed, QBism locates the quantum state at the "subjective" end of the spectrum of probabilistic assertions, namely, those which are assumed to differ *a priori* among the given set of possible agents. By contrast, Healey, chooses to locate the quantum state in the "normative" component of quantum theory, assuming that there is a single quantum state that represents a norm of "best advice" for the given set of agents. Thus, on Healey's account, the quantum state is a set of probabilistic assertions that is "objective" in the sense of being invariant across the relevant set of agents. There is still much work to be done in comparing and contrasting different pragmatist approaches to quantum theory; an overview can be found in Healey (2017).

Inferentialism therefore not only can provide the missing piece that is needed to combine Husserl's phenomenology of probabilities with QBism's subjective decision-theoretic approach, but might also provide a unifying framework in which to compare and contrast QBism with other "pragmatic" interpretations of quantum theory including that of Healey. A fuller investigation of these themes must be left to future work.

5.8 Conclusion

We conclude by summarizing our findings. QBism and phenomenology have many points in common, and indeed a common goal: to articulate a metaphysics of "experienced outcomes" or "perceived phenomena" which are neither figments of the observer's mind, nor phantoms in a world beyond all observation. However it is by no means obvious that phenomenologists and QBists see eye to eye on every issue. We have examined the question of the existence of unobservable entities in science from both points of view, and compared their approaches. We found that QBism distinguishes between system-like entities and probability-like entities, and adheres to the Prolongation Thesis, which renders unobservable system-like entities potentially observable, contrary to Wiltsche's objection

(Wiltsche, 2012). On the other hand, in order to maintain its claim that probability-like entities cannot be said to physically exist, QBism should embrace Wiltsche's "Originality Thesis of Justification" over Berghofer's counter-proposed "Criterion of Justification", as the latter opens the door to the possible physical existence of quantum states and hidden variables.

Our rejection of Berghofer's criterion raised the question of what exactly are probability-like entities, if not things that physically exist. Here we encountered the difficulty that phenomenologists have said very little about the interpretation of probability. We argued that Husserl's account of probabilities as "qualities of judgement" and "weighted presumptions" could be made compatible with QBism if we interpret Husserl's usage of the word "objective" to describe features of probability theory as being equivalent to QBism's usage of the word "normative" to describe the invariant nature of these rules among rational agents.

Furthermore we pointed out that although Husserl and QBism provide seemingly opposing accounts of the origin of the probability values, with Husserl appealing to the reasons that "speak for" a probability assignment, and QBism appealing to the gambling commitments that stem from a probability assignment, a reconciliation can be found in Brandom's inferentialism, if we regard probabilities as assertions which entail both supporting reasons and consequent commitments. This leads us to interpret probability-like entities as collections of assertions made by an agent, which stem from the agent's prior probabilities and past experience, and serve to co-ordinate their future actions, within a context of normative rules. In this manner it may be possible to make sense of QBism's distinction between system-like and probability-like unobservable entities, and explain QBism's claim that system-like entities can be observable in principle (with an appropriate intracorporeal measuring instrument), while probability-like entities such as quantum states cannot.

Acknowledgement

This work was supported in part by the John E. Fetzer Memorial Trust.

Notes

1 See Fuchs and Schack (2013), Fuchs and Schack (2014), Fuchs et al. (2014), Fuchs (2010, 2016, 2017), DeBrota and Stacey (2019), and Fuchs and Stacey (2019).

2 A more accurate term would be "indexed to the agent", since QBism does not insist that "agent" refers to a person (DeBrota and Stacey, 2019, p. 23).

3 Note that in light of the preceding discussion, the "observability in principle" of something must be understood as being relative to a particular set of possible agents.

4 Note that the agreement of two agents' beliefs in one or more aspects would not imply that the agents hold the same belief, any more than the identical appearances of two apples would imply that they are the same apple.
5 This technical term – givenness in originary presentive intuition – is here understood as referring to X's being an object of the agent's sense perception.
6 Note that one is not forbidden to *believe* that such an entity physically exists – just that one's belief cannot be justified on "epistemic" grounds. The question of what other grounds there might be for believing in something is left open.
7 There may already be a precedent for this idea in the literature: the quantum de Finetti theorem (Caves et al., 2002), which is a cornerstone of QBism, may be thought of as proving that if an agent believes that a set of systems is "exchangeable", they are thereby entitled to make a probability assignment that has a particular mathematical form.

References

Barad, K. (2007). *Meeting the Universe Halfway: Quantum Physics and the Entanglement of Matter and Meaning*. Duke University Press, Durham, NC.

Berghofer, P. (2017). Transcendental phenomenology and unobservable entities. *Perspectives*, 7, 1–13.

Bernardo, J. M., and Smith, A. F. M. (1994). *Bayesian Theory*. Wiley, Chichester.

Brandom, R. (2000). *Articulating Reasons: An Introduction to Inferentialism*. Harvard University Press, Cambridge, MA.

Caves, C. M., Fuchs, C. A., and Schack, R. (2002). Unknown quantum states: The quantum de Finetti representation. *Journal of Mathematical Physics*, 43(9), 4537–4559.

Crease, R. P., and Sares, J. (2021). Interview with Physicist Christopher Fuchs. *Continental Philosophy Review*, 54, 1–21.

de Finetti, B. (2017). *Theory of Probability* (A. Machi and A. Smith Trans.). Wiley, New York (Original work published 1974).

de La Tremblaye, L. (2020). QBism from a Phenomenological Point of View: Husserl and QBism. In Wiltsche, H. and Berghofer, P. (Eds.), *Phenomenological Approaches to Physics. Synthese Library (Studies in Epistemology, Logic, Methodology, and Philosophy of Science)*, 429, 243–260. Springer, Cham.

DeBrota, J. B., Fuchs, C. A., and Schack, R. (2020). Respecting one's fellow: QBism's analysis of Wigner's friend. *Foundations of Physics*, 50, 1859–1874.

DeBrota, J. B., and Stacey, B. C. (2019). FAQBism. https://doi.org/10.48550/arXiv.1810.13401

Eliot, G. (1965). *Middlemarch*. Penguin Books, London (Original work published 1871).

Fuchs, C. A. (2010). QBism, the Perimeter of Quantum Bayesianism. https://doi.org/10.48550/arXiv.1003.5209

Fuchs, C. A. (2016). On Participatory Realism. In Durham, I. T. and Rickles, D. (Eds.), *Information and Interaction: Eddington, Wheeler, and the Limits of Knowledge*, 113–134. Springer, Berlin.

Fuchs, C. A. (2017). Notwithstanding Bohr, the reasons for QBism. *Mind and Matter*, 15(2), 245–300.

Fuchs, C. A., Mermin, N. D., and Schack, R. (2014). An introduction to QBism with an application to the locality of quantum mechanics. *American Journal of Physics*, 82(8), 749–754.
Fuchs, C. A., and Schack, R. (2013). Quantum-Bayesian coherence. *Reviews of Modern Physics*, 85, 1693–1715.
Fuchs, C. A., and Schack, R. (2014). QBism and the Greeks: Why a quantum state does not represent an element of physical reality. *Physica Scripta*, 90(1), 015104.
Fuchs, C. A., and Stacey, B. C. (2019). QBism: Quantum Theory as a Hero's Handbook. In Rasel, E. M., Schleich, W. P., and Wölk, S. (Eds.), *Proceedings of the International School of Physics "Enrico Fermi" Course 197 – Foundations of Quantum Physics*, 133–202. IOS Press, Amsterdam; Società Italiana di Fisica, Bologna.
Healey, R. (2012). Quantum theory: A pragmatist approach. *British Journal for the Philosophy of Science*, 63(4), 729–771.
Healey, R. (2017). Quantum-Bayesian and Pragmatist Views of Quantum Theory. In Zalta, E. N. (Ed.), *The Stanford Encyclopedia of Philosophy* (Spring 2017 ed.). Metaphysics Research Lab, Stanford University, Palo Alto, CA.
Husserl, E. (2019). *Logic and General Theory of Science, Lectures 1917/18 with Supplementary Texts from the First Version of 1910/11*, volume 15 of *Husserliana: Edmund Husserl – Collected Works* (C. Ortiz Hill Trans.). Springer International Publishing, Cham (Switzerland). Husserliana XXX (Original work published 1918).
Lobo, C. (2019). Husserl's logic of probability: An attempt to introduce in philosophy the concept of 'intensive' possibility. *Meta: Research in Hermeneutics, Phenomenology, and Practical Philosophy*, 11, 501–546.
Merleau-Ponty, M. (1964). In Edie, J. M. (Ed.), *The Primacy of Perception: And Other Essays on Phenomenological Psychology, the Philosophy of Art, History and Politics* (W. Cobb Trans.).Northwestern University Press, Evanston (Original work published 1947–1961).
Merleau-Ponty, M. (1968). In Lefort, C. (Ed.), *The Visible and the Invisible* (A. Lingis Trans.). Northwestern University Press, Evanston (Original work published 1964).
Merleau-Ponty, M. (2012). *Phenomenology of Perception* (D. Landes Trans.). Routledge, London (Original work published 1945).
Pienaar, J. (2020). Extending the agent in QBism. *Foundations of Physics*, 50, 1894–1920.
Pienaar, J. (2021). QBism and relational quantum mechanics compared. *Foundations of Physics*, 51(5), 96.
Todes, S. (2001). *Body and World*. The MIT Press, Cambridge, MA.
Wiltsche, H. (2012). What is wrong with Husserl's scientific anti-realism? *Inquiry*, 55, 105–130.
Zahavi, D. (2018). *Phenomenology: The Basics*. Routledge, Oxford.

Part II

From Phenomenology to QBism

6 On the Consilience between QBism and Phenomenology

Hans Christian von Baeyer

Tout notre raisonnement se réduit à céder au sentiment.
(All our reasoning boils down to yielding to feeling.)
(Blaise Pascal)

6.1 Introduction

The study of the *interpretation of quantum mechanics* (IQM) is poised to enter a new epoch. The story of IQM began with the creation of quantum mechanics itself in 1925/1926, but its appeal as a research topic was usurped by the theory's unprecedented successes in explaining experiments and suggesting new devices. Around the year 2000, amid the rapid development of *quantum information theory*, novel information-based IQMs appeared. Among them was QBism, which here serves as an exemplar of a whole class of different approaches. The fundamental novelties of QBism (Fuchs 2017) were its substitution of personalist Bayesianism for frequentist probability and its recognition that quantum measurement outcomes were not available "for all the world to see," but rather of the nature of personal experience. Now, 20 years on, the third epoch in the IQM story focuses on the fundamental role of perception, as opposed to concept formation, in theoretical physics (both quantum and classical). Since phenomenology is, crudely speaking, the study of perception, students of IQM should broaden their horizons to include this philosophical and psychological topic – as indeed many of the founders of quantum mechanics thought a century ago. For popularizers of physics, including myself, this poses a triple challenge. First, we must learn a modicum of the language of philosophy. Then we must urge our physicist colleagues to temper their disdain for philosophy. Finally, we must marshal the most persuasive philosophical arguments in support of the relevance of phenomenology for QBism. Building on the pioneering work published in this volume and its predecessor (Wiltsche & Berghofer 2020), as well as the largely overlooked

DOI: 10.4324/9781003259008-8

work of the American philosopher Samuel Todes (1927–1994), I suggest steps toward meeting this challenge.

6.2 Seeking Consilience between QBism and Phenomenology

At the beginning of the 21st century, the late biologist Edward O. Wilson re-introduced the felicitous 19th-century word *consilience* into the vocabulary of the two-cultures debate. He set the stage in lofty terms: "The greatest enterprise of the mind has always been and always will be the attempted linkage of the sciences and humanities" (Wilson 1998, p. 8). The Oxford English Dictionary, for its part, defines consilience as "Agreement between the approaches to a topic of different academic subjects, especially [of] science and the humanities." QBism was developed by physicists, and has attracted the attention of philosophers, particularly phenomenologists. This volume, together with its predecessor, offers a timely contribution to the effort of finding consilience between the two approaches.

Perhaps the most profound attempt to explore the connection between philosophy and physics, "between soul and matter," was proposed by Wolfgang Pauli under the rubric *Psychophysics* (Gieser 2005, p. VII). Sadly, he got distracted from the project and died, much too early, before he got back to it. Another pioneer of quantum mechanics, Erwin Schrödinger, placed the origin of the clash between science and the humanities close to the beginning of Western philosophy, two and a half millennia ago, by quoting Democritus. After describing the Greek atomic hypothesis, Schrödinger continued:

> Yet Democritus at the same time realized that the naked intellectual construction which in his world-picture had supplanted the actual world of light and color, sound and fragrance, sweetness, bitterness and beauty, was actually based on nothing but the sense perceptions themselves which had ostensibly vanished from it. In fragment D125, taken from Galen and discovered only about fifty years ago, he introduces the intellect (dianoia) in a contest with the senses (aesthesis). The former says: 'Ostensibly there is color, ostensibly sweetness, ostensibly bitterness, actually only atoms and the void'; to which the senses retort: 'Poor intellect, do you hope to defeat us while from us you borrow your evidence? Your victory is your defeat.' *You simply cannot put it more briefly and clearly.*
>
> (Schrödinger 1996, p. 32. My emphasis.)

The whimsical Fragment D125 is a reminder of the obvious and fundamental realization that everything that enters our individual minds does so via the senses. Just as Descartes's *cogito* symbolizes the logical relationship between thought and existence, Democritus's little gem encapsules the paradoxical links between physics and phenomenology. The resolution of the paradox lies in consilience.

In modern times, Democritus's complaint against the intellect has been sharpened by science. Many sense experiences have indeed been linked to well-understood physical concepts: color to light frequency, warmth to molecular motion, sound to material waves, pressure to inter-atomic forces, etc. But a bewildering variety of optical illusions reminds us that materialistic "explanations" of sense experiences are much more problematic than many physicists – especially teachers – like to believe. Furthermore, conditions such as *synesthesia*, the involuntary confusion of different senses, which has no basis in physics, hint at the entangled map of cognitive pathways that awaits its future cartographers.

An accessible, concrete example of the power of consilience between philosophy and science is the story of the *Molyneux Question* (Degenaar & Lokhorst 2021), which was debated by the philosophers Locke, Berkeley, Hume, and others, before turning up recently in a lab at MIT and an eye clinic in New Delhi (Held et al. 2011). The question concerns the perception of the shape of a solid object, a feat that can be achieved by two different senses. The concept of shape *per se* does not seem to be very problematic. Nevertheless, in 1688, William Molyneux wrote to Locke, asking, in effect: "Would a blind subject, on suddenly gaining sight, be able to recognize an object previously known only by touch?" For centuries, even the phrasing of this question provoked controversy. Today, sketchy experimental evidence supports Locke's answer, which was NO. However, after a recovery period that could be as short as a few days, the newly sighted were able to reconcile what they saw with what they had felt earlier. For 333 years Molyneux's trenchant question has continued to fascinate philosophers and scientists alike.

In the case of the Molyneux Question, the arc of the story pointed from philosophy to science; with respect to QBism, the arrow points in the opposite direction: Initiated by physicists, QBism is now attracting the attention of philosophers. The two communities don't always speak the same language. In the mind/matter discussion, for example, here are some of the dichotomies in use:

Greek atomists:	Intellect	/	Senses
The lay public:	Thinking	/	Feeling
Physicists:	Theory	/	Experiment
Phenomenologists:	Conception	/	Perception

Physicists tend to focus on the third line, philosophers on the fourth. Rather than favoring one definition over another, in the spirit of consilience, it will be necessary to develop a language comfortable to both communities. Theory and Conception, for example, refer to the **Imagination,** Experiment and Perception to human **Experience.** The meanings of the terms, however, overlap substantially. Painstaking analysis of abstract concepts, a core business of philosophy, has already borne fruit for QBists: It was the examination of the meaning of *probability* that inspired Quantum Bayesianism, the precursor of QBism. Conversely, QBism has contributed to the unraveling of the philosophical problem of the meaning of time (Mermin 2013). Here, I am not recommending any specific nomenclature, but suggesting that clarifying the terms of the conversation among QBists and Phenomenologists is an essential first step in promoting understanding.

6.3 Samuel Todes: A Forgotten Philosopher of the Human Body

In 1963, Samuel Todes (Arias 2011) submitted his doctoral dissertation in philosophy, titled *The Human Body as Material Subject of the World*, to Harvard University. Although he successfully continued his academic career, his thesis (Todes 1990) was not published until almost 30 years later as part of a series of distinguished Harvard dissertations, including those of luminaries such as Goodman, Davidson, Putnam, and Quine. It would be another decade before it appeared in a more accessible form under the less daunting title *Body and World* (Todes 2001). In keeping with the simplification of this title, I will avoid the controversial technical terms *subject body* and *body subject* (Cf., for example Sheets-Johnstone 2020, p. 2).

In an essay dated 1993, a year before his death, Todes summarized his radical program essentially as follows: All sense we can make, whether of what is actually perceivable, or possible as conceivable, draws on our sense of our [active] body, and is unintelligible without reference to it (Todes 2001, p. 268).

In his book, Todes points to what he calls a common, fundamental error: *the misinterpretation of the human body as merely a material thing in the world* (Todes 2001, p. 88). Instead, he suggests that "The human body is not only a material thing found in the midst of other material things in the world, but it is also, and moreover thereby, that material thing whose capacity to move itself generates and defines the whole world of human experience in which any material thing, including itself, can be found."

To illustrate his take on the relationship between perception and conception, he used the metaphor of a building: "There are two levels of objective experience: the ground floor of perceptually objective experience; and the upper storey of imaginatively objective experience (of which theoretical understanding is one kind), which presupposes for its objectivity (i.e. for its dependability as living quarters) that the ground floor unto which it is built is itself on firm foundations" (Todes 2001, p. 100). The firm foundation is our individual sense of our own living body.

As physicist, I would suggest the related metaphor of a Cartesian frame of reference; a foundation is, after all, under the frame of a house. Furthermore, Todes's own word *reference* in the summary quoted above brings to mind the idea of a reference frame. In my image, an understanding of the world in terms of abstract concepts resembles the graph of a mathematical function, which only makes sense with respect to a frame of reference whose origin, axes, and scale must be specified before the function itself. Todes proposes that the reference frame is defined by the perceptions of the human body in a manner I will try to illustrate.

Todes's choice of the body as the source of an immutable certainty on which to build a philosophy is a surprising as it is bold. Nevertheless, it makes sense to me! Descartes's *cogito*, for example, worked for him, but although I can learn *what* he thought, I have no way of knowing *how* he thought. But I think I am safe in assuming that his body's perceptive mechanisms were similar to mine. Todes explains:

> Modern evolutionary biology supports the contention that mankind has not evolved over historical time – that is, in the last five or ten thousand years – and makes it plausible that the species-specific character of the human body may condition all human experience within some broad but definite limits. A phenomenological analysis, I will argue, justifies our current biological belief by demonstrating how our sense of our body plays a basic formative role in our making sense of everything. And not only in a culturally bound way. For historical knowledge, travel, reading, and imagination show that we can make *some* sense of different cultures. But the level at which our sense of our body enters into our sense-making is such that experience shorn of that body-sense is *unintelligible*. We cannot imagine having any sense at all of things of that type.
>
> I will try to show this for perceptual objects, which *actually* present themselves to us, and for ideas or concepts, which present to ourselves as forms of *possibility*.
>
> (Todes 2001, p. 263. Italics in the original.)

Before I provide concrete examples of what Todes is suggesting, I want to mention why his proposal appeals to me. As the problems of the world multiply and intensify, science is not only an important tool for defining them but also one of the most reliable sources of solutions. Before such solutions can be implemented, however, scientists must first persuade politicians and the general public to accept the findings of science. To this end, a growing community of writers, reporters, and teachers is using all available means of communication in a grand enterprise of *public understanding of science*. Unfortunately, their task, which is mine too, grows increasingly difficult as science reaches unimaginable realms of vastness, minuteness, and complexity. Sometimes I despair of ever reaching the elusive "informed public."

The first obstacle is scale. Quantum mechanics, for example, which is poised to reach into our lives in the immediate future via quantum computing and quantum cryptography, possibly even quantum currency, owes its unfortunate reputation for weirdness in large part to the unimaginable chasm between the typical dimensions of our world and those of the quantum world. If QBism, with the help of phenomenology, is to become a household word, how can we talk about it in accessible language? If the nature of fundamental concepts such as space, time, reality, probability, causality, and consciousness is to be explained in terms of quantum mechanics, phenomenology, and neuroscience, how does the average citizen have a chance to understand?

I hope Todes can help. He writes: "... the level at which our sense of our body enters into our sense-making is such that experience shorn of that body-sense is *unintelligible*," and, as quoted above, "All sense we can make, whether of what is actually perceivable, or possible as conceivable, draws on our sense of our [active] body, and is unintelligible without reference to it." Since we all have similar bodies, we are justified in assuming that others have similar body-senses as well so that we can use the human body and its movements as the foundation for the construction of our worldview – confident that it will have some claim on universality.

6.4 Space according to Todes

In High School, I loved Euclidean Geometry. The triumph of seeing before my eyes the steps of a successful proof gave me a little high – the conclusiveness of the proof satisfied my thirst for certainty. But for me, Euclidean geometry has one major drawback. Most proofs require the construction of "auxiliary lines," straight lines or circles that needed to be chosen cleverly and used as stepping stones before they were erased again from the original drawing. I never found a systematic method for the choice of these lines: You just had to be clever or lucky. Going into an exam, I worried whether I would be able to find them.

A year after Euclid, as we called it then, we took Analytic Geometry. Here, the game is to translate all points and lines into algebraic equations, which are then solved by well-established routines. No auxiliary lines here – no guessing – no need for luck. What a relief! The method requires the construction of a frame of reference, consisting of an origin and two or three mutually perpendicular axes, which serve as the fixed stage on which the game of geometry plays out. I have always thought about the three axes of that Cartesian frame of reference as being interchangeable. In three dimensions, there is a subtlety about labeling the three positive directions (a property called chirality or handedness), but otherwise the identification of which one is x, y, or z is arbitrary. When I think of space, I imagine a Cartesian frame with numerical values attached to its points.

I was, therefore, surprised to learn from Todes that our bodies experience the three directions in three different ways. Our left and right sides (along the y-axis, say) are mirror images, at least for many parts of our bodies. Their obvious differences and similarities provide the intuitive foundations for concepts of symmetry, parity, chirality, and reversibility, as well as perennial conundrums such as "Why does a mirror reverse right and left but not up and down?" and "How do you define left and right, over the telephone to someone who doesn't know the words?" In sharp contrast, front and back (along the x-axis) lack mirror symmetry. Thereby the body acquires a direction, a natural forward arrow that is missing from the y-axis. Since the description and experience of motion require the specification of a direction, Todes associates the intuition of back-to-front with motion, and motion, in turn, with time (Todes 2001, p. 49). The third dimension, in the vertical direction, along the z-axis, is perceived via gravity. It too has a direction, but unlike the body's direction, that direction is constant and global, rather than local and changeable at will. Thus, the three axes of a Cartesian coordinate system, while mathematically identical, are differentiated by the human body.

For Todes, the perception of the spatial attribute of solid objects called *extension* is related to the sense of motion of the active, living body. In his critique of Hume, he writes: "Hume holds that we are aware of extension through sight and touch, but he never understands that we may become aware of extension through movement" (Todes 2001, p. 48). The word "may" is puzzling because Todes then proceeds to show that body movement is actually necessary for the perception of shape, as Molyneux's question reveals. Simply "staring fixedly or resting our hand inertly on an object" do not reveal its shape or extension in three dimensions. As Todes puts it, "In movement alone we are not restricted to being passively aware of objects, in the way Hume thinks characteristic of all our experience."

6.5 Time according to Todes and Mermin

An early demonstration of consilience between QBism and Phenomenology was David Mermin's solution of the problem of the NOW (Mermin 2013). Before that paper, the present moment, familiar though it is to everyone, played no role in physics. Mermin showed that the principal difficulty was not the relativity of simultaneity, but the compulsive exclusion of subjectivity from science. As noted, Democritus had already suspected that this attitude would lead to frustration. With the emphasis of both QBism and phenomenology on a first-person perspective, the problem of the NOW melts away.

NOW, according to both Mermin and Todes, is simply the moment in time at which the anticipation and prediction of a future event turns into memory of the past experience of that event. Since each of us is capable of distinguishing between those two modes of cognition, that moment is not only well defined but also surely of surpassing importance.

Mermin re-ignited the discussion among physicists and phenomenologists about the ancient question "What is Time?" (I am, of course, not referring to sophisticated ideas like string theory, loop quantum gravity, and quantum foam, which concern the *nature* of time. This essay is about its *meaning*.) Exactly half a century earlier, Todes had come to similar conclusions by a different route.

The simple, paradigmatic story Todes tells as an illustration of my perception of time is that of myself approaching, passing, and leaving behind an object (Todes 2001, p. 120). Initially, the object, say a girl, is in front of me. My direction of motion is forward, as defined by the asymmetry of my body. After the encounter, I have *passed* the girl, and she is now in my *past*. So, Todes claims, "the front-back body distinction makes possible the *passage* of time." (He points out that the *passed-past* homonymy is no mere linguistic coincidence: The verb and the noun are both *passé* in French, and *vergangen-Vergangenheit* in German (Todes 2001, p. 300).) In this way, the asymmetry of my body combines with my ability to move to produce my sense of the passage of time. Building on this humble foundation, Todes devotes additional analysis to other attributes of time, such as irreversibility, continuity, and the metaphor of flow.

6.6 Time in Popular Understanding

According to one of the several different versions, Ernest Rutherford said: "A theory that you can't explain to a bartender is probably no damn good." If you want to avoid offending bartenders, substitute an 11-year-old. It turns out that the concept of the NOW, according to Todes and Mermin, passes this test. I know, because I was such a child.

In my book about QBism, written for the public, one chapter is devoted to Mermin's treatment of the NOW (von Baeyer 2016). It tells the story of an experiment I performed at about age 11, when I was living in Switzerland with my family. What's remarkable is the precision with which my experience matched the scenario described by Todes to explain how we perceive time. Sitting in a train from Basel to Zurich one day, I knew that we would soon pass a fake castle which fascinated me, and later turned out to be a brewery. Anticipating the passage, I concentrated on the exact moment of closest approach, and, when it arrived, shouted: NOW. I never forgot that moment 72 years ago. I had stopped time in its course.

Note the ingredients of the story. The "event" mentioned by Todes was my body passing the brewery. My motion was supplied by the train. The fixed forward orientation of my body was required by my need to concentrate on what was coming. My mental image of the brewery represented the future, until, in refreshed form, it became the past.

Decades later, I repeated the experiment with groups of school and university students. The re-enactments differed from my original experiment in two details. For one thing, we experienced at first-hand what Mermin called *The NOW of Many People*. In our relativistic world, such a collective event makes sense only when the participants are physically close together. Our common instant occurred when we shouted NOW in unison. As soon as we dispersed, our common NOW divided into so many personal, unrelated NOWs. Second, we were not moving, as I had been on the train. For bodily movement, we substituted a different kind of directed, continuous change. The change required for the perception of time was provided by our collective count-down: ten, nine, eight ... NOW. Afterwards, we discussed that NOW almost as if it were a material thing – an atom of time. What happened, according to Todes's scheme, was that we raised the bodily perception of motion from the ground floor of his metaphorical building to the upper story of conception: We *imagined* the approach of the NOW.

Some students remembered the event years later. For me, however, the re-enactments lacked the spontaneity of the original and soon faded from memory. What I do remember, though, is the eerie, powerful sensation of the spatial approach of the NOW. I could almost see that moment approaching, like the front of a green, Swiss locomotive. I could almost feel the vibration of the floor. It was exhilarating, and my pupils loved it.

6.7 Conclusion

Among interpretations of quantum mechanics, QBism shows with special urgency that physics, in order to offer a satisfying approach to the nature of reality, must reconnect with philosophy to reach consilience on the meaning of words such as space, time, certainty, reality, belief, truth, and probability. Phenomenology is the branch of philosophy that is currently a candidate for partnering with physics to achieve this goal.

The phenomenologist Samuel Todes suggested that the human body, with its free will, its senses, and its moveability, provides the foundation for defining and understanding all human experiences. In defense of this radical proposal, he offered a number of observations: The body is universal, unchanging, and accessible. It combines mind and matter in a way that may be difficult to disentangle, but that all of us experience from the moment of birth.

In this essay, I add a modest but potentially useful observation. For the purpose of public understanding of science, considering the living human body as the origin of the frame of reference for coming to terms with the world may help to overcome the impediments imposed by the unimaginably small and large scales of most of the material universe, as well as by the complexity of modern science. Accepting even parts of Todes's phenomenology may benefit not only physicists and philosophers but popularizers of science as well.

References

Arias, D. (2011). *Samuel Todes Blog*. http://samueltodes.blogspot.com/2011/12/who-was-samuel-todes.html

Degenaar, M. & Lokhorst, G.-J. (2021). Molyneux's Problem. *The Stanford Encyclopedia of Philosophy*. https://plato.stanford.edu/archives/win2021/entries/molyneux-problem/

Fuchs, C. A. (2017). Notwithstanding Bohr, the Reasons for QBism. *Mind and Matter 15*(2), 245–300. https://arxiv.org/abs/1705.03483

Gieser, S. (2005). *The Innermost Kernel: Depth Psychology and Quantum Physics. Wolfgang Pauli's Dialog with C.G. Jung*. Berlin: Springer.

Held, R., Ostrovsky, Y. & de Gelder, B. et al. (2011). The Newly Sighted Fail to Match Seen with Felt. *Nature Neuroscience 14*, 551–553. https://doi.org/10.1038/nn.2795

Mermin, N. D. (2013). *QBism as CBism: Solving the Problem of 'the Now'*. https://arxiv.org/abs/1312.7825

Schrödinger, E. (1996). *Nature and the Greeks and Science and Humanism*. Cambridge, UK: Cambridge University Press.

Sheets-Johnstone, M. (2020). The Body Subject: Being True to the Truths of Experience. *The Journal of Speculative Philosophy*, *34*(1), 1–29.

Todes, S. (1990). *The Human Body as Material Subject of the World*. New York, NY: Garland Publishing.

Todes, S. (2001). *Body and World*. Cambridge, MA: MIT Press.

von Baeyer, H. C. (2016). *QBism – The Future of Quantum Physics*. Cambridge, MA: Harvard University Press.

Wilson, E. O. (1998). *Consilience – The Unity of Knowledge*. New York, NY: Knopf.

Wiltsche, H. A. & Berghofer, P. (2020). *Phenomenological Approaches to Physics (eBook)*. Cham: Springer Nature, Switzerland AG. https://doi.org/10.1007/978-3-030-46973-3

7 QBism

Realism about What?

Thomas Ryckman

> But whatever you do, *PLEASE* call QBism "*any damned thing but*" non-realist!
>
> (Fuchs, 2017, p. 126)

7.1 Not "realism" in Any Traditional Sense

QBism[1] is an interpretation of quantum mechanics according to which probabilities are regarded as personal degrees of belief, operationalized as bets made by a (coherent) agent (user of quantum theory) concerning contents of future experiences, among them possible measurement outcomes. Correspondingly, quantum states, symbolized as usual by wave functions ψ, are not agent-independent and do not represent "elements of physical reality" (as in ontic views) but simply give expression to an agent's personal probability assignments, reflecting his or her subjective degrees of belief regarding future contents of experience. Now under familiar construal, an *interpretation* of quantum mechanics purports to tell us what the world is like according to quantum mechanics. And QBism considers itself in accord with a customary self-conception of the scientific enterprise whose aim is to say *how the world really is*. Bristling at any instrumentalist or anti-realist construal of QBism, Chris Fuchs insists that QBism shares the usual realist ambitions of physical theory:

> For 20 years … I have pleaded with the community to understand my efforts at understanding quantum mechanics as being part of a realist program, i.e., as an attempt to say something about what the world is like, how it is put together, and what's the stuff in it.
>
> (Fuchs, 2017, p. 117)

To "say something about what the world is like" is minimally to say something about the world external to the agent (the "objective world" in the sense of realism); QBists surely wish to march under the banner

DOI: 10.4324/9781003259008-9

of realism. But such affirmations stand beside others that appear rather distant from any realist conception of physical theory. For QBists also proclaim

> ... quantum mechanics itself does not deal directly with the objective world; it deals with the experiences of that objective world that belong to whatever particular agent is making use of the quantum theory.
>
> (Fuchs, Mermin, and Schack, 2014, p. 750)

This tells us that QBism has a more ambitious agenda than to be just another interpretation of quantum mechanics. It is a vastly broader outlook on science itself, a radically first-person perspective underlying the entire scientific enterprise, one *inter alia* reminiscent of J.S. Mill, W. James, Russell, Carnap, and P.W. Bridgman. QBists regard individual personal experience (not to be understood as immediate data of sense) as the bedrock upon which the whole edifice of science, including probabilistic inference, is built up. Certainly, this is far more empiricism than nearly all scientists and most philosophers will condone; it requires us, as Mermin (2019, p. 2) observes, "to think about science in a radically unfamiliar way". To be sure, locating the origins of empiricism in experiences of an individual mind or consciousness is nothing more than empiricism consistently carried through, even to the paradoxical dead end of solipsism. Accordingly, recent philosophical critics, among them Brown (2019) and Earman (2019) worry that lurking under QBism's philosophical hood is an ogre of solipsism. Such contentions, in my view, are best to be understood as rhetorical pleas for philosophical clarity rather than substantive criticisms. QBism's empiricist strictures, however, do suggest it feels at home at or near the intersection of several known tendencies or movements: a broadly empiricist quadrant bounded by Hume, Mill, and Mach; the "radical empiricism" of William James (an explicitly metaphysical strand of American pragmatism tinctured with a dose of late 19th-century monism due to Mach, but trademarked by Russell as "neutral monism"); and finally, an assortment of Delphic pronouncements by visionary physicist John A. Wheeler. The latter, inspiration for bundling QBism into a "participatory realism" package, appears loosely related to empiricism and can be construed as the physicist's corollary to the voluntarist metaphysics of W. James. QBists recognize the centrifugal tendencies of its diverse philosophical progenitors, occasionally gesturing toward a future happy reconciliation, perhaps refuge, within the hazy metaphysical projects of William James and other pragmatists. Still, to avoid even rhetorical insinuations of solipsism, QBism no less than its empiricist and pragmatist precursors, must address a fundamental epistemological problem: How it is that, beginning from bedrock empiricism, *viz.*, "experiences ... belong(ing) to whatever particular agent ... making use of the quantum theory", can plausibly realist conclusions can be reached regarding "that objective world" out there, affirmations regarding "what the world is like, how it is put together, and what's the stuff in it". And just here we find ourselves in a frustrating void.

A recent endeavor to clarify these murky philosophical waters, an attempt somewhat uncomfortably welcomed by QBists, underscores QBism's apparent affinity to, or parallelism with, phenomenology, whether in the guise of past masters like Husserl and Merleau-Ponty, or more contemporary proponents of an "eco-phenomenology". This is not surprising as phenomenology begins with a first-person *Ansatz*, i.e., "pure experience". In this light, QBism's empiricist strictures resonate with Husserl's assessment of transcendental phenomenology as "the ultimate fulfillment ... of English empiricist philosophy". But then must QBism take the next step and embrace a corresponding transcendental phenomenological methodology, i.e., pursue an "investigat(ation of) the transcendental phenomenological 'origins' ... the origins of objectivity in transcendental subjectivity"? (Husserl, 1956, p. 382). M. Bitbol and L. de la Tremblaye (2024) propose QBism indeed should take this route, advocating adoption of "an uncompromising phenomenological attitude". Naturally, Bitbol and de la Tremblaye identify formidable hurdles along this road, also noting that QBist melodies are simultaneously sung in two disparate and disharmonious registers, in the sonorous voice of traditional physical theory, implicitly realist and explicitly within the natural attitude of working scientists, and in the dulcet tones of first-person experience wherein phenomenologists distinctly hear notes of "transcendental constitution of objectivity". Certainly, if QBism assumes an "uncompromising phenomenological attitude" it thereby abandons naturalism and quashes its realist ambitions, for any interpretational pursuit in the familiar sense is then to be understood as merely a posture within the natural attitude of the practicing physicist, signifying only "the surreptitious substitution of the mathematically substructed world of idealities for the only real world, the one that is actually given through perception, that is ever experienced and experienceable..." (Husserl, 1970, pp. 48–49). The mandate to seek the origins of objectivity in transcendental subjectivity precludes realist affirmations altogether; they must be reduced to assertions of an "as if" realism.[2] A Solomonic solution might be found. It needn't demand unconditional surrender of QBism's realist aims despite admittedly close affinities between QBism and phenomenology. For phenomenological investigation should be, and has been, directed to the realist claims of physical theories. Phenomenological analysis does not, like positivism, seek to reduce realist assertions to first person experience but rather only to reveal how realist expressions within physical theory are to be understood "transcendentally", i.e., how their meaning is constituted from the source of all meaning, personal experience.[3] As will be discussed below, this remains an important role for phenomenology within the philosophical project of illuminating QBism. Yet phenomenological inquiry need not preclude understanding QBism as an *interpretation* of quantum mechanics, one purporting to tell us what that theory does affirm about a physical world external to the experiencing agent. Accordingly, to view QBism as an interpretation of quantum mechanics, we should turn from the uncompromising phenomenological path in favor of a path of compromise, in recognition of the fact that QBism's realist ambitions, no less than those of any other physical theory or interpretation of quantum mechanics, require a philosophically critical articulation, as well as phenomenological clarification, on their own terms. In what follows, I shall attempt to do that, drawing from a figure QBists themselves regard as a *sine qua non*, Niels Bohr.

7.2 In Bohr's Footsteps?

> Without Bohr, QBism would be nothing. But QBism is not Bohr.
>
> (Fuchs, 2018, p. 1)

Interpreters of Bohr's philosophy do not find ready agreement on what precisely Bohr's philosophy is, or might be.[4] Nevertheless, Fuchs views QBism as a direct evolutionary offshoot, at least along a line of descent from Bohr through Wheeler:

> the … accumulated ingredients for what ultimately became QBism all originated in one form or another through Wheeler, and Wheeler through Bohr.
>
> (Fuchs, 2018, p. 2)

In the remaining sections I shall try to identify neglected or dismissed ingredients of Bohr's philosophy that are of possible relevance to QBism's realism without pretending to resolve internecine interpretational disputes regarding Bohr's philosophy as a whole. My focus will be to elucidate philosophical, more precisely, epistemological, resources available within Bohr's philosophical reflections, resources that I regard as arguably sufficient to fill lacunae in QBism's philosophical self-portrait and to make sense of a realism suited to both Bohr and QBism.

The most prominent of these lacunae was already mentioned, the absent but apparently epistemologically necessary argument showing how QBism transits from first-person agential experience, from "quantum measurement outcomes [that] *just are* personal experiences for the agent gambling upon them" (Fuchs, 2018, p. 1), to what is widely regarded as a principal accomplishment of natural science, i.e., producing objective statements about nature. Certainly, such objective statements need not be understood in the usual realist sense of a match between representation and states of affairs in the world, in Rortyan terms, a "mirroring" between statement and reality. But what sense of "objective" might be implied in QBism's many affirmations about "that objective world" of which the agent using quantum theory has personal experience? The question must be addressed within the broad confines of QBism's non-representationalist epistemology wherein the wave function $|\psi\rangle$ is not to be regarded as descriptive of, or representing, the physical state (in its full potentiality, prior to preparation or measurement) of a quantum system but simply as symbolic expression of an agent's degree of belief concerning future experiences.[5] Non-representationalism regarding the wave function is both a core premise and a principal attraction of QBism, integral to its resolution of the familiar paradoxes of quantum theory – wave function "collapse", non-locality, Wigner's friend, etc. (see further below). QBism holds these paradoxes arise only within a flawed epistemology of representationalism, viewing wave functions as mathematical expressions standing for microphysical states of affairs, and so mirroring (in an implied sense of isomorphism) aspects of quantum reality. On the contrary, the QBist non-representational view is radically local in the sense it is *personal*, the wave function merely is an expression for a state of belief in the agent's head concerning expectations of future experiences, given experiences of past ones.[6] Along similar lines, QBism affirms that "the general structure of quantum theory" (by which is meant the Born rule, Schrödinger equation, Kochen-Specker and other no go theorems, violation of Bell inequalities, etc.) also does not represent "the real" but nonetheless is "correlated with the real" (i.e., presumably with catalogs of protocols of measurement outcomes, as experienced and recorded by agents) as normative rules guiding users of quantum theory.

QBism's non-representational account of the wave function is one of its principal philosophical attractions, as is the fact that it is an existence proof that quantum theory, as de Finetti believed but did not demonstrate, can perfectly well abide a radically personalist conception of probability. Moreover, QBism's agent-centered first-person conception of science has both empiricist purity and illustrious precursors.[7] However, in rejecting representationalism, QBists have not secured a clear meaning for its use of the term "objective" and indeed the concept of objectivity. Admittedly, the concept is not be understood as implying the correspondence theory of truth, linking mathematical expressions and the "world as it really is". But what purchase does the term have within QBism's non-representationalism? On occasion QBism aligns its non-representationalism with ostensibly similar views of Richard Rorty, archenemy of the conception of epistemology as an inquiry into the conditions justifying knowledge claims. The worry (as will be argued below) is to secure guard rails ensuring that embrace of non-representationalism does not lead QBism over the cliff of Rortyan relativism, not merely to anti-representationalism in epistemology but to anti-epistemology. Going "full Rorty" entails that accuracy of representation is replaced with a denial of the idea that justification has anything to do with truth or objectivity, that belief is contrasted not with purported fact but only to other beliefs. In full Rortyan mode, science is refashioned as another conversation, merely "one genre of literature" (Rorty, 1980, p. xliii). But the needed guard rails can be found within Bohr's non (not anti)-representationalist conception of the wave function, based as it is on his understanding of the epistemological significance of the quantum postulate. Enmeshed as he was within decades of dialectic with Einstein, Bohr believed that physics' use of mathematical symbolism in descriptions of phenomena precludes any need for empiricist obeisance to the first-person experience of the individual observer or experimenter (Bohr, 1954, p. 68). Like today's QBists, Bohr was also a non-representationalist about the wave function; similarly, his views (more precisely, doctrines purported to be Bohr's) have been tarred and feathered with accusations of instrumentalism and anti-realism.[8] With the prominent exception of Einstein, such charges were leveled largely following Bohr's death due to the reverence in which Bohr was held by the broad community of physicists. In retrospect, these criticisms stemmed from a collective failure to come to grips with Bohr's admittedly intricate if not inchoate reasons for why the quantum postulate implies non-representationalism about the wave function. What the critics never quite grasped is that Bohr had fashioned a moderate realism within these non-representationalist confines, an epistemological achievement in itself not widely recognized. I suggest that within Bohr's philosophical writings one can find, or perhaps reconstruct, a moderately realist account of quantum theory from which QBism might draw with profit.

7.3 Bohr in Brief

> Our present QM formalism is a peculiar mixture describing in part laws of Nature and in part incomplete human information about Nature – all scrambled up together by Bohr into an omelette that nobody has seen how to unscramble. Yet we think that the unscrambling is a prerequisite for any further advance in basic physical theory ...
>
> (Jaynes, 1990, p. 387)

E.T. Jaynes and other "objective" Bayesians seek to put robust constraints on an agent's opinions prior to specific inquiry. For Jaynes, an agent's priors in using and reflecting upon quantum theory are to be constrained by a conviction that "all of Einstein's thinking – in particular the EPR argument – remains valid today, when we take into account its ontological character" (*ibid*). I suspect that for Jaynes the "ontological character" of Einstein's thinking is pithily expressed by the (in)famous EPR "Criterion of Reality"[9]; accordingly, I suggest the alleged "omelette" attributed to Bohr is not Bohr's but stems from Jaynes' robust prior, a conviction that any concept of "objectivity" is to be exposited ontologically, as in the Criterion of Reality. A closer consideration of Bohr's treatment of subjectivity and objectivity in quantum theory shows why dropping it is crucial to any mental unscrambling.

Quantum mechanics, according to Bohr, provides "an epistemological lesson with bearings on problems far beyond the domain of physical science" (Bohr, 1954, pp. 68–69). The lesson is contained in drawing out the consequences of the "quantum postulate" ("symbolized by Planck's quantum of action") that Bohr regarded as expressing the "essence" of quantum theory (Bohr, 1927, p. 53). The postulate is encapsulated in two relations in which Planck's constant appears fundamentally, linking quantities (energy-time, momentum-space) that Bohr will term "complementary",

$$E = hf \quad \text{and} \quad p = h\sigma := h = \frac{1}{\lambda}\hat{n},$$

where λ is the wavelength and $\hat{n}$ a unit vector in the direction of momentum. The relations state that a determinate energy is associated with a definite frequency, and that a given value of momentum is associated with a particular wavelength. Then it is immediately clear from the Heisenberg uncertainties that a definite value of either energy or of momentum on the left-hand side gives rise to a massive underdetermination for the value of the quantity on the other, e.g., a definite momentum is associated with a wave of infinite extent. Two thoughts already come to mind. First of all, Bohr, allegedly an anti-realist or instrumentalist about quantum theory, appears to be a realist about Planck's constant coupling both sides in these relations. Second, what bearing does Bohr's "quantum postulate" have on the EPR "Criterion of Reality", supposedly the epitome of the ontological character of Einstein's thinking? The answer is readily apparent: an ontological "picture" of the physical world in which individual entities are theoretically assigned "states" tacitly assumed to completely describe localized real states of affairs in which the entities possess definite valued, pre-existing properties (properties whose obtaining explains what a measurement outcome would yield) is no longer possible. And if this is so, what is to be done? Jettisoning the ontological picture, Bohr's answer to this question in couched in the language of measurement, observation, and interaction. It consists of giving principled reasons why it is necessary to transit from the realist notion of objectivity presumed in the rejected ontological picture (i.e., correct state assignment of pre-existing values to properties independently of measurement), to a novel notion of objectivity, or, as Bohr preferred, of "objective descriptions of experience" in the domain of atomic physics. Bohr's overriding concern is with "the problem of how objectivity may be retained during the growth of experience beyond the events of daily life" (Bohr, 1954, p. 67). His solution, in light of the non-representationalism about the quantum formalism implied by the quantum postulate, is to secure the concept of objectivity through a quasi-transcendental argument, illuminating the conditions of possibility of unambiguous communication. Descriptions of phenomena are objective because they can be, and in fact are, unambiguously communicated. Properly understood, the quantum postulate mandates a profound transformation of representationalist assumptions undergirding the presumed notion of objective description of classical physics. It demonstrates that the descriptive framework of classical physics, sketched above, is too narrow to encompass atomic phenomena; it also shows precisely why and where the representational shoe doesn't fit. Furthermore, by revealing the need for a fluid distinction between subject and object, it teaches "an epistemological lesson with bearings on problems far beyond the domain of physical science".

7.4 "Suspended in Language"

> We are suspended in language. Our task is to communicate experience and ideas to others. We must strive continuously to extend the scope of our description, but in such a way that our messages do not thereby lose their objective or unambiguous character.
>
> (N. Bohr)[10]

"Objective description", presupposes or as Bohr put it, "demands the contraposition of subject and object" (Bohr, 1960, p. 14). The "differentiation between subject and object" underlies "the general limits of man's capacity to create concepts" (Bohr, 1929, p. 96). It is an indispensable distinction, though usually only implicit, in ordinary language usage. Familiar speech acts presuppose it and are straight-forwardly understood: A subject refers, speaks "about", attributes properties to objects. The distinction has a simple grammatical subject-predicate form; properties are attributed to objects by a speaker within a common perceptual reality; canonically, the attention of interlocutors is directed to a commonly perceptible particular object situated in a commonly shared space and time. If I speak of the color of a particular birch tree, I can identify the tree by its position in relation to readily observable macroscopic states of affairs, "in the northwest corner of the yard" or "behind the garage". This paradigmatic use of everyday language in a commonly shared space and time is the basis for a representationalist (or, "pictorial" as Bohr preferred) understanding of predication or attribution, i.e., a semantics of word-world relations supposed necessary for understanding and a conception of truth of assertions as accuracy of representation.[11] Bohr stressed that the meanings of kinematical and dynamical concepts "position", "direction of motion", "cause", "effect", etc. are grounded in this ordinary usage. In any particular speech situation the meaning of these terms are always contextually bound yet they retain a universally understood significance as pertaining to objects in space and time, at rest or undergoing change in motion on account of perceptible or inferable causes; by default, such objects with their pre-existing properties are regarded as isolated or autonomous, without need to consider means or methods of observation or knowability.[12]

Classical physics simply assumes the fixed subject/object distinction of everyday language with its word-world semantics. On the other hand, the language of classical physics is replete with mathematical symbols; these can be unambiguously defined, and this allows circumvention of mention of any particular conscious subject/observer attributing properties, an implicit reference that "infiltrates"[13] everyday language usage. As Bohr also observed, the kinematical and dynamical terms of classical physics are idealizations. The idealization is already evident in the notion of "observation" taken over from everyday usage; a position can be attributed to a classical particle without suggesting interaction with a device that might measure position with respect to a given reference body. While implicitly following ordinary language usage, the concept of position is quantified through assigning an object coordinate values in a given frame. Attribution of definite position to an object no longer invokes relation to the space and time of a specific macroscopic states of affairs, but locates the object within a particular frame of spatio-temporal reference (an abstract concept), assigning it a particular *n*-tuple term in that frame.[14] The representational or referential character of ordinary language is retained except that the word-world relation has been idealized to the form of abstract term-world. Attribution of properties is idealized in two ways, through precise symbolic term-object correlation, and without need of reference to the circumstances or agency through which attributions of definite values are knowable. Employing these idealizations, classical physics retains the concept of an object as an autonomous bearer of determinative properties. In fact, the classical concept of observation implies a firm distinction between subject and object. Assignment of a physical state to an object or system is understood to provide a consistent and complete list of possessed properties;

observation or measurement simply reveals values of pre-existing properties which, if changing in time, nonetheless have definite values at any given time, without regard to measurement or observation. The classical concept of observation gives rise to a corresponding realist notion of objective description, e.g., Einstein's.[15] The default understanding in classical physics of objectivity, and of objective description, appears realist and representationalist, resting upon correct attribution of definite values to pre-existing properties. Objective description, in classical physics has, presumptively, the representational sense taken over from everyday language, of attribution of properties corresponding with an antecedently determinate reality. But objectivity in classical physics can also have another sense, often, but not necessarily, regarded as a consequence of the first. In virtue of eliminating mention to any "individual subjective judgment",[16] the expressions of classical physics, expressed by mathematical symbols well-defined in Hamiltonian or other formalisms, can be unambiguously communicated. The former "pictorial" understanding of representationalist description can be retained in circumstances in which the quantum of action can be neglected.[17] Where it cannot, the latter notion of objectivity alone remains.

7.5 Redefining "phenomenon"

> The main point ... is the distinction between the *objects* under investigation and the *measuring instruments* which serve to define, in classical terms, the conditions under which the phenomena appear.
>
> (Bohr, 1949, p. 50)

The quantum postulate has shown that the representationalist meaning of objective description customary from ordinary life and commonly assumed in classical physics fails for atomic physics; it can no longer be upheld on account of "the impossibility of distinguishing in our customary way between physical phenomena and their observation". The impossibility of strict separation between phenomena and means of observation reveals a "failure of our [customary] forms of perception", a failure connected with forced surrender of the pictorial property realist semantics adequate for descriptions of experience in everyday life and in classical physics as well as the rigid subject/object distinction this semantics assumes. But as the human capacity to form concepts relies on the differentiation between subject and object,[18] the distinction must be drawn. In effect, Bohr argued that there must be a reconciliation of two seemingly incompatible postulates, one stemming from the quantum postulate, the other from the requirement of objective description. The first states that the "wholeness" of the quantum postulate has shown an "*impossibility of any sharp separation between the behavior of atomic objects and the interaction with the measuring instruments which serve to define the conditions under which the phenomena appear*" (Bohr, 1949, pp. 39–40; original emphasis). The second postulate, apparently antagonistic, is nonetheless to require "a *fundamental distinction between the measuring apparatus and the objects under investigation*" (Bohr, 1958a, p. 3). The latter distinction is required for attribution of properties to objects and so for objective description in a sense to be further specified.

Reconciliation is attained by requiring that the term "phenomenon" be redefined in the quantum context such that description of phenomena is "exclusively to refer to the observations obtained under specified circumstances, including an account of the whole experimental arrangement" (Bohr, 1949, p. 64). Extension of classical kinematical and dynamical concepts to describe experience in quantum physics mandates, via the aforementioned "impossibility", reference to the conditions of epistemic access to the quantum world. These concepts are well-defined only in the context given by a specific experimental set up, state preparation, choice of agencies of measurement for a particular observable, etc., in virtue of which attribution of specific values of these properties to atomic objects can be ascertained. In the quantum context, kinematic and dynamical terms are but homonyms of their robust counterparts in classical physics that, even if idealized, convey "classical pictures" of property attribution to objects considered as independent of any epistemic access. But in quantum mechanics, in view of the revised notion of "phenomenon", these classical terms acquire a restricted, highly contextual, meaning defined within specific experimental conditions suited for outputting a definite value the corresponding classical kinematical or dynamical property. Complementarity insists that a classical kinematical or dynamical concept be defined in the context of a distinct. and mutually exclusive, type of experimental set-up suitable for measuring or obtaining information about this or that kinematical or dynamical property. This is precisely the basis of Bohr's (1935) response to Einstein. Seemingly predicated as particular values of properties of microphysical systems revealed by measurement, the classical concepts have a completely restricted meaning; the terms expressing them are definable only in contexts in which the phenomenon described includes the entire experimental set up. Classical concepts thus lose any visualizable or picturable character familiar from property attributions in classical physics. But they are integral to the description of atomic phenomena, of "what has been learned", a core component of objective description, without reference to any individual observer, and as such, it can be communicated unambiguously.[19] Unambiguous communication is essential; expressed in Wheelerian terms, "no elementary quantum phenomenon is a phenomenon until it is a registered ('observed,' indelibly recorded) phenomenon".[20]

7.6 Realism, without Representation

> There is no quantum world. There is only an abstract quantum physical description. It is wrong to think that the task of physics is to find out how nature is. Physics concerns what we can say about nature.[21]

Whether or not Bohr actually made these statements reported by Petersen, it follows from Bohr's revision of the term "phenomenon" that wave functions only indirectly represent quantum systems; the very idea of a quantum state, supposed representational target of the wave function, has been replaced by a notion that has necessary reference "to the observations obtained under specified circumstances, including an account of the whole experimental arrangement". As consequence, property attribution to phenomena in the quantum domain is completely different than in descriptions of objects of classical mechanics; attributions of particular properties, e.g., q or p, to quantum objects, must be understood contextually. Bohr's use of indirect representation can be illustrated by considering the problem of quantum measurement in von Neumann's account. The representationalist presupposition of von Neumann, that the measurement state is the tensor product superposition $|S_i\rangle \otimes |M_i\rangle \otimes |O_i\rangle$ is denied.[22] Bohr of course holds that objective description, by definition, omits reference to any individual observer S; at most, implicit reference to an experimenter or observer occurs only in the particular choice of observable to measure and to the corresponding experimental set up in which measurement apparatus M is a suitable component. But in general, superposition expressions are to be understood as indifferently symbolizing many completely different, possibly mutually exclusive, phenomena, of each of which an unambiguous account must include a description of all relevant features of the experimental arrangement. By itself, the superposition is merely part of "a mathematical formalism ... adapted to the measuring results obtainable in the new fields of observation" (Bohr, 1946, p. 125). It is an expression without empirical meaning; it is mere abbreviation or "expectation catalogue" (Schrödinger, 1935, p. 157). With component indices running from 1 to n, it abstractly symbolizes all possible experimental set ups for the observable physical quantities, each with an associated as yet unknown outcome.[23] It is an objective expression in only the sense that it is unambiguously definable and so communicable within the linear algebra of Hilbert space. It is not the objective description of any "phenomenon" in Bohr's sense which pertains exclusively to a definite result within a specific measurement set up.

A prerequisite for any realist theory of the microphysical world must allow that observation and measurement indeed reveals information about the quantal aspects of the atomic objects of interest. Bohr of course grants that quantum theory does produce such information but insists that it is produced only indirectly.[24] Attribution of a given value of a classical kinematic or dynamical quantity to target system of interest involves "an essential ambiguity" since such attribution necessarily refers to the experimental conditions by which that information is obtained.[25] Property attribution can be made only after the microphysical object system has "made contact with physical reality", i.e., the laboratory set up and apparatus.[26] The value of the property attributed to a quantum object as result of a measurement cannot be regarded as an intrinsic property of the object itself but only of a behavior manifested observationally when the object is placed in a given experimental situation and a result recorded. As such, the concept pertains to the entire phenomenon, of which the object is but a non-separable component: The concept "indirectly represents" the object. The term "indirect representation" is perhaps not a happy one, but its intent is to flag that a moderate realism can be retained when jettisoning a metaphysical notion of *the real* as comprised of objects "out there" possessing at all times definite values of properties, values that are exactly revealed by measurement or observation. "Objectivity", on the metaphysical view of realism, is getting the representation of such objects correct, without stain of human subjectivity. Objects indirectly represented are always objects situated within conditions of epistemic accessibility consistent with the quantum postulate. Indirect representation is non-representational in that the quantum postulate denies that in the domain of atomic physics the metaphysically realist conception of object is, or can be, the target of physical description, i.e., can be an object of which the quantum formalism provides an exactly representative expression.

7.7 "Suspended in Language" Reprise

It seems almost paradoxical that for Bohr ordinary language usage lies at both the beginning and at the end of the sea-change from a representationalist concept of objectivity to a non-representationalist one. At the beginning since, "suspended" as we are in language, ordinary language is as a matter of fact understood to be an often necessary vehicle for unambiguous communication of information describing our experiences. It is as well the *fons et origo* of meaning for kinematical and dynamical concepts, meanings idealized and made more precise in classical physics. At the end, we are still "suspended" though now our language is supplemented with classical terms, the kinematical and dynamical terms bearing the restricted meaning they have acquired in the light of the revised notion of "phenomenon". This language is not only sufficient but also necessary for objective description of phenomena in the domain of atomic physics. If we reconsider Bohr's insistence that "*however far the phenomena transcend the scope of classical physical explanation, the account of all evidence must be expressed in classical terms*" (Bohr, 1949, p. 39), we may recall that Bohr deemed the epistemological problem posed by the quantum postulate as that of addressing "how objectivity may be retained during the growth of experience beyond the events of daily life". Bohr's philosophical reflections hazily trace a step-wise progression of how the growth in human experience, ranging from everyday to classical and then atomic physics, poses continual challenges for what can be considered the objective description of experience. In everyday experience, paradigmatically perceptual experience, objective description resides in presumed correct representations of objects, although in ordinary usage there is always implicit reference to the individual subjective judgment of the speaker or describer. In classical physics, the representationalist "correct description" presuppositions of ordinary language are inherited but improved upon since descriptions now employ unambiguously communicable mathematical symbols permitting elimination of subjective reference to any individual observer or perceiver. Finally objective description of experience in the domain of quantum phenomena drops altogether any sense of representational correctness with its fixed subject/object divide in favor of a non-representationalist sense compatible with the quantum postulate. The macroscopic character of experimental set ups as well as the recorded results of observation and measurement are describable and communicable within ordinary language as supplemented with classical concepts of "space", "energy", "momentum" defined within specific concrete contexts of attribution. The classical kinematical and dynamical concepts, in contexts of specific experimental set ups, are attributed to quantum systems with a meaning restricted to recorded results of interaction of these systems with the agencies of measurement. In the absence of representationalism, objective description of experience rests upon, and is secured by, a quasi-transcendental requirement of the conditions of possibility of "unambiguous communication".[27] These conditions Bohr located in the descriptive resources of language. What can be communicated unambiguously about quantum phenomena must be communicated in ordinary language, supplemented as needed with the idealized concepts of classical physics. The universality of ordinary language and the pre-existence of classical kinematical and dynamical concepts, now with their restricted meanings, are conditions of possibility of unambiguous communication about quantum phenomena.[28]

7.8 Obstacles or Problems for QBism?

Both explicitly and implicitly, QBists regard important components of their conception as posing insurmountable objections to any assimilation to Bohr's views. I suggest when Bohr's position is understood as outlined above, these objections can be countered, either by allowing Bohr to help himself to the subjective notion of probability, or by clarifying misconceptions about Bohr.

We agree completely that "QBism is not Bohr". On the other hand, we have argued that Bohr's philosophical reflections on quantum theory contain an account of objectivity that QBism requires in any case, and might usefully employ to connect its realist aspirations with its insistence on the primacy of personal experience. With the above presentation of Bohr's views in mind, the major QBist objections to Bohr can be answered. The principal objection appears to be an insistence that the "pure experience" of the user or agent of quantum theory connects somehow to "the real" and that Bohr has obliterated this connection by employing a concept of "detached observer". Paraphrasing William James, Fuchs has written,

> Importantly James saw his notion of "pure experience" as the ultimate building block of reality – i.e., like Wheeler of Bohr's "phenomenon".
>
> (Fuchs, 2017, p. 127)

and

> The greatest and most difficult task will be ... to return to what Bohr tried to jump prematurely with his "classically described agencies of observation" and to *disambiguate* the "pure experience" notion from the agents gambling upon them.
>
> (Fuchs, 2018, p. 18)

James' "pure experience" is admittedly resistant to philosophical analysis (Wild, 1969) but parallels can be found in Husserl's theory of intentionality, where the *reele* contents of an intentional act (as opposed to the act's *ideal* contents, which are sharable) are exclusively *individual* moments, non-repeatable and non-sharable (Husserl, 1982, §§ 76–77). Indeed, if the above presentation of Bohr is anywhere near correct, the philosophical project can be encapsulated precisely in Bohr's attempt to show how the growth of human experience into atomic physics, requires transformation of the concept of objective description once quantum postulate forces surrender of a representationalist sense. As we have seen, Bohr found unambiguous communication to be the touchstone of the new concept of objectivity. In fact, if we have understood Wheeler correctly, the mantra (which Wheeler took to paraphrase Bohr), "no elementary quantum phenomenon is a phenomenon until it is a registered ('observed,' indelibly recorded) phenomenon" paraphrases the same thought. What is not unambiguously communicable is not, in science, to be considered objective.

A closely related QBist objection to Bohr stems from Bohr's use of the term "detached observer", and it can be answered in a similar fashion. Pauli (1954) had used the term in print in a different sense at roughly the same time. He objected that Bohr's term is similar in meaning to Einstein's "ideal of the detached observer", allowing unrestricted application of concepts of space, time, cause and effect, in describing "real states of affairs", individual localizable physical objects with their inherent pre-existing properties "out there" independent of human thought and action. Pauli argued, of course, this could not be the case in quantum theory. But Bohr's "detached observer" is not Einstein's. Complementarity is deemed a rational generalization of the unrestricted application of concepts in descriptions of phenomena by the detached observer. The representationalist sense of objectivity lying behind Einstein's "ideal of the detached observer" might have been rejected for classical physics on purely philosophical grounds. But after the quantum postulate it is simply no longer in play. Bohr essentially stated this to Pauli in a letter of March, 1955:

> ... we have in quantum physics attained the same goal [i.e., Einstein's] by recognizing that we are always speaking of well-defined observations obtained under specified experimental conditions. These conditions can be communicated to everyone who also can convince himself of the factual character of the observations ...[29]

The sole meaningful role allotted to Bohr's "detached observer" is to fashion objective descriptions of experimental results that are unambiguously communicated to similarly situated physicists possessing relevant background and experience. Arguably, this is a role sorely missing an actor in QBism.

Finally, Bohr did not hold, nor as far as we know, was he even aware of de Finettian personal probability. But for de Finetti personalist probability was broadly conceived as one's degree of expectation that some particular event will happen, a subjective conception operationally quantifiable, hence communicable, in one's behavior, real or potential, with respect to that event. This is de Finetti's framework for all inductive reasoning, clothed in psychological richness, competing inclinations, and cognitive abilities to weigh factors of considerably different characters. The arid QBist conception of an "agent", on the other hand, is taken from probability theory as a primitive and irreducible notion. A QBist agent is an abstract function updating expectations when given specific data (Fuchs and Schack, 2014, p. 2). With de Finetti's richer notion of "agent", is there any conclusive reason why Bohr could not have adopted subjective probability?

QBists allow that a wave function is a symbolic expression going proxy for a conscious experience of an agent. This is all that Bohr requires.

Symbolic expressions admit of unambiguous definition and by this fact need make "no explicit reference to any individual observer". Probabilities refer to communicable information gained in sense experience from particular macroscopic devices suited to prepare and examine microphysical systems. These are probabilities of specified outcomes of measuring devices under specified conditions. Bohr's own conception of probability is empirical and statistical, based on relative distribution of distinct measurement outcomes on repetition under the same experimental set-up.[30] Moreover, as with QBists, conditional probability is fundamental for Bohr since Born rule probabilities always refer to outcomes obtained for experimental set ups that are appropriate apparatus for measurement of the corresponding observable. Complementarity always lies in the background of the use of conditional probabilities for specific measurement outcomes, for it proscribes what kind of measurements cannot be made, given others are.[31]

Mermin states that the radically personalist conception of probability of QBism is a fundamental difference with Bohr, and more generally, from "orthodox ('Copenhagen') thinking about quantum foundations". The orthodoxy overlooks the QBist postulate that "*individual* personal experience is at the foundation of the story *each* of us tells about the world", and in fact seeks to replace "*private personal* experience" by "impersonal features of a common 'classical' external world" (Mermin, 2019, p. 4). In QBism on the other hand, "the role of measurement is played for [a particular agent] by *any action whatever* that [the agent] take(s) on the world, whether or not with the help of a 'measurement apparatus'". Mermin indeed allows that "the outcome of an action, being an experience, is subjective" (Mermin, 2019, p. 7). And Mermin recognizes the need of supplementing QBism's first-person perspective with an appeal to linguistic and other communicative means in sketching how measurement outcomes, understood as the personal response of an agent ("user", Mermin), can be considered intersubjective. Language is the ladder for climbing out of the prison of solipsism; once the outcome of an action can be represented in language, it can be communicated to others:

> for a QBist, language – ordinary or technical – is even more essential than it was for Bohr ... the only way one can try to represent to others one's private personal experiences.
>
> (Mermin, 2019, p. 13)

Yet Mermin has a vastly ecumenical view of what language can be, for the term embraces not only spoken and written language but "gestures, touches, hugs, and so on". In fact, "language" is defined functionally: It is "any way you have to induce experiences in me, through which you can try to give me some sense of the context of experiences of your own" (Mermin, 2019, p. 5). Here is the problem with which we began this essay, the huge fissure between QBism's insistence on the primacy of first-person experience and its purport to make objective claims about the external world. It is difficult to see how Mermin's loose appeal to "language" as a means of gaining intersubjectivity can more than gesture at the problem. In point of fact, the gesture might even appear superfluous in the face of personalist affirmations as "What is real for an agent rests entirely on what that agent experiences;" (Fuchs, Mermin, and Schack, 2014, p. 750) "*QBism don't do third person*;" (Fuchs, 2018, p. 22) "reality is *more* than any third-person perspective can capture"; and "I *do* think we have a kind of direct evidence of 'the real'. It is in the very notion of experience itself" (Fuchs, 2017, pp. 113 and 118). These statements tie "reality" (however construed) to personal experience, without so much as a suggestion of other agents or the need of any medium of communication. Besides opening the door to spurious charges of solipsism, they leave QBism no apparent way to fashion that merger of opinion at the core of any Bayesian account of objectivity. From the third-person standpoint of scientific statement, an epistemological conundrum ensues if both cognizing subject and object of cognition have been subsumed under the broad tent of agential personal experience, even as that experience might be supplemented with instrumental protheses.[32] While denying, with Bohr, that the wave function represents an element reality external to and independently of the agent (Fuchs and Schack, 2015, p. 2) we are in the dark regarding how the personal experience of the agent when assigning a state to a quantum system results in communication of information about that quantum system to a wider world. QBism's non-representationalism has left an epistemological void.

7.9 Conclusion: Realism Enough?

> [T]he proper meaning of the words "physical reality" must refer to the content of an objective account of physical experience.
>
> (N. Bohr)[33]

The quantum formalism informs us about the quantum world but not directly. Laboratory practice is where "physical reality" is to be found, in the carefully controlled contact of microphysical entities and processes with macroscopic apparatus that provide the necessary context for definition of the physical quantities attributable to microentities. Bohr effectively construes objectivity not as accuracy of representation but via intersubjectivity, as objective description of phenomena unambiguously communicated to others. In this regard, Bohr adopts a notion of objectivity quite similar to positions Rorty termed "epistemological behaviorism", identifying it with Dewey, the late Wittgenstein, W. Sellars and Quine.

> To be a behaviorist in epistemology is ... to look at the normal scientific discourse bifocally, both as patterns adapted for various historical reasons and in the achievement of objective truth, where "objective truth" is no more and no less than the best idea we currently have about how to explain what is going on.
>
> (Rorty, 1979, p. 385)

Rorty's construal of "epistemological behaviorism" is deemed a public antidote to the traditional practice-transcending private quest for legitimation of knowledge statements characteristic of modern epistemology since Descartes. But perhaps only post-moderns have elected to follow Rorty's rejection of the epistemological project *in toto*, by agreeing that "justification is not a matter of a special relation between ideas (or words) and objects, but of conversation, of social practice" (Rorty, 1979, p. 170). What is common to the "epistemological behaviorists", as well as to Bohr, is a renunciation of the representational presuppositions of traditional epistemology, rooted in a presupposition that a rigid non-contextual line can and must be drawn between what is subjective and what is objective, between what is given and what is added by mind, what is mere appearance and what is real; in short, all reject representationalist accounts of objectivity. Dewey, Sellars, and Wittgenstein all sought to fashion a "antifoundational" alternative to the modern "quest for certainty" enshrined in representationalist criteria of objectivity, truth, and rationality, one insisting that knowledge of nature is of a mind-independent reality, expressed in terms that directly refer and in statements true, or approximately so, because they accurately represent that reality. Bohr had no similar antifoundational agenda; his rejection of representationalism stems entirely from the implications of the quantum postulate for the subject-object relation concerning the objective description of quantum phenomena. Many interpreters of Bohr regard his non-representationalism to imply an anti-realist or instrumentalist stance toward quantum theory. This is not correct. Non-representationalism, occasionally termed "renunciation of visualization", is a necessary epistemological step when fashioning objective description of the physical world beyond the objectifying methods employed in classical physics. Such knowledge as we have of the microphysical world is empirically based and fallible, it can only be projected from the interpretation of our experiences and is therefore indeterminate, but it is nonetheless knowledge of the real.

Acknowledgments

Many thanks to the participants, both live and virtual, at Linköping in June, 2022. I am especially grateful for discussions of Bohr with Zach Hall at Stanford.

Notes

1 Throughout this chapter, the term "QBism" refers to views held by "QBists". But as the term "Copenhagen Interpretation" does not designate a unique doctrine, neither does "QBism". One might even question whether QBism purports to be an *interpretation* of quantum theory, after all, Mermin (2019, pp. 1–2) describes it as a "new view of physics" and "a view of science … anathema to most physicists", while Fuchs and Schack (2014, p. 1) speak of the "QBist conception of quantum theory". However, other QBist texts, e.g., Fuchs and Schack (2013, p. 1693), Fuchs, Mermin, and Schack (2014, p. 749), Fuchs and Schack (2015, p. 2) refer to QBism as an "interpretation of quantum mechanics" or "interpretation of quantum theory".

2 Remark of Michel Bitbol, Linköping, June 8, 2022.

3 See, e.g., H. Weyl's phenomenological clarification of Einstein's gravitational theory through the lens of a "purely infinitesimal geometry" (Ryckman, 2005) or F. London and E. Bauer's phenomenological treatment of the quantum measurement problem (see French, 2023).

4 For surveys of the wide spectrum of views about Bohr's philosophy, see Holton (1988), Kaiser (1994), Folse (1995), and Faye and Folse (2017).

5 Regarding a symbolic expression as going proxy for a conscious experience of an agent is, unwittingly, a Bohrian move; see further below.

6 QBism, following de Finetti, holds that the primitive notion of probability is both subjective and conditional; see Fuchs and Schack (2013) and further below.

7 Mermin (2019) advances CBism, extending the non-representationalist understanding to statements of classical physics.

8 The descriptivism of Kirchhoff and, more explicitly, Duhem's "symbolic" conception of the formalism of a physical theory are significant non-representationalist precursors of Bohr (see Ryckman 2017, pp. 288–289). But these philosophical views were chosen in opposition to a metaphysics of realism. Non-representationalism about the quantum formalism, for Bohr, is not an option: The quantum postulate *mandates* non-representationalism.

9 Einstein, Podolsky, and Rosen (1935, p. 777): "*If, without in any way disturbing a system, we can predict with certainty (i.e., with probability equal to unity) the value of a physical quantity, then there exists an element of reality corresponding to this physical quantity*". Original italics.

10 As reported by Petersen (1963, p. 10).

11 A semantics of "the cat is on the mat" variety, in which terms pick out and are associated with readily perceptible objects and relations, the ontology of what philosophers call "middle-sized dry goods". For an extensive discussion, see Wilson (2006) and (2017).

12 Bohr (1958b, p. 59): "The description of ordinary experience presupposes the unrestricted divisibility of the course of the phenomena in space and time and the linking of all steps in an unbroken chain in terms of cause and effect. Ultimately, this viewpoint rests on the fineness of our senses, which for perception demands an interaction with the objects under investigation so small that

in ordinary circumstances it is without appreciable influence on the course of events. In the edifice of classical physics, this situation finds its idealized expression in the assumption that the interaction between the object and the tools of observation can be neglected, or, at any rate, compensated for".

13 Bohr's term (1954, p. 68).

14 Such descriptions are also suited to the demands of relativity since they need make "no explicit reference to any individual observer" (Bohr, 1958a, p. 3).

15 Einstein (1953, p. 6): "There is such a thing as the 'real' state of a physical system existing independently of any measurement or observation that in principle can be described by the means of expression in physics".

16 Bohr (1960, p. 10): "... our task must be to account for [human] experience in a manner independent of individual subjective judgment and therefore objective in the sense that it can be unambiguously communicated in the common human language".

17 Bohr (1958a, p. 2): "The pictorial description of classical physics represents an idealization valid only for phenomena in the analysis of which all actions involved are sufficiently large to permit the neglect of the quantum".

18 Bohr (1929, p. 96): "... a close connection exists between the failure of our forms of perception, which is founded on the impossibility of a strict separation of phenomena and means of observation, and the general limits of man's capacity to create concepts which have their roots in our differentiation between subject and object".

19 Bohr (1958a, p. 3): "The description of atomic phenomena has ... a perfectly objective character, in the sense that no explicit reference is made to any individual observer and that therefore ... no ambiguity is involved in the communication of information".

20 Wheeler, as quoted by A. Ananthaswamy, https://blogs.scientificamerican.com/observations/what-does-quantum-theory-actually-tell-us-about-reality/

21 Bohr, as quoted in Petersen (1963, p. 12).

22 The tensor product of state vectors for *subject S*, *measuring apparatus M*, and *object system O*.

23 Note that this analysis of superposed states provides Bohr's answer to the "Wigner's friend" paradox; see the discussion of quantum measurement below.

24 Bohr (1958a, p. 4): "... the quantal features of the phenomenon are revealed in the information about the atomic objects derived from the observations".

25 Bohr (1949, p. 40): "... an essential ambiguity is involved in ascribing conventional physical attributes to atomic objects, as is at once evident ... where we have to do with contrasting pictures, each referring to an essential aspect of empirical evidence".

26 I am indebted here to Dieks (2017).

27 Bohr (1954, p. 67): "Every scientist ... is constantly confronted with the problem of objective description of experience, by which we mean unambiguous communication. Our basic tool is, of course, plain language which serves the needs of practical life and social intercourse".

28 Commenting on Rutherford's "whole attitude", that "the aim of experimentation is to put questions to nature", Bohr (1958b, p. 59) noted that if "the inquiry may augment common knowledge" it is "an obvious demand that the recording of observations as well as the construction and handling of the apparatus, necessary for the definition of the experimental conditions, be described in plain language", a demand "amply satisfied" in investigation of quantum phenomena.

29 Bohr to Pauli, 2 March 1955, in Bohr (1999, p. 568).

30 Bohr (1958a, pp. 3–4): The laws of quantum mechanics are "of an essentially statistical type" since "(t)he very fact that repetition of the same experiment [including in description all relevant features of the experimental arrangement] in general yields different recordings pertaining to the object, immediately implies that that a comprehensive account of experience in this field must be expressed by statistical laws".

31 Osnaghi (2017, pp. 159–160): "… only in *some* experimental contexts, namely, the contexts in which a certain property is measured, can the question of whether a given quantum system has that property meaningfully be asked, then, *provided that the simultaneous occurrence of some experimental contexts is precluded*, it becomes possible to devise models which can account for all the statistical correlations that are actually recordable".

32 Pineaar (2021) calls this the "prolongation thesis".

33 Unpublished "Post Scriptum", dated 13 August 1957; Niels Bohr Archive: Bohr Manuscripts, reel 22; copies in AHQP-AIP. Cited in Kaiser (1994, p. 136, note 39).

References

Bitbol, M. & de la Tremblaye, L. (2024). QBism: An Eco-Phenomenology of Quantum Physics. This volume.

Bohr, N. (1927). The Quantum Postulate and the Recent Development of Quantum Theory. [Como Lecture] As reprinted in *The Philosophical Writings of Niels Bohr*, volume I *Atomic Theory and the Description of Nature*. Woodbridge, CT: Ox Bow Press, 1987 Reprint edition, pp. 52–91.

Bohr, N. (1929). The Quantum of Action and the Description of Nature. As reprinted in *The Philosophical Writings of Niels Bohr*, volume I *Atomic Theory and the Description of Nature*. Woodbridge, CT: Ox Bow Press, 1987 Reprint edition, pp. 92–101.

Bohr, N. (1935). Can Quantum-Mechanical Description of Physical Reality Be Considered Complete? As reprinted in J. Faye & H. Folse (Eds.), *The Philosophical Writings of Niels Bohr*, volume IV *Causality and Complementarity*. Woodbridge, CT: Ox Bow Press, 1998, pp. 73–82.

Bohr, N. (1946). On the Problem of Measurement in Atomic Physics. As reprinted in J. Faye & H. Folse (Eds.), *The Philosophical Writings of Niels Bohr*, volume IV *Causality and Complementarity*. Woodbridge, CT: Ox Bow Press, 1998, pp. 122–125.

Bohr, N. (1949). Discussion with Einstein on Epistemological Problems in Atomic Physics. As Reprinted in *The Philosophical Writings of Niels Bohr*, volume II *Essays 1933–1957 on Atomic Physics and Human Knowledge*. Woodbridge, CT: Ox Bow Press, 1987 Reprint edition, pp. 32–66.

Bohr, N. (1954). The Unity of Knowledge. As reprinted in *The Philosophical Writings of Niels Bohr*, volume II *Essays 1933–1957 on Atomic Physics and Human Knowledge*. Woodbridge, CT: Ox Bow Press, 1987 Reprint edition, pp. 67–82.

Bohr, N. (1958a). Quantum Physics and Philosophy: Causality and Complementarity. As reprinted in *The Philosophical Writings of Niels Bohr*, volume III *Essays 1958–1962 on Atomic Physics and Human Knowledge*. Woodbridge, CT: Ox Bow Press, 1987 Reprint edition, pp. 1–8.

Bohr, N. (1958b). The Rutherford Memorial Lecture. As reprinted in *The Philosophical Writings of Niels Bohr*, volume III *Essays 1958–1962 on Atomic Physics and Human Knowledge*. Woodbridge, CT: Ox Bow Press, 1987 Reprint edition, pp. 30–73.

Bohr, N. (1960). The Unity of Human Knowledge. As reprinted in *The Philosophical Writings of Niels Bohr*, volume III *Essays 1958–1962 on Atomic Physics and Human Knowledge*. Woodbridge, CT: Ox Bow Press, 1987 Reprint edition, pp. 8–16.

Bohr, N. (1999). *Niels Bohr, Collected Works*, volume 10. Part V, Selected Correspondence. Amsterdam: Elsevier.

Brown, H. R. (2019). The Reality of the Wavefunction: Old Arguments and New. In A. Cordero (Ed.), *Philosophers Look at Quantum Mechanics*. Cham: Springer, pp. 63–86.

Dieks, D. (2017). Niels Bohr and the Formalism of Quantum Mechanics. In J. Faye & H. Folse (Eds.), *Niels Bohr and the Philosophy of Physics: Twenty-First Century Perspectives*. London, UK: Bloomsbury Academic, pp. 303–33.

Earman, J. (2019). Quantum Bayesianism Assessed. *The Monist* 102 (4), 403–423.

Einstein, A. (1953). Einleitende Bemerkungen über Grundbegriffe. In *Louis de Broglie: Physicien et Penseur, collection dirigée par André George*. Paris: Éditions Albin Michel, pp. 4–15.

Einstein, A., Podolsky, B. & Rosen, N. (1935). Can Quantum-Mechanical Description of Physical Reality Be Considered Complete? *Physical Review* 47, 777–780.

Faye, J. & Folse, H. (2017). (Eds.), *Niels Bohr and the Philosophy of Physics: Twenty-First-Century Perspectives*. London, UK: Bloomsbury Academic.

Folse, H. (1995). Essay Review: Niels Bohr and the Construction of a New Philosophy. *Studies in the History and Philosophy of Modern Physics* 26(1), 107–116.

French, S. (2023). *Cutting the Chain of Correlations: Reviving a Phenomenological Approach to Quantum Mechanics*. New York: Oxford University Press.

Fuchs, C. A. & Schack, R. (2013). Quantum-Bayesian Coherence. *Reviews of Modern Physics* 85, 1693–1715

Fuchs, C. A. & Schack, R. (2015). QBism and the Greeks. *Physica Scripta* 90, 015104, 1–6.

Fuchs, C. A. (2017). On Participatory Realism. In I. Durham and D. Rickles (Eds.), *Information and Interaction: Eddington, Wheeler, and the Limits of Knowledge* (pp. 113–134). Cham: Springer.

Fuchs, C. A. (2018). *Notwithstanding Bohr, the Reasons for QBism*. arXiv:1705.03483v2 [quant-ph] 11 Nov 2018.

Fuchs, C. A. & Schack, R. (2014). *Quantum Measurement and the Paulian Idea*. arXiv:1412.4209v2 [quant-ph] 16 Dec. 2014.

Fuchs, C. A., Merman, N. D. & Schack, R. (2014). An Introduction to QBism with an Application to the Locality of Quantum Mechanics. *American Journal of Physics* 82(8), 749–754.

Holton, G. (1988). The Roots of Complementarity. In *Thematic Origins of Scientific Thought*. Revised Edition. Cambridge, MA: Harvard University Press, pp. 99–145.

Husserl, E. (1956). Zur Auseinandersetzung meiner transzendentalen Phänomnologie mit Kants Transzendentalphilosophie, ca. 1908, one of many *Beilagen* first published in *Erste Philosophie, Erster Teil.* R. Boehm (Hrsg.). The Hague: Nijhoff, p. 382.

Husserl, E. (1970). *The Crisis of the European Sciences and Transcendental Phenomenology.* Evanston, IL: Northwestern University Press. Translation by D. Carr of original German ed., first published in 1954.

Husserl, E. (1982). *Ideas Pertaining to a Pure Phenomenology and to a Phenomenological Philosophy. First Book: General Introduction to a Pure Phenomenology.* Translated by F. Kersten from original German 1913 edition. The Hague: Martinus Nijhoff.

Jaynes, E. T. (1990). Probability in Quantum Theory. In W. Zurek (Ed.), *Complexity, Entropy and the Physics of Information.* Boulder, CO: Westview Press, pp. 381–402.

Kaiser, D. (1994). Bringing the Human Actors Back on Stage: the Personal Context of the Einstein-Bohr Debate. *British Journal for the History of Science* 27, 129–152.

Mermin, N. D. (2019). Making Better Sense of Quantum Mechanics. *Reports on Progress in Physics* 82, 1–16.

Osnaghi, S. (2017). Complementarity as a Route to Inferentialism. In J. Faye and H. Folse (Eds.), *Niels Bohr and the Philosophy of Physics: Twenty-First-Century Perspectives.* London, UK: Bloomsbury Academic, pp. 155–178.

Pauli, W. Jr. (1954). Probability and Physics. *Dialectica* 8, 112–124.

Petersen, A. (1963). The Philosophy of Niels Bohr. *Bulletin of the Atomic Scientists* (September), pp. 8–14.

Pineaar, J. (2021). Extending the Agent in QBism. *Foundations of Physics* 50, 1894–1920.

Rorty, R. (1979). *Philosophy and the Mirror of Nature.* Princeton, NJ: Princeton University Press.

Rorty, R. (1980). *Consequences of Pragmatism.* Minneapolis, MN: University of Minnesota Press.

Ryckman, T. A. (2005). *The Reign of Relativity: Philosophy in Physics 1915–1925.* New York, NY: Oxford University Press. (Oxford Studies in the Philosophy of Science.)

Ryckman, T. A. (2017). *Einstein.* New York, NY and London, UK: Routledge. (Routledge Philosophers.)

Schrödinger, E. (1935). The Present Situation in Quantum Mechanics. As translated by J.D. Trimmer in J. Wheeler & W. Zurek (Eds.), *Quantum Theory and Measurement.* Princeton, NJ: Princeton University Press, 1983, pp. 152–67

Wild, J. (1969). *The Radical Empiricism of William James.* Garden City, NJ: Doubleday.

Wilson, M. (2006). *Wandering Significance. An Essay on Conceptual Behavior.* New York, NY: Oxford University Press.

Wilson, M. (2017). *Physics Avoidance. Essays in Conceptual Strategy.* New York, NY: Oxford University Press.

8 QBism

An Eco-Phenomenology of Quantum Physics[1]

Michel Bitbol and Laura de la Tremblaye

8.1 Introduction

Since its inception, QBism has been torn apart between two seemingly conflicting ontological inclinations.

QBism has inherited from Asher Peres a strong instrumentalist commitment (Fuchs & Peres 2000). But QBism is simultaneously developing an original realist research program (Fuchs 2017).

QBism holds a conception of quantum theoretical statements that embraces "two levels of personalism" (Fuchs & Stacey 2020), about probabilities and about possible experimental outcomes, thereby ascribing them a decidedly first-personal status. But QBism is also looking for aspects of quantum theoretical formalism that would "take advantage from a third-person perspective" (Fuchs 2019). It is striving to find an objective rationale for the most efficient subjective betting strategies, and especially to extract an objective component from the procedures used for combining subjective probabilities.

Even more strikingly, it is essential to the QBist dissolution of celebrated quantum enigmas, such as the Wigner friend "paradox", that measurement outcomes be equated with *lived experiences,* rather than with properties of macroscopic experimental devices (Fuchs et al. 2014). So much so that, in this context at least, discourse about the entities of the world (including laboratory furniture) tends to be set aside in favor of an insistent reference to their manifestation in (and *qua*) the personal experience of scientists. Yet, QBism defines the phenomenon as something new that *really happens in the world* (Fuchs 2002) when two of its material fragments meet: the "physical systems" belonging to the external world on the one hand, and the body of the agent (together with its experimental instruments construed as prostheses) on the other hand.

The latter reified conception of the phenomenon and its production prompted a cogent criticism from Hervé Zwirn (2021). If experimental outcomes only occur as (personal) experiences, he writes, one cannot

DOI: 10.4324/9781003259008-10

consider that something new happens *in the external world* as a result of experiments.

A proper reply to Zwirn's criticism is certainly not beyond the reach of QBism. It just requires two moves.

The first move consists in carefully restating the nature of the "something new" produced by measurements. This restatement can indeed be found in some QBist writings: "A measurement does not, as the term unfortunately suggests, reveal a pre-existing state of affairs. It is an action on the world by an agent that results in the creation of an outcome—a new experience for that agent" (Fuchs et al. 2014). According to QBism, what is newly produced by a measurement, what happens through it, is nothing else than an agent's *experience*.

The second indispensable move is that it is only if one suspends the traditional dualist distinction between experience and world, and/or if one tends to construe experience as the basic stuff of the world (Section 8.3), that a new experience is *ipso facto* a new event of the world.

But this is not yet enough to dispel Zwirn's qualms. For, although the latter non-dualist ontology is by no means unfamiliar to QBists, it is far from being consistently adopted in the bulk of their literature. In the same paper where an outcome is equated with an experience (and especially in the quoted sentence), the QBist account of a measurement process persistently relies on a dual (and object-like) description of its protagonists: an agent with body and prostheses on the one side, and the acted-upon world on the other side. Here, the usual description of a world made of a multiplicity of "physical systems" external to the agent and to one another, is maintained throughout: "Acting as an agent, Alice can use the formalism of quantum mechanics to model any physical system external to herself. QBism directs her to treat all such external systems on the same footing, whether they be atoms, enormous molecules, macroscopic crystals, beam splitters, Stern-Gerlach magnets, or even agents other than Alice" (Fuchs et al. 2014).

This language raises a serious problem of conceptual consistency. It features once again what Ryckman (2023) rightly calls "the huge fissure between QBism's insistence on the primacy of first-person experience and its purport to make objective claims about the external world". If QBists acknowledge that "What is real for an agent, rests entirely on what that agent experiences" (Fuchs et al. 2014), they should not posit *a priori* the real-world-out-there as a pre-existent being whose interaction with the agent gives rise to an experienced outcome. The so-called "external" world made of various "physical systems" should be seen as a problem, not as a given; its concept should be elaborated out of a careful analysis of experience, rather than taken for granted before any such analysis.

Satisfying the latter requirement is one of the major tasks of phenomenology. Starting all over again, from a reflection about lived experience, in order to clarify the meaning we ascribe to the entities of our "natural" ontological attitude, such as macroscopic bodies, microscopic physical systems, fellow agents, or "external world", has been attempted by Husserl under the (Kantian) label "constitution of objectivity". Husserl here advocated a two-step procedure: first, suspend beliefs and preconceptions about the world; and then show how (and to what extent) those beliefs and preconceptions can be justified on the basis of the reflected-upon experience. In his own terms, "I must lose the world by the *epoché*, in order to regain it by a universal self-examination" (Husserl 1960, 157 [183]).

This is a daring endeavor indeed, one that has been painstakingly pursued by Husserl during his whole life, and has discouraged Carnap (1967) despite a remarkable effort in this direction (De la Tremblaye & Bitbol 2022). Moreover, as a consequence of this endeavor, Husserl has been accused by some of his own students of having (apparently[2]) missed the fact that experience is immediately suffused by a sense of embodiment, community, and being-in-the-world, before any complex operation of constitution of this world (and the bodies in it) has been performed. But it is only at the cost of adopting this kind of uncompromising phenomenological attitude throughout, that QBism can be protected against the inconsistency of patching together a common-sense view of the world and a phenomenological construal of measurement outcomes. It is only at this cost that the QBist dissolution of quantum "paradoxes" can be defended against acute criticisms such as Zwirn's. It is only at this cost that the two conflicting views, ontologies, and standpoints that coexist in QBism can be reconciled.

Things are not entirely settled at this point however. Assuming that we have indeed approached quantum mechanics with this thoroughly phenomenological attitude, we still have a challenge to meet. How can one rescue the most uncontroversial aspect of the realist intuition of QBism (Section 8.4), despite the methodologically idealist option of Husserl's phenomenology? How can one overcome the standard dualist epistemology that is the standard presupposition of scientific realism, and yet retain a sense that "something" is beyond our direct control? These questions will be addressed by borrowing from the French tradition of the phenomenology of embodiment and belonging, from Merleau-Ponty and Henry to Barbaras and Bégout. In particular, we will see in Sections 8.5 and 8.6 of this chapter that Barbaras' (2016, 2019) "cosmological phenomenology", and Bégout's (2021) concept of "Eco-Phenomenology", remarkably fit with the difficult philosophical synthesis QBism is half-consciously heading to.

8.2 Naturalized Versus Transcendental Theories of Knowledge: Another Name for the Dilemma of QBism

As a preliminary, let us notice that the QBist discrepancy between its instrumentalist and realist inclinations, or between its experiential and interactional concepts of measurement, is one among many expressions of a fundamental strain that haunted epistemology throughout its history. It is the tension between an actor's and spectator's standpoint on cognition, between a first-person and a third-person approach of the process of knowing, between a transcendental/normative and a naturalized/descriptive theory of knowledge. At first sight these two types of conceptions of knowledge are antinomic; but they can be shown to be complementary and even synergic under appropriate conditions (Bitbol 2010, chapter 7).

A transcendental/normative epistemology aims to formulate rules to achieve a very high goal in knowledge: universal *a priori* certainty in its Kantian version, or domain-bound self-consistency in several neo-Kantian versions such as Cassirer's (Friedman 2009). It pretends to establish the validity of the body of evidence offered in support of a proposition or a theory, and to ensure the (absolute or context-dependent) *truth* of such propositions or theories. Since a transcendental/normative epistemology purports nothing less than to justify in advance the capacity of science to reach some sort of truth, it is not (and should not be) scientific. A transcendental/normative epistemology does not offer any scientific representation or description of the process of scientific knowledge: it is *a priori* rather that *a posteriori*; it precedes (and *should* precede) the material, the contents and the methods of the science it pretends to found. A transcendental/normative epistemology adopts the standpoint of an actor who wishes to build a science and who makes a preliminary reflection on the conditions of possibility of her own projects and methods, before a mature science has provided her with a description of these very methods from the standpoint of a spectator. The sought pre-scientific normative foundation of science is then supposed to be an absolute starting point which testifies to its own validity. This self-obvious starting point includes lived experience, rules of inference, anhypothetical[3] principles (such as the principle of non-contradiction), etc.

By contrast, a naturalized/descriptive theory of knowledge is concerned with the actual formation of beliefs in concrete organisms or artifacts. It sets out to describe the interaction between the (objectified) knowing subjects and the world, as well as the subsequent modifications of the subjects' beliefs about the surrounding world, from the standpoint of a spectator of the knowledge process. Deprived of any self-justifying source of certainty, naturalized epistemology declares that it has nothing better to do than adopt the methods of the science of nature it aims to elucidate (Quine 1969). It thus offers a scientific representation of science from the standpoint of a (scientific) researcher taking scientific research as her object of study. As a result, a naturalizing epistemologist no longer dreams of some ground firmer than science, on which science can be founded; she acts to defend science from within, against the doubts it has about itself (Quine 1974).

Let's notice at this stage that the actor's standpoint is associated with a normative approach of knowledge, whereas the spectator's standpoint implies a descriptive approach of knowledge, including the knowledge of knowledge. Let's also remark that, from the actor's standpoint, it would be absurd to presuppose the objects and properties of a mature science before science is adequately founded. Even more strikingly (according to Kant), it would be wrong to presuppose the objects and properties of *ordinary language and common sense* before they are appropriately founded in a procedure of "constitution of objectivity", performed by due ordering and selection of fractions of the actor's experience. Instead, from the spectator's standpoint, the entities and categories of ordinary knowledge and science are used unproblematically for the description of everything, including the very procedures by which ordinary and scientific knowledge are acquired.

The orientations of the two types of epistemologies are so antinomic that they tend to accuse each other of being self-defeating. Advocates of a naturalized/descriptive epistemology usually reproach transcendental/normative epistemologists to posit norms in the name of a superior kind of knowledge, whereas this alleged knowledge has never been submitted to the stringent tests of scientific knowledge. According to naturalists, there is no better procedure than extracting the scientific methods *a posteriori*, from the very science of nature that has obtained success by letting them emerge in the course of its development; for positing these methods *a priori* would put us at risk of arbitrariness. Conversely, supporters of a transcendental/normative epistemology accuse naturalized/descriptive epistemologists of vicious circularity. How can they pretend to found science on a scientific description of itself, without committing a *petitio principii*? How can they not see that, by affording a scientific, purely *factual*, description of science, they weaken the traditional ideal of a science capable of approaching *truth*? For, does truth not reach beyond the domain of facts to attain the domain of norms and values?

One of the most vocal critics of naturalism along this line was Husserl (1973, 16). According to him, if one accepts (say) a scientific evolutionist account of the acquisition of knowledge, it turns out that no kind of knowledge (not even science itself!) will have any significance other than adaptative. And therefore, *evolutionism itself is not true, but only adaptative.*

Beyond this antagonism, the two kinds of epistemologies (naturalist/descriptive and transcendental/normative) need one another; and if they become mindful of this need, they are likely to borrow features from each other.

A naturalist/descriptive epistemology can benefit from recognizing the normative element in its own approach. After all, a naturalist/descriptive epistemology *prescribes normatively* to use scientific methods, and to adopt, in its description of any process of knowledge, the ontology of scientific theories.

Conversely, a transcendental/normative epistemology is *de facto* dependent of the science it is meant to found. Just consider the way Kant (1781) derived his normative framework of knowledge by a regressive (transcendental) deduction starting from the strong credentials of the mathematics and physics of his time, and then deriving their condition of possibility.

Moreover, a transcendental/normative epistemology can benefit from representing the knowledge process by means of a naturalized description of it, while ascribing this representation no status other than heuristic. A transcendental/normative epistemologist can indeed exploit the isomorphism between her theory and a naturalized theory of knowledge, to take the first steps of her conception, and to illustrate it.

Here again, we may find a good example in Kant himself, who progressively converted an earlier naturalized/descriptive epistemology of relational knowledge into his later transcendental/normative relativizing epistemology.

In his first works, written before 1770, Kant had not yet formulated his transcendental epistemology, and he expressed himself in a characteristic mixture of metaphysical and scientific styles. A central thesis of Kant (1755), was that the *relations* between "substances" are irreducible to (sets of) their immutable intrinsic *properties*. Indeed, he argued against Leibniz' concept of "windowless monads", if this were not the case, relations would be epiphenomenal to properties, and there would be no basis for real *changes* in natural "substances". Applying this relational thesis about nature to the particular relation between the knower and the known, Kant was led to the conclusion that the knower cannot disclose the intrinsic properties of things (substances), but only their *dispositions* to *relate* to her. This is the naturalized/descriptive source of a well-known principle of Kant's transcendental/normative epistemology: the principle that knowledge is not meant to be faithful to the thing-in-itself; that knowledge can only pre-organize the phenomenal byproduct of our interaction with the thing-in-itself, so that we can think and act *as if* these phenomena were objects independent of us.

The latter example is especially relevant to QBism and its own struggle between a normative and a naturalized approach of microphysical knowledge, between the actor's and spectator's standpoint on microphysics. In QBism as in Kant, one starts with a naturalized picture of knowledge as an interaction between an agent (ordinary human being or scientist, endowed with her hands and sensory organs or with her measurement devices) and entities out there (the external "physical systems", substances, or things-in-themselves). In QBism as in Kant, this interaction is supposed to yield an experienced phenomenon that is not just a copy of the (hypothetical) "primary" qualities of the thing out there, but an emergent byproduct of the relation with it, i.e., a so-called "secondary" quality. In QBism as in Kant, also, the standpoint of experience is eventually adopted, and the whole process of knowledge is tentatively seen anew from there.

But Kant became more and more consistent in his choice of the first-person agent's standpoint, and he tended less and less to think of the "thing-in-itself" literally as an intrinsically existent "thing" facing the knower. In his subsequent writings, the thing-in-itself names both an inarticulate starting point, and an inaccessible horizon of knowledge: it refers both to the formless "*Grund*" (Ground) of phenomena, and to a "regulative ideal" of research. Kant then progressively dismissed the naturalized scaffolding of his own earlier thought, and sticked more and more consistently to the first-person, experiential and normative, standpoint of his mature epistemology.

It seems to us that this is precisely the result QBism should strive to obtain, if it is to become fully self-consistent: considering the whole process of microphysical knowledge from the standpoint of the actor, the knower, the experiencer, and putting at rest those representational scaffoldings that involve agents-with-prostheses acting upon physical systems pre-given out there. In other terms, it seems to us that QBism should progressively decrease the import of its pragmatist representations, and bring to completion its phenomenological inclination (to the point where even pragmatist pictures are given a phenomenological meaning). For, if it does *not* do so, it exposes itself to the risk of self-"fissuring" mentioned in the introduction.

Several subsequent naturalized theories of cognition followed Kant's pattern, and they can be equally inspiring for QBists. They include Jakob Von Uexküll's (2010) "biosemiotics", and James Gibson's (1986) "Ecology of perception".

Von Uexküll's central concept is that of "*Umwelt*" (environment) *qua* different from "*Welt*" (world). The *Umwelt* of an animal, or of an animal species, is a set of features that co-emerge with the activity of the animal, and are *significant* for the basic concerns and the survival of this animal. But, from the first-person standpoint of the animal, what *we* call its *Umwelt* is (mis)taken for The World (*Welt*); and the significant features that co-emerge with the activity of the animal are (mis)taken for objects of The World. In other terms, Von Uexküll invites us to distinguish between:

i. the apparent world (*Umwelt)* of an animal, aroused by its activity, and then seen from its first-person standpoint;
ii. the "real underlying world" (*Welt*) seen from a third-person standpoint, which, through its interaction with the animal, lets the latter's *Umwelt* emerge.

Now, can we say something about this "real underlying world"; can we characterize it further? Von Uexküll makes clear that what we call "the real world", what we represent as something that interacts with animals to give rise to their *Umwelt,* is nothing else than *our own Umwelt*, namely the correlate of the most advanced epistemic activity of the human species. Just as animals do, we tend to (mis)take our own (collective) *Umwelt* for *THE World*. Our naturalized, third-person-singular, description of knowledge, once again turns out to be nothing else than the misleading form taken by an unnoticed first-person-plural approach. This being granted, we can conclude that Von Uexküll's concept of "*Umwelt"* finally yields the dissolution of the very anti-concept of "*Welt* (real world)" by opposition to which it was initially defined. Similarly, Kant gave such a convincing demonstration that we tend to mistake our objects for things-in-themselves, that the very concept of thing-in-itself became suspect of being one more expression of (a sophisticated version of) this very same mistake, called the "transcendental illusion". This is why the "thing-in-itself" was progressively set aside by Kant (2013, 2014) as a "regulative ideal" or an open problem for reason, and later eagerly criticized by the majority of the post-Kantian German philosophers (Cassirer 1969).

Let us now come to James Gibson. His original concept is that of "Affordance". An affordance of the environment is what this environment offers (affords) a living being, namely a manifold *disposition* to satisfy its needs, to preserve its vulnerabilities, and to fit with its capabilities. In the same way as the elements of Von Uexküll's *Umwelt,* Gibson's affordances emerge from the activity of the living being in *what there is*.

Rom Harré's (2006, 2014) Gibsonian philosophy of physics is especially interesting for us, since it has breathed new life into the concept of affordance, as a powerful way to express Bohr's interactional conception of quantum phenomena. According to Harré's definition, "affordances are dispositions that are created in the interplay between an agent and the possibilities that the target of the agent's activity makes available". This enables one to make sense (e.g.) of the (in)famous wave-particle duality without creating an ontological chimera, since by using "a different apparatus, the experimenter can get the *World* to afford interference phenomena with the same starting point as the experiment that afforded particles" (Harré 2013). But what about this "World" that plays a capital role, together with agents and apparatuses, in the onset of affordances; and what about our prudent "what there is"? Just as Von Uexküll's, Harré's concept of "World" is bound to remain problematic, since any characterization we can ascribe to it, is derived from the affordances we arouse in it by our experimental activities. What we usually call (and describe as) *The* World thus identifies to *Our* world. This is the reason why Harré (1997) coined a new, somehow onomatopoeic, term to refer to his highly indeterminate concept of a truly "independent" world: *the Glub*. Harré thinks of his Glub as a reservoir of dispositions, or even as a second-order disposition: a low-level disposition to let experimentally or biologically relevant higher-level dispositions (affordances) come up. A seemingly third-personal concept of "world", named "the Glub", is thereby reconstructed regressively out of the first-person-plural *Umwelt(s)* of mankind.

The latent role of the first-person standpoint, in all these naturalized epistemologies that pretend to give a third-person description of the process of cognition, is striking, and full of lessons for QBism. But another kind of naturalized epistemology would be even more efficient to defuse the conflict between the two QBist ontological orientations: a naturalized, third-person, epistemology of Kant-Uexküll-Gibson kind, that be nevertheless designed from the outset as a pathway toward the first-person standpoint of phenomenology. This is precisely the case of *Enaction* (or *Enactivism*), a theory of cognition initially formulated by Francisco Varela, with the collaboration of Evan Thompson and Eleanor Rosch (Varela et al. 2017). It was derived from its biological forerunner, the "autopoietic" theory of cognition, which also involved a circulation between the third and the first-person approach of knowledge (Maturana & Varela 1980).

According to the enactive theory, cognition is neither tantamount to a passive reproduction of some external reality by a subject, nor to a mere projection of subjective operations onto this reality. It rather arises from an intermediate level that stands between the knowing subject and the reality-to-be-known: the *activity* of the subject of cognition embedded in her environment. By her activity, the subject selects, and retroactively alters, the features to which she is sensitive. By this combination of selection from, and feed-back on, her environment, the subject (the agent) determines, and even molds, her own specific *Umwelt*. Conversely, the molded *Umwelt* exerts a pressure on the cognitive organization of the subject-agent. Here, one can say that the subject and the environment mutually *constitute* each other. And cognition is accordingly construed as a process of co-emergence of the knower and the known (Bitbol & Luisi 2004).

Thus far, the theory of enaction has been presented as a naturalized, third-person, epistemology; an epistemology of the productive interaction between agent and environment, as seen by a spectator of such process. But, according to its complete agenda, the theory of enaction promotes a displacement from the third to the first-person standpoint; it seeks a self-revelation of the enactive nature of knowledge in the lived experience of the enacting subject; and it then aims to reformulate the whole problem of knowledge in terms of the first-person experience of knowing subjects.

Eleanor Rosch has successfully clarified this point in the preface to the second edition of *The Embodied Mind* (Varela et al. 2017). There, she establishes a clear-cut distinction between phase 1 enaction and phase 2 enaction. Phase 1 enaction refers to a description of the process of sense-making achieved by an agent in her effort to define features of her environment that are available as targets of her actions, and relevant to the maintenance of her organism. This is the third-person aspect of enaction, which involves a duality of subject and world, together with a description of the co-emergence of the form of the subject and the form of her world (her *Umwelt*). Phase 2 enaction, instead, investigates *what it is like* (in the first person) to realize that one is caught in such co-emergence of the two poles of the process of knowledge; and, more precisely, *what it is like* to partake of a process of knowledge that is based neither on the firm ground of a pre-formed external world, nor on the firm ground of a pre-formative cognizant *ego*. It turns out that "phase 2 enaction is a non-dual mode of knowing that allows for *a direct experience of groundlessness*" (Meling 2021). An experience of one's significant targets of action, together with the dizzying awareness that these experienced targets are not grounded on any intrinsic properties of inner or outer entities, is the non-dual first-person aspect of enaction.

Retrospectively, it appears that Phase 1 enaction, namely the naturalistic description of a co-emergence of the subject's knowledge and the structure of its environment, is just a mental tool to figure out why we have the kind of experience outlined in Phase 2 enaction. *Actually, Phase 1 enaction can by no means be* more than a mental tool. For, someone who believes that the (Phase 1) enactive naturalistic description of the subject-world interaction is a *true representation of a real process (of cognition) unfolding out there*, would be caught in a self-contradiction. Indeed, believing this, would mean considering that our knowledge of the process of knowledge is not enacted, but rather obtained quasi-passively like a "mirror of nature" (Rorty 1981). Believing this, means that we do not take the concept of enaction seriously enough to apply its consequences to itself. In other terms, taking enaction at face value, immediately undermines the third-person dualistic *picture* of an enactive transaction between subject and world. This picture is decidedly nothing more than an ancillary tool for imaginative minds.

Such reasoning, in which the consequences of a naturalized picture of knowledge *qua* interaction turn against the very acceptability of the interactive picture and of naturalization in general, can easily be extended to QBism. Just as enactivism, QBism makes use of a naturalized picture of quantum knowledge *qua* interaction. To draw this picture, QBism imposes a "conceptual split" of the world into "one part treated as an agent" and another part treated "as a kind of reagent" (the "quantum system") (Fuchs 2010). Then, an "agent's taking an action on the quantum system" results in "a unique creation within the previously existing universe" (Fuchs 2010). The problem is that, in virtue of its own consequences, this picture cannot be a *faithful representation* of the process of quantum knowledge. Indeed, in the framework of QBism, if an agent wants to know quantum knowledge, the only strategy she can use is the one she uses to know any other process or system: she must "take an action" on this quantum (epistemic) process or system, and trigger "a unique creation" that inextricably combines her own contribution with the contribution of this "reagent". Such unique creation being usually called a *phenomenon*, one is concerned in this case by the phenomenon of *quantum knowledge*. But then, by such procedure, the agent has not disclosed the (alleged) true intrinsic nature of quantum knowledge. She has only co-produced an experienced novelty: some phenomenon-of-quantum-knowledge. This is enough to show that the QBist's dual image of an agent "really" acting on a "real" quantum system is self-defeating (or self-dissolving). What remains of it is just the *experience* of partly unexpected, and therefore creative, phenomena.

The previous reflections are not without consequences on the QBist defense against the charge of solipsism. According to Chris Fuchs (2010), "Two points are decisive in distinguishing this picture of quantum measurement from a kind of solipsism: (1) The conceptual split of agent and external quantum system .... (2) Once the agent chooses an action ... to take, ... the actual outcome is not a product of his whim and fancy". Point (1) has just been proved to be self-defeating. Therefore, only point (2) remains to avoid integral solipsism. The only sign that the individual agent is not "alone" (*solus ipse*) is "that the consequences of measurement actions are beyond the agent's control; (that) the world can *surprise* the agent" (DeBrota et al. 2020a). If a phenomenon surprises us to a certain extent, this indicates that we have not entirely manufactured it.

Beware, at this point. Surprise is not sufficient to prove that subject and object, agent and quantum system, are two ontologically different poles of the world. Surprise is also an experience; it is a compound experience that arises from a discrepancy between experienced expectations and experienced outcomes. Far from representing a break in the fabric of the experience, surprise thus partakes of its continuity. But it plays a highly non-trivial role in it.

To begin with, remind that, from Husserl's standpoint, experience is made up of two poles – potentiality and effectivity. And notice that this is remarkably parallel to QBist thought, in which the agent makes bets about the potential outcomes of future measurements, and then observes some effective outcome as soon as she can testify that the measurement has been performed. But then, how would surprise not be a challenge to the phenomenologist? If there is only room for anticipated and actual phenomena in the fabric of experience, how can we accommodate surprise within experience itself? And how can we avoid to jump immediately to the conclusion that surprise proves the existence of an outside world beyond any possible experience? If surprise were to fit in the standard minimal framework of Husserlian phenomenology alluded to above, even it should be anticipated, even it should be deemed possible to a certain extent. This would immediately deprive surprise of its own definition: that of a phenomenon that escapes any anticipation.

But let's take a closer look at what phenomenology has to tell us on this precise point. In the phenomenological theory of perception developed by Husserl, there is room for what he calls "the disappointment of expectations". Here, a true surprise represents "the tipping of the possible into the impossible" (Serban 2016, 89). It is true that, for there to be a surprise, there must be the disappointment of an expectation, a frustration, or to put it another way, a *non-fulfilled intention*. But what, in the shock of the event, appeared impossible, immediately afterwards acquires the value of an open, yet indeterminate, possibility. A surprise should then be construed phenomenologically as a "defeat of possibility motivated by open possibility". "Its emergence shows [...] that the 'impossible' is nothing other than a more open possibility" (Serban 2016, 89). According to Husserl, surprise then represents a lived condition for the unfolding of ever more open possibilities within lived experience itself. As for its standard interpretation as evidence of the impact of something that transcends experience, Husserl brackets it by the *Epoché* that prepares the ground for phenomenological inquiry. At most, he makes such metaphysical interpretation conditional on an internal analysis of its credentials in experience.

To recapitulate, in a phenomenological framework, an unexpected phenomenon is both a disappointment (*Enttäuschung*) and the metamorphosis of the impossible into an expanded field of open possibilities. Surprise just recasts the lived sense of the possible, by broadening it to an indeterminate extent. Reciprocally, we should understand that "[...] experience [...] consists of a horizon of 'open possibilities' (Husserl 1972: 108) which, precisely, creates room for the emergence of surprise" (Serban 2016, 179). Surprise therefore does not go against prior motivated possibilities; it adds to them a larger domain of open, indeterminate, possibilities; it constitutes a space for experienced indeterminate possibilities somewhere in between motivated anticipation and recognized actuality.

From what has just been said, we tend to infer that the lived experience of agents is the unique domain in which every challenge to QBism (including the challenge of the solipsistic *reductio ad absurdum*) should be addressed. Even the desire to reach something of a world deemed to be "external", arises in experience. Even its credibility must be evaluated on the basis of experience. Even the project of disclosing an aspect of the quantum formalism that would "take advantage from a third-person perspective" is bound to find its justifications in the first-person perspective.

8.3 Experience First and Foremost: The Phenomenological Ghost that Haunts QBism

The pre-physical, and not physically definable, concept of *experience* plays a pivotal role in QBism. In one of the most emblematic presentations of QBism (Fuchs et al. 2014), the word "experience" appears 58 times, initially in relation to Bohr's claim that physics is just a set of "methods for ordering and surveying human experience" (Bohr 1987, 10), and then in the more specific context of a discussion of the status of experimental outcomes. And that' not all. In a collection of written debates between discussants of QBism (Fuchs 2015), the word "experience" appears ... 595 times; and it serves as the pivotal theme of several of these dialogues.

Such intensive use of the word "experience" (that is preferred to "consciousness", presumably in view of the strong reflective connotation of the latter), is one of the main features that bring QBism dangerously (or happily) close to phenomenology. Another phenomenology-like feature of QBism is its conception of both microscopic *and* mesoscopic entities as "bundles of expectations", made of (i) a central perceptive or experimental nucleus of experiences and (ii) a "horizon" of anticipated experiences (De la Tremblaye 2020). Our task in this section will be to clarify the role played by the concept of experience in the QBist approach of quantum mechanics, and to evaluate the strength and limits of the phenomenological commitment of QBism.

Some QBist sentences are among the stronger statements of the role of experience that can be found in the literature about the interpretation of quantum mechanics. Thus, "according to QBism, quantum mechanics is a tool anyone can use to evaluate, on the basis of one's past experience, one's probabilistic expectations for one's subsequent experience" (Fuchs et al. 2014). Here, there is no question of a previous knowledge of the world, and no question of writing down some "state" of the world, but only of a probabilistic connection between two successive *experiences*. This is how the age-old conundrums of quantum mechanics (or rather of its clumsy interpretations) are immediately defused.

First of all, "The notorious 'collapse of the wave-function' is nothing but the updating of an agent's state assignment on the basis of her experience". No mysterious influence of gravity is required to impose the "spontaneous collapse", or "objective reduction", of the "state of physical systems". No baroque multiverse, and no cumbersome emergence of classical univocity from quantum plurivocity, are required either.

Besides, in QBism, the quantum "paradoxes" that involve the *comparison* between several observer's/agent's outcomes are put to the decisive test of their formulation within a single agent's experience. This is the case of the comparison between the outcomes and memories of Alice and Bob, in the Einstein, Podolsky, Rosen correlations; and this is also the case of the outcomes and predictions of Wigner and his friend, in Wigner's friend paradox. The universal key to the dissolution of this family of "paradoxes" is the thesis that "Bob's answer is created for Alice *only when it enters her experience*" (Fuchs et al. 2014). This crucial thesis extends a general principle of QBism to the issue of intersubjectivity: the principle that phenomena are created in the experience of each agent by the meeting between this agent and a reagent, rather than observed passively by the agent.

This being granted, the alleged discrepancies between the descriptions various observers give of the "state" of "physical systems", are immediately defused. Indeed, such discrepancies can only be certified from an external standpoint, and they vanish whenever a process leading to an agent's lived experience is completed. They only arise as the illusory by-product of a "view from nowhere" of the experimental processes, and they are dissolved within a "view from somewhere". No non-local influences are then needed to account for the *experienced* quantum correlations (Fuchs et al. 2014; Smerlak & Rovelli 2007). No spurious action of consciousness on the physical world is required either (Bitbol 2022), to avoid the apparent contradiction between Wigner's friend *observing* a definite experimental result, and Wigner's *representing* his friend as being in a state of superposition.

Such clarification is obtained at what a majority of (realist) physicists would consider excessive cost. Some of them (those who were educated in the Copenhagen interpretation) would concede to Bohr that the experience accounted for by quantum mechanics is an experience of macroscopic events displayed and recorded by quasi-classical apparatuses. This Bohrian doctrine at least gives them the sense of something real *of which* observers get an experience: the events of the macro-world. But QBists resist this remnant of standard realism that takes the form of macro-realism. They essentially agree with other interpreters of quantum mechanics (Brukner 2020) that no observer-independent notion of experimental outcomes can be maintained; and they go as far as declaring that experimental outcomes only make sense *qua* experiences of agents-observers. We have seen that QBists have good reasons to do so. But other reasons can be adduced at this point. In particular, if they maintained the Bohrian notion that an experimental outcome is a macroscopic event described by way of classical concepts, they would be caught into the endless debate about the locus of the quantum/classical boundary within the measurement chain.

At the end of this operation of defusing any ontological commitment about events that (allegedly) first happens out there, and are then secondarily noticed by an observer, QBists willy-nilly end up with a phenomenological *tabula rasa*. Here, no pre-given world is supposed, no pre-given events, processes, or objects, but a flux of experience organized into a network of quasi-invariants that can be dealt with *as if* they were appearances of intrinsically existent objects and properties. Not only "What is real for an agent rests entirely on what that agent experiences"; but "*Any user's own experience constitutes all of the raw material out of which she constructs her world*" (Fuchs et al. 2014). The latter sentence is a short and (perhaps too) dry statement of the previously sketched Kantian and phenomenological procedure called the "constitution of objectivity".

Among the members of the QBist circle, Jacques Pienaar is probably the author who is most willing to bite the bullet and endorse a fully phenomenological approach. His endorsement is bold, and it is expressed in a few powerful sentences. To begin with, "in QBism, *an element of reality is an experience*" (Pienaar 2020). The latter statement is tantamount to adopt a phenomenological ontology, in the strongest sense of the word. It is analogous, e.g., to Eugen Fink's concise definition of a central doctrine of (uncompromising) phenomenology: "(Phenomenology) merely claims that Being is identical to the phenomenon" (Fink 1994, 120). In the latter sentence, "phenomenon" is the phenomenon of phenomenology, not the phenomenon of classical physics; it means an appearance in experience, not a macroscopic event occurring in a laboratory. Then, just after his decisive ontological claim, Pienaar goes on: "(An experience) contains as a fundamental *internal structure* a pairing of an experiencing subject with an experienced object; such experiences are called *Events*". Here, the split between agent and system, between subject and object, is no longer taken as a pre-phenomenological assumption aimed at preserving something of the usual concept of an "external" world; it is itself a phenomenological structure that turns the concept of an external world into a problem of "constitution" for phenomenology. And the term "event" is redefined accordingly as an *experience* of pairing.

By the way, this represents an unambiguous rejection of the first strand of the standard QBist defense against the accusation solipsism, which implied maintaining a naturalized description of the transaction between an agent and an external reagent called the "physical system". To make things even clearer, Pienaar insists that "since the Agent and World represent *internal aspects of Events*, one should be careful not to think of the Agent and the World as being *causes* of the Events". In the former sentence, the naturalized picture of an agent acting on "physical systems" of the world, thereby triggering (causing) objective events to occur in the world, is entirely replaced with a phenomenological redefinition of Events *qua* experiences, and of Agent and World *qua* internal structures of experience. Here, the very distinction between agent and some outer world relies on an inner feature of experience. It relies on the difference we make *within experience* between voluntary acts and partly unexpected outcomes (Pienaar 2020).

To sum up, our experience includes *everything*, including our aiming for a transcendent whole called "the world", and for transcendent entities called the "objects of the world". "QBism ... acknowledges that an agent's experience encompasses both subjective and transcendent elements (i.e., relating to a world beyond the agent)" (Pienaar 2021). Accordingly, the objects of the world, which include what physicists call "physical microsystems", are endowed with a purely phenomenological meaning: "Phenomenologists emphasize that every object given in perception is given within a certain context or 'horizon', against which it takes on certain significance. This significance might include the object's being a concrete instance of some theoretical abstraction, or embodying some formal mathematical model" (Pienaar 2021). For a phenomenologist, and for a QBist as well, an object or a physical system is (or should be) nothing more than that: a nucleus of perception(s) surrounded by a fuzzy background together with a more or less (mathematically) formalized horizon of expectations, and whose becoming is constantly monitored within the agent's experience.

As we have seen previously, even Fuchs comes (dangerously?) close to such a fully "constitutive" conception of the world. And this is apparently confirmed when he quotes approvingly a post-kantian idealist view advocated by Schrödinger (1951): "In *Nature and the Greeks,* (Schrödinger) takes a QBist view of science more generally, and hardly even mentions quantum mechanics. He stresses that, because *everything any of us knows about the world is constructed out of his or her individual private experience,* it can be unwise to rely on a picture of the physical world from which personal experience has been explicitly excluded, as it has been from physical science" (Fuchs et al. 2014). Even the idea of a fully phenomenological ontology, such as the one offered by Pienaar, has been a permanent temptation for Fuchs. This can be seen in two emails Christopher Fuchs wrote to Manuel Bächtold in January and June 2009: "I try to view these 'pure experiences' as the active monads of the world, similar to James, and similar to John Wheeler with his 'elementary acts of observer-participancy' being the building blocks of the world". Even more directly, Fuchs claims that "What I am aiming for is a pluralistic ontology of something like '*pure experience*'" (Fuchs 2015, 1661, 1738). And he justifies this quest by a personal conviction: "I *do* think we have a kind of direct evidence of 'the real'. It is in the very notion of experience itself" (Fuchs 2017, 113 and 118).

Yet, when Robert Crease (Crease & Sares 2021) tried to push Fuchs all the way down in this direction, the answer was an expression of reluctance, and a renewed realist act of faith. "The starting point for me is that there is a world external to any agent", Fuchs replied. This is very much unlike some of his former statements (and unlike most of Pienaar's statements), in which the concept of an external world sounded like an *ending point* based on an ontology (or just an incontrovertible fact) of "pure experience", rather than a metaphysical starting point. In the wake of this realist act of faith, Fuchs insisted on his agreement with the implicitly realist presupposition of pragmatist thinkers. Such pragmatist presupposition can be stated in two steps. According to the first step, "the starting point of Deweyian pragmatism is that *there is a world out there* for each of us" (in Crease & Sares 2021). Yet, according to the second step, borrowed from F.C.S. Schiller, "The actual situation is of course a case of interaction, a process of cognition in which the 'subject' and the 'object' determine each other, and both 'we' and 'reality' are involved, and, we might add, evolved"; so much so that "it is meaningless to inquire into *nature as it is in itself*" (Fuchs 2015, 1366). By this second step, pragmatism can be considered as a major philosophical source of the semantic, enactive, ecological, naturalized theories of cognition we presented in the previous section. Just as these naturalized theories of cognition, pragmatism retains a (more or less nuanced) form of realist approach to the world. Crease then concluded that Fuchs' construal of QBism "remains in the natural attitude by adhering to the idea of a world that pre-exists and exists independently of the subject". According to Crease, the pragmatist basis of QBism is definitely averse to the phenomenological attitude, since a phenomenologist would necessarily ask: "how do we *know* that quantum mechanics refers to anything 'external' in the world, beyond our own experience of using quantum mechanics?" (Crease & Sares 2021).

Now, there is a momentous difference between the two documented steps of the pragmatist presupposition borrowed by QBism. If taken in isolation, the first step may easily be mistaken for plain "external realism", namely for the crude belief that the world "has a determinate nature which the knowing reveals but does not affect". It is only through the second step that this misunderstanding is retrospectively dispelled, and that one acknowledges that, in pragmatism, "the 'determinate nature of reality' does not subsist 'outside' or 'beyond' the process of knowing it".

But, as we saw in the previous section, this second step of the pragmatist presupposition doesn't content itself with downplaying the naive realist interpretation of the first step. The enactive component of pragmatism surreptitiously undermines its realist component; the enactive component of pragmatism dissolves any realist remnant under the pressure of its very consequences. For when, at the first step, we say "there is a world out there", do we not tacitly assume that this real world is something determinate, located in some determinate place (outside), even before we have undertaken a process of knowledge of it? And is this assumption not irreconcilable with the lesson of the second step, namely that "the Real is *nil,* as unknown: it is only potentially real" (F.C.S. Schiller, quoted by Fuchs 2015, 1359)? In this case, how can we even speak of such *potentiality* as if it were completely independent of the gestures of actualization an agent performs on it, and independent of their manifest consequence in/*qua* lived *experiences?*

The pragmatist philosophy borrowed from William James here repeatedly collides against itself, and insistently calls for another aspect of William James thought: his doctrine of "pure experience" as the primordial stuff of what there is, his *radical empiricism* (James 1912), his proto-phenomenology. If a form of complementarity of the pragmatist and (proto-) phenomenological sides of James' thought is to be achieved, this can only happen within a phenomenological framework. And this observation also applies to the pragmatist and phenomenological aspects of QBism.

To outline the strategy we'll use to articulate these two Jamesian lines of thinking, the shortest way is to comment on a biological picture offered by Fuchs as an analogue of the QBist view of quantum knowledge. "We are like euglenas—tiny single-cell organisms with little tails—swimming in the big environment surrounding us. The tail is a tool the organism uses to move in the direction of better nutrients. We are not much different than that" (in Crease & Sares 2021). The analogy is sound, and it closely fits with the spirit of pragmatist, autopoietic and enactive theories of cognition. Since the world (*Welt*) of the euglena is an ocean, this little organism is immersed in it instead of facing it. The euglena then co-defines its *Umwelt* made of opportunities and threats, nutrients and predators, by its very behavior within the oceanic world. The euglena has no map of its environment, but it can count on a repertoire of stereotyped conducts that anticipate the affordances it is likely to meet. In other terms, instead of relying on some exhaustive description of the world for its survival, it contents itself with a genetically encoded "user's manual" for coping with the *Umwelt* it lets emerge as it continuously acts (by swimming).

As stunning as it may sound, we, human beings capable of formulating and using quantum mechanics, "are not much different than that". We co-define our physical *Umwelt* made of opportunities and obstacles, by our technological interventions. We have no pre-defined map of the world (*Welt*), but (within our *Umwelt*) we can count on an integrated system of probabilistic anticipations that obey conditions of Dutch-Book coherence. Instead of relying on some theoretical description of the world as it is independently of us, we content ourselves with a mathematically encoded "user's manual" for coping with the *Umwelt* of phenomena that our agency lets emerge.

But let us push the analogy even further, beyond what Fuchs stated explicitly. The euglena does not reflect on its own cognitive behavior; it does not have a representation of its transactions with its environment; it cannot disengage from its own life and see it from outside. As far as we can tell (but who are we to tell this, beyond our "heterophenomenological" interpretation of a living being's behavior?), the euglena just has a primitive experience of what it is like to be immersed in its oceanic environment, and to cope moment after moment with the reactions of this bath.

Here again, *we are not much different than that,* despite our grand claims about the superiority of the human intellect.

What do we mean? Are we not able to reflect on our own condition? Are we not capable of providing a *representation* of ourselves while we engage in our epistemic endeavor, unlike the euglena? Yes and No. We are indeed capable of providing pictures of our knowledge process: this is precisely what naturalized theories of cognition claim to do. However, we have also understood in the previous section that, at the end of the day, any such picture is nothing more than a mental instrument for our effort of orientation in a technologically co-defined *Umwelt,* without any pretension to resemble the (putative) *Welt.* In other terms, any such picture is an aid *in* the process of knowledge, not a faithful imitation *of* some alleged object of knowledge (not even of the process of knowledge taken as a meta-object). Just as the euglena, we are entirely immersed in the only habitat we ever had: the first-person experience of being there, acting, and coping moment after moment with what we make occur by our actions.

With respect to the euglena, our knowledge has both a disadvantage and an advantage. Our disadvantage is that, precisely because we elaborate pictures of our knowledge process, pictures of ourselves in an *imagined* world-out-there, we tend to make a major mistake that the euglena cannot even figure out: the mistake of looking, beyond our *Umwelt*, for something else called *Welt*, or thing-in-itself. Our advantage with respect to the euglena, is that human experience is likely to be populated by many more intellectual tools than the algae's: it includes mental patterns and fictions that allow us to universalize our biological strategies of coping. But knowing, in us as in the euglena, is still experience throughout; experience first and foremost. The pragmatist component of the QBist approach is bound to be embedded into its radical empiricist or phenomenological component.

Embedding the pragmatist component of (quantum) knowledge into its radical empiricist component. Embedding the (realist) claim of transcendence of the world into the immanence of *lived experience*. Embedding what is beyond the agent's control into the horizon of possibilities of her *present experience*. Embedding the notion that what occurs is "(*not*) plastic to our every demand" (Fuchs 2015, 1359), into the stringent rules of the actions that we *experience* as promising. Such is our program of thorough phenomenologization of QBism for the subsequent sections. But this program will be prepared, in the next section, by an analysis of the carefully weighed realist demand of QBism.

8.4 Participatory Realism from the Standpoint of the Participator: Preliminary Steps

QBism walks on the thin line that separates a wholly subjectivist construal of most quantum theoretical symbols, and a rejection of flat solipsism. It walks on the thin line that separates the conviction that phenomena are (co-)created by us, and the observation that phenomena can nevertheless surprise and resist us *as if* they were provided by something completely external to us. The metaphysical response of QBism to the challenge of reconciling these seemingly antagonist tendencies has been coined "participatory realism" (Fuchs 2017). It develops what we may now call "the euglena approach to knowledge", into a doctrine that comes remarkably close to the pragmatist and enactivist epistemologies. According to such doctrine, the oceanic reality in which we are immersed remains irrepresentable from a strictly third-person (extra-oceanic) standpoint; but it *constrains* our predictions of its reactions, and *imposes normative rules* to our actions within it, if they are to become efficient on it.

This being granted, even the first (disputable) point of the QBist's defense against the charge of solipsism, i.e., the split between agent and world, and the idea of their mutual interaction, becomes elegantly subordinated to the oceanic image of participatory realism. To see this, it suffices to read one of Fuchs' (2015, 26) most lucid criticisms of the subject-object, agent-world, dualities, which was borrowed from a specialist of science studies, Karen Barad. "Within a given context, classical descriptive concepts can be used to describe phenomena, our *intra-actions* within nature (we use the term 'intra-action' to emphasize the lack of a natural object-instrument distinction, in contrast to 'inter-action', which implies that there are two separate entities) …". As a consequence, "Our characterizations do not signify properties of objects but rather describe the *intra-action* as it is marked by a particular constructed cut chosen by the experimenter". These sentences clearly suggest that the agent-world split is not given out there, but rather defined (or "constructed") by the experimenter in the course of her activity. They also imply that is no "inter-action" *between* the entities on the two sides of the split, but an "intra-action" *within* the unsplit oceanic reality of which we partake. *Inter-action* is a mediated or unmediated collision between two things, whereas *intra-action* is a budding or a surge out of a single, initially undifferentiated, *continuum*. Barad further writes: "According to my *agential realist* ontology, or rather ethico-onto-epistemology ..., 'individuals' do not preexist as such but rather materialize in *intra-action*. That is, *intra-action* goes to the question of the making of differences, of 'individuals,' rather than assuming their independent or prior existence" (Barad 2007; Kleinman 2012).

We have seen earlier that Fuchs often puts aside this advanced *intra-active* concept of a measurement, and replaces it by other formulations that tacitly presuppose the division of the world into a plurality of objects. For instance, in Fuchs (2015, 64), the split between the "measuring system" and the "measured system" is still posited unproblematically. But we should definitely consider such standard characterization of the measurement process as an unfortunate remnant of a common prejudice, that lags behind the cutting edge of QBist research. For, unlike the old-fashioned *inter-active* concept of measurement, the *intra-active* concept is the only one that does full justice to participatory realism.

Now, can we take advantage of the surprises that impinge on us, of the constraints that are felt by us, or of the form of the normative rules that maximize our technological success, to extract some information about this "oceanic reality" we are exploring from the midst of it? More specifically, can we go beyond the internal constraint imposed on our system of bets by the clause of Dutch-Book coherence, and identify an external or interfacial constraint that might perhaps teach us something indirectly about the "oceanic reality"? QBists tend to incline toward a positive and carefully argued answer to these questions. This quest for indirect expressions of the transcendent "oceanic reality" in the immanence of our experimental endeavor, is probably the core of their participatory realist research program. According to QBists, the nature of the transcendent "oceanic reality" presumably manifests by way of a non-trivial determination imposed to the normative rules that govern our most successful predictions and actions within it. And we can therefore learn a lot by deciphering this cryptical signal.

One of the few places in which such faint signal can be detected, is the Born rule. In its ordinary form, the Born rule allows one to calculate the probability of some experimental outcome, from the state vector or the density operator of the "physical system" on which the experiment is performed. In its QBist form, the Born rule connects probabilities bearing on two successive experiments; more specifically, it connects the probability an agent assigns to some outcome in the second experiment, to the conditional probability of obtaining this latter outcome O_2 *if* the first experiment resulted in a certain outcome O_1. Now, in both forms (the standard form and the QBist form), the Born rule can be derived from two classes of assumptions: (1) assumptions bearing on the internal coherence of a set of probability assignments and (2) assumptions that seems to go beyond any clause of internal coherence.

In its standard form, the Born rule has been derived out of two such assumptions, initially by Paulette Destouches-Février (1946, 1951), and later on by Andrew Gleason (Caves et al. 2004; Pitowsky 1997). The original derivation of Paulette Destouches-Février starts from the *contextuality* of quantum phenomena (what she calls the "*subjective*" character of quantum phenomena), together with the mutual incompatibility of certain pairs of experimental contexts corresponding to conjugate variables (Bitbol 1996a, 2014). Then, she takes a further step by wondering what is the condition for maintaining the *probabilistic* nature of the prediction of quantum phenomena, throughout the variety of mutually incompatible experimental contexts (this is what we may call a clause of *non*-contextuality of probabilities). And she finds that the Born rule is the only one that makes the probabilistic nature of predictions, especially their additivity to 1, invariant across the various experimental contexts. Indeed, adopting the rule of squared amplitudes (i.e., of squared projections of state vectors on the eigen-directions of the observables corresponding to each experimental context), allows one to apply the Pythagoras theorem to these projections. The sum of the squared amplitudes is then equal to the square of the norm of the state vector, which is constant across time, and equal to 1. The squared amplitudes required by the Born rule then automatically obey a basic condition (Kolmogorov's second axiom) for their probabilistic nature.

Now, what can we say about the *meaning* of Paulette Destouches-Février's two assumptions? The assumption of *non-contextuality of probabilities* is a clause of robustness and coherence of the system of probabilistic valuations across the manifold experimental situations. It is purely internal to the procedure of agent's prediction (or gambling). In other terms, it is a type (2) assumption. Instead, the assumption of *contextuality of phenomena* is of type (1) since it reaches beyond the internal rules of coherence agents impose to their predictions. It is tempting to see the contextuality of phenomena as a fact of the world, though an essentially negative one: the fact that the world *cannot* be neatly separated into an observing system and an observed system. This fact, in turn, may be derived from the fact that "the quantum of action" (as Bohr would have said) takes on a nonzero value: the value h of the Planck constant. Later on, the *amount* of the quantum contextuality of phenomena, gauged by the Tsirelson bound, has been taken as a further fact of the world (Aerts & Sozzo 2014).

The QBist approach retains the general strategy of deriving the Born rule from an assumption of internal coherence of predictions, plus a constraint that may perhaps manifest the contact of agents with some "transcendence". But it differs from Destouches-Février's and Gleason's approach on two points. First, QBism tends to avoid the misleading concept of "quantum state" entirely, and stick to the concept of probability throughout. Second, when it deals with the issue of contextual phenomena, QBism refuses to posit in advance (or even to uniquely derive, as Paulette Destouches-Février did after Jean-Louis Destouches 1942) the standard structure of observables in a Hilbert space, together with the commutation relations between them. Here is what Fuchs wrote one of us (MB) on 04/10/2021 in connection to this: "I am resistant to previous foundational efforts (even heroic ones like that of Destouches-Février) that take noncommutativity or contextuality (as it is usually posed) as their starting points. I instead want to see noncommutativity and all the other things come from this most QBist-conducive starting point: A statement that 'unperformed experiments have no results' (Peres 1978)".

The statement that "unperformed experiments have no results" is clearly less determinate than an axiomatically imposed contextuality plus noncommutativity between pre-defined observables. It introduces additional intellectual degrees of freedom, that can be exploited to explore a larger space of derivations of formal elements of quantum theory such as the Born Rule. It also expresses a fundamental option recently taken by QBism (see Stacey 2021): identify *two* levels of personalism in quantum predictions, instead of just one; assume not only personalism in probabilities, but also personalism in the choice of observables. "Both the probabilities *and* the choice of POVM (Positive Operator-Valued Measure) elements are personal judgments, expressions from the agent's own mesh of beliefs cashed out as gambling commitments" (Fuchs & Stacey 2020). But not even the second level of personal judgment is arbitrary; not even it, is immune from normative constraints. Indeed, for the set of POVMs to be "informationally complete", i.e., for the probabilities of their experimental values to be sufficient to generate a state vector, quantum theory requires that its cardinal be equal to the square of the number of dimensions of the corresponding Hilbert space. It is precisely this exponent 2 (which appears in the *square* of the number of dimensions of the Hilbert space) that, according to QBists, is the mark of the quantum nature of the "systems" and environment, in the normative structures of quantum formalism. Here is their conclusion about the status of the Born rule (DeBrota et al. 2020b): "The Born rule … can be viewed as a normative constraint on an agent's probability assignments. It is a normative constraint above and beyond the standard rules of probability theory. On their own, the rules of probability theory do not tell an agent how their probabilities for one experiment (Experiment Two) should constrain their assignments to another slightly different experiment in which one of the measurements is missing (Experiment One). To make this connection requires some extra empirically motivated assumptions about the physics relevant to these two experiments. We identified a set of such assumptions, the first three of which represent general assertions about physical systems and are compatible with both classical and quantum systems, while the last (the rule of squared dimension) represents a minimal requirement for believing the systems to be essentially quantum in spirit if not letter". In other terms, to generate the Born rule, one needs two classes of assumptions. The first one is plain Dutch-Book coherence (the internal constraint): it is a type (1) assumption. Instead, the second assumption is "a little more", namely an external constraint, a constraint that can be suspected to express the "physics" of the environment under investigation: it is a type (2) assumption.

But are we thus allowed to think that the second class of assumptions represent "assertions *about physical systems*"? Are we allowed to suppose that the second class of assumptions are *about something completely separate* from the agent that makes normative use of them to guide her bets? That the norms of our agentivity into the "oceanic reality" are not arbitrary, that they are strongly constrained by a factor that exceeds the mere necessity of internal coherence of the set of our gambles, does not entail that this constraint tells us anything about what we are exploring *independently of the activity of exploration*. What is revealed by the type (2) constraints imposed on the norms of our bets and actions, cannot be entirely disentangled from our betting and acting. They inform us that we are not dreaming reality, but they disclose no feature that can be said to belong to a putative reality *entirely distinct from us*. The transcendence has put its recognizable mark on the norms of our actions, but it has not been cut from its roots in the immanence of our lives.

Let us recapitulate. The participatory realist picture of the knowledge process allows one to dissolve the usual "external realist" picture. But it also has enough resources to achieve the dissolution of its own (quasi-dualist) narrative. Acknowledging that we are radically immersed in what we intend to know, acknowledging that we fully "participate" in it, implies that a faithful account of knowledge can be obtained only by adopting the standpoint of the knower-participator. But approaching knowledge from the standpoint of the knower, focusing on the direct experience of what it is like to know, turns the picture of an agent-immersed-in-oceanic-reality into a mere propaedeutics to a radical change of priorities, from a third-person to a first-person stance. In other terms, what the participatory realist conception willy-nilly pushes us to develop, is a thoroughly *phenomenological* approach of knowledge, and in particular of quantum physical knowledge; just as the enactive theory of cognition was a propaedeutics to a phenomenological reconstruction of it.

This being granted, the challenge we now have to meet is to accommodate the realist intuition of "participatory realism" within a phenomenological conception that has the reputation of being thoroughly idealist.[4] Think of Husserl's insistence that the field of *pure experience* is *"the All of absolute being"*, unlike the *objects* of experience, that, since they are only given by incomplete adumbrations, can at most "claim being". Think also of Husserl's thesis that "Nature cannot be the condition for the existence of consciousness, since *nature itself turns out to be a correlate of consciousness*" (Husserl 1983, 116 §51). Is this not one of the boldest statements of idealism in the history of philosophy? Husserl acknowledged his own idealist inclination, but with a momentous nuance. In phenomenology, he wrote, "Idealism is not a metaphysical construction ... but the only possible and absolute truth ... of an *ego* recollecting on itself, on its own doing and its own capacity to give meaning" (Husserl 2007, 48). This is a statement of what we may call Husserl's *methodological* idealism: any claim about the existence of something, ultimately derives its credibility from the contents and structure of experience; as for the "real" existence/inexistence of anything independently of such source of credibility, this is just (once again) a matter of metaphysical speculation.

Even more importantly (for us), Husserl's root-*ego*, together with its lived experience, is not necessarily to be construed as an isolated abstract point facing its object-like intentional correlates. Husserl's reference to the "ego's" *doing*, and capacity to give meaning, may be read as an indirect suggestion that the *ego* should be construed as an embodied and participating agent. This is exactly what he did in the second volume of his *Ideas* (Husserl 1982) and in his *Crisis of the European Sciences* (Husserl 1989). But in what follows, we'll mostly rely on the lineage of French phenomenologists who extrapolated the latter tendencies in Husserl's pioneering work, by pursuing a thorough exploration of the *embodiment* of experience.

Before pursuing the inquiry in this direction, however, we have to acknowledge an obstacle that has hindered it until now. Relying on phenomenology is usually perceived as superfluous or cumbersome by the community of physicists (see Berghofer & Wiltsche 2020; Crease 2020; French 2020). The opponents to QBism are even likely to get the feeling that this amount of ("continental"!) philosophical intricacies is a *reductio ad absurdum* of any interpretation of quantum mechanics that would require it. As for QBist physicists themselves, they may prefer to retain a form of discourse that does not depart too much from what can be easily understood by their colleagues.

But we definitely think the phenomenological option is not just a luxury in the present situation. If we wish to retain the QBist *experiential* dissolution of quantum paradoxes, without remaining caught into a conceptual "omelette" (Jaynes 1990) that mixes up realism, pragmatism, and a touch of phenomenology, the only viable approach consists in trying to make sense of every ingredient of the "omelette" within a unified framework. And since Phenomenology is the only contemporary philosophical research program that does not turn lived experience into some ghostly epiphenomenon, and that takes instead experience as its absolute starting point, we claim that it is the only unified framework suitable for making sense of QBism. At any rate, the attempt is worth making, because it represents one of our best hopes to get out of the quagmire of quantum "paradoxes". If it proves successful, this will confirm that nothing less than a "philosophical revolution" (Healey 2017) is required if we want to make sense of quantum physics. Perhaps an even more radical revolution than Healey himself would be ready to accomplish.

8.5 From Embodiment to En-Worldment

How do we know something we partake of? How can we grasp a reality in which we participate? How can we contemplate that with which we have no distance? What does "knowing" mean, when what is to be known is simultaneously presented *qua* experience, *and* capable (under certain conditions) of experiencing? The paradigmatic ground for testing this list of questions is *our own body*, of which we partake, and which is both experienced and experiencing. No wonder that the most recent phenomenological approaches of a participatory epistemology stem directly from the phenomenology of *embodiment* first developed by Edmund Husserl at the end of his career, and then by Maurice Merleau-Ponty.

To start with, which features distinguish my own body (that I partake of, to the point of *being it* in a way) from standard material bodies (that I can only perceive in outer space)? Four major distinctive features have been identified and described by Husserl and Merleau-Ponty.

The first one is that our own body is the site *from which* any standard material body is perceived and prehended.

The second one is the subject-object reversibility, or double-faced nature, of our own body, that differs from the purely object-status of standard material bodies.

The third one is that our own body immediately moves at will, whereas standard material bodies can only be moved mediately.

And the fourth one is the exceptional mode of presentation in experience of our own body, to a large extent different from the perspectival presentation that characterizes standard material bodies.

About the first feature, Husserl (1982, 223) writes: "Things appear under such and such a facet; and in this mode of appearance is included … the relation with a 'here' and with its fundamental directions. … The own-body then possesses this distinctive trait, unique in its kind, that it carries within itself the *point zero* of all orientations". Notice that this fundamental feature of our own body is revealed as a mirror effect of the mode of presentation of standard material bodies. At any given moment, standard material bodies (or "things") are presented only by one facet or one profile; therefore, one is bound to acknowledge the situatedness of "that wherefrom the facet is seen".

Beware at this point: according to Husserl, such mode of presentation by facets or profiles is essential to the concept of a thing, rather than just accidentally connected with our particular relation with it. Indeed (unlike a realist), a phenomenologist does not say that things have several facets simultaneously available in space, and that (unfortunately) we discover them as we go along instead of seeing them all at once. Instead, a phenomenologist considers that the mode of donation by incomplete facets, and the expectation that we have of their completion, is an unsurpassable constituent of the concept of a thing.

But this difference between two interpretations of the facet-presentation of things is irrelevant to the main lesson we are drawing from it. In both cases the perspectival presentation of what we tend to consider as "things" implies that they are perceived *from* a certain site, *relative to* a certain standpoint. It also implies that varying such perspective and standpoint will give rise to the manifold facets of those putative objects.

The intricacies arise when the phenomenological status of our own body is at stake. On the one hand, we can perceive parts of our own body in the space of standard material bodies' presentations; these parts of our own body are perceived perspectivally, almost in the same way as standard material bodies. But on the other hand, the "here" of point zero, which is the core part of our own body, remains an unperceived perceiving origin of perspectives. Yet, this initial distinction between perceived and perceiving parts of our own body must be refined. When carefully attended to and analyzed, even the parts of our own body that can be perceived in the standard thing-like way manifest a capacity to be perceiving.

This introduces us to the second, most crucial, feature of the bulk of the own-body: its double-faced, perceived-perceiving, ability. Such remarkable characterization of the own-body *in* and *qua* lived experience was first formulated by Husserl. When he did so, Husserl distinguished the case of the visual own-body from the case of the tactile own-body. According to him, the visual own-body is pure seeing, since the site of seeing (the eye) is not directly seen. On the contrary, the tactile own-body is both touching and touched. Indeed, whenever we touch something with one of our hands, we simultaneously (or alternatively) feel this hand's being affected by this gesture. Moreover, we can also feel this hand with the other hand. "What I call my seen own-body, is not a seeing-seen, unlike my body *qua* touched body which is a touching-touched" (Husserl 1982, 211).

Unlike Husserl, Merleau-Ponty subsequently downplayed the difference between the visual and tactile modalities, and he emphasized the primacy of the perceiving over the perceived in both of them: "My visual body is certainly an object as far as its parts far removed from my head are concerned, but as we come nearer to the eyes, it becomes divorced from objects .... It is no different, in spite of what may appear to be the case, with my tactile body, for if I can, with my left hand, feel my right hand as it touches an object, the right hand as an object is not the right hand as it touches: the first is a system of bones, muscles and flesh brought down at a point of space, whereas the second shoots through space like a rocket to reveal the external object in its place. In so far as it sees or touches the world, my body can therefore be neither seen nor touched" (Merleau-Ponty 2002, 105). To sum up, the own-body can be both perceiving and perceived, but what can be perceived is by no means the perceiving itself. The fact of perceiving is the core of one's intimate participation in the own-body, whereas the fact of being perceived manifests an act of (necessarily incomplete) distancing from oneself.

Now, when I bring back my attention closer and closer to this core, what I find is ... *nothing:* what I find is just "a quasi-space to which (I) have no access", writes Merleau-Ponty (2002, 105). This is the site of the perceiving. But this is also the site of the first impulse to act. What we call our "will" is a source of motion situated in this inaccessible quasi-space, in this extended point zero to which we identify. We are thus introduced to the third distinctive feature of the own-body: "(our own-body) is the organ of will; it is the one and only object which can be set in motion spontaneously and immediately by my will, and the one and only means to produce the movement of other things" (Husserl 1982, 215). The limits of my body are the limits of those changes that are sensed as immediately originated in the point zero to which I identify. Beyond these limits, changes are experienced either as mediately produced by my body, or as occurring independently of my will altogether.

That the core of our participation in our own body bears the negative characteristics of being un-perceived, un-accessed, un-traceable, has further consequences.

Although its core is nothing more than a blind spot in the field of our experience, our own body remains stubbornly present to us at every moment. This is the fourth feature by which our own body radically departs from the mode of presentation of standard material bodies: its holistic, permanent, and unanalyzed presence, that contrasts with the partial, transient, and sharp presentation of "things". "An object is an object only in so far as it can be moved away from me, and ultimately disappear from my field of vision. Its presence is such that it entails a possible absence. Now, the permanence of my own body is entirely different in kind: it is not at the extremity of some indefinite exploration; it defies exploration and is always presented to me from the same angle. … The presence and absence of external objects are only variations within (the) field of (its) primordial presence" (Merleau-Ponty 2002, 104). I can explore an object that is detached from me, see it from various angles, and even get away from it, to the point of making it disappear. But my own body, the body I intimately partake of, is a compulsory presence self-perceived from a single angle, not to say from no angle at all. Exploring (parts of) my body from various angles becomes possible only at the cost of its *self-splitting* into the core and non-core fractions of it.

This remarkable situation of *embodiment,* of inextricable participation to our own body, has been progressively extended by Merleau-Ponty to our participation to the *world.* We may accordingly call the broadened participation he suggests a case of "*en-worldment*". Such extension from body to world has been sketched first in Merleau-Ponty's *Phenomenology of Perception*, and then developed, with more dramatic metaphysical undertones, in his *Visible and Invisible*. In Bitbol (2020a, 2021) Merleau-Ponty's extrapolation of the characteristics of the own-body to the characteristics of the world, and its consequences for a phenomenological understanding of quantum mechanics (Merleau-Ponty 1995) were already pointed out. Here, we will briefly come back to the forerunners of the concept of en-worldment in Merleau-Ponty's *Phenomenology of Perception,* before we ponder the post-Merleau-Pontian development of a full-blown "cosmo-phenomenology".

First of all, let us notice that Merleau-Ponty's bold generalization, which extends our embodied mode of being to the cosmos as a whole (thus taking the momentous step from embodiment to en-worldment), is a special case of a non-standard mode of reasoning that was used by a few heroic thinkers of the past when they were trying to meet the challenge of the "hard problem" of consciousness by coming back to its very source. The standard method of scientific knowledge consists in deriving singular cases from a general proposition, after the general proposition has been extracted (or assumed) as an invariant of a set of particular situations. But some thinkers soon realized that, in virtue of its very definition, such scientific method is powerless to tackle the issue of phenomenal consciousness, of pure experience, that is, of this exceptional feature which occurs only in the first person singular. They then decided to turn the method upside-down. Despite the apparent logical invalidity of such move, and against the standard direction of inference, they attempted to derive general propositions about consciousness from a singular first-person observation. The reason they had to do so, is that adopting the first-person standpoint is the only way to avoid overlooking lived experience in an inquiry about consciousness.

A good example of this reverted approach (Bitbol 2020b) was given by Pierre Teilhard de Chardin, a French paleontologist and thinker of the middle of the twentieth century. According to him, the proper way to disclose the significance of phenomenal consciousness, or pure experience, is precisely to "*discover the universal under the exceptional*". Teilhard de Chardin's argument in favor of this non-conventional strategy is both simple and strange: "Consciousness appears with complete evidence *only* in humans, we were tempted to say; so it is an isolated case of no interest to science. Consciousness appears with *evidence* in us humans, one should rather say; therefore, *seen in this single flash, it has a cosmic extension*" (Teilhard de Chardin 1955, 52). The exceptional fact of our own lived experience cries out for ascribing it a universal import.

At any rate, in those cases where acquaintance is the most appropriate form of knowledge, a singular piece of evidence is the only available basis for inquiry. Accordingly, the usual neglect of isolated facts in a science that strives too quickly and too abstractly for universality, turns out to be the most challenging of the obstacles it poses to the investigation of such cases. Now, embodiment is among the situations in which the only way to know (part of) the world is by being acquainted with it. No wonder that Merleau-Ponty took the singular evidence of being embodied as our strongest basis for any further inquiry about the world: a basis that cries out, here again, for generalization. Several features initially ascribed to the body were then found by him to characterize the world as a whole. More precisely, they were found by him to characterize the world *qua* capable of including living bodies.

This being granted, Merleau-Ponty granted the inaccessibility to knowledge of our body's core a cosmic counterpart, and a cosmic significance: "(Our own body) is not merely one object among the rest, which has the peculiarity of resisting reflection and remaining, so to speak, stuck to the subject. Obscurity spreads to the perceived world in its entirety" (Merleau-Ponty 2002, 232). In so far as the own-body has a knowing but unknown face, the entire world that encompasses this body is haunted by this *lacuna*. The world as a whole must be ascribed a knowing but unknown core, by extension of this own-body that partake of it. "There can be no question of describing perception itself as one of the facts thrown up in the world, since we can never fill up, in the picture of the world, that gap which we ourselves are, and by which it comes into existence for someone, since perception is the 'flaw' in this 'great diamond'" (Merleau-Ponty 2002, 241). The separation between the perceiving and the perceived we impose within our body by varying the focus of our attention, indeed extends to the entire world. According to Merleau-Ponty, it is not correct to say that our body is the locus of the perceiving, whereas the whole world is the perceived. One should rather say that the world *qua* capable of including our own body, imposes on itself a variable limit between its perceiving side and its perceived side. This is what Merleau-Ponty stated more explicitly in his *Visible and Invisible,* when he gave the name "flesh" to the stuff that has the double-faced quality of being both perceiving and perceived, and when he finally declared that "the world *is* flesh" (Merleau-Ponty 1964, 182).

This reflection of Merleau-Ponty culminated in the thesis that perception is not an apprehension of the world by a subject, but the consequence of a self-splitting of the world into subject-like side and an object-like side. In line with this thesis, we access the world in two ways: the object way, and the own-body way. On the one hand, in accordance with the object way, we access the world facet after facet, and not all at once; this sequential unfolding expresses the successive steps of the self-splitting of the world within our own moving body. On the other hand, in accordance with the own-body way, the world manifests as a massive, permanent and indivisible presence perceived from no angle at all. "From the very start I am in communication with one Being, and one only, a vast individual from which my own experiences are taken, and which persists on the horizon of my life as the distant roar of a great city provides the background to everything we do in it" (Merleau-Ponty 2002, 382). The keyword, here, is "background". The world is not facing me; it mostly manifests as the "vast" background of all my life. The unity of the world is not obtained after a process of synthesis of its manifold facets, as one does for objects; it precedes any effort to theorize the cosmos, as a permanent ambiance of everything we do in it. "The world has its unity, although the mind may not have succeeded in inter-relating its facets and in integrating them into the conception of a geometrized projection" (Merleau-Ponty 2002, 181).

Beware at this point, however. To repeat, the world according to Merleau-Ponty is no substantial entity, no metaphysically "external" being that exceeds experience. It should rather be understood as an experienced guiding thread of the flux of lived experience. In-the-world-ness arises as an experience of being immersed, as an experience of partaking of a process greater than our individual lives, as an experience of counting on something sturdy beyond our own failures to master it. "The natural world is the horizon of all horizons, the style of all possible styles, which guarantees for my experiences a given, not a willed, unity underlying all the disruptions of my personal and historical life" (Merleau-Ponty 2002, 385).

A style, like an atmosphere or an ambiance, is no *object* of experience; it suffuses experience, just as a *basso continuo* suffuses the melody; and it serves as a prospective and retrospective binder to fill in the temporal gaps of sequences of experienced contents. As for a horizon, it is an experience of indefinite promise, that relies on a generalized anticipation of future experiences. From a phenomenological standpoint, an object is composed of a perceptive nucleus surrounded by a horizon of expectations. And, since the world promises the discovery of a host of unseen objects, it surrounds the present experience of seeing objects with a higher-order horizon of expectations. *Qua* horizon of horizons, the world is what we *experience as if* it were "the inexhaustible reservoir out of which things are drawn" (Merleau-Ponty 2002, 401). This fits with what has been said earlier, provided we remember that a "thing", in this phenomenological context, is itself to be understood as nothing else than a series of experiences whose endless and sometimes unexpected unfolding makes it *look* transcendent.

To sum up, here, neither the nature of the world nor the nature of its objects is fundamentally different from the nature of lived experience, although they both transcend any local, momentaneous, individual content of experience.

Such is the starting point of Renaud Barbaras, in his attempt to draw the ultimate consequences of Merleau-Ponty's concept of the *world's flesh*: that "*there is nothing more in the world than its appearing*" (Barbaras 2019, 83). Barbaras borrows this uncompromising conception of phenomenology from Jan Patočka, and this serves as a preparation of his inquiry about the phenomenological concept of "world". Barbaras considers, after Patočka, that the phenomenological *epoché* must be pushed to a point of completion, beyond what Husserl and Merleau-Ponty did. The suspension of judgments and interpretations must be pressed to a point where none of the old divisions of epistemology subsists, not even the difference maintained by Husserl between consciousness (with its flux of lived experiences) and its intentional objects, a residue of the difference between subject and object. After this relentless *epoché*, nothing else remains than "the 'there is' as such, in its neutrality between subjective and objective" (Barbaras 2019, 83): a "there is" equated with "there appears".

Among the concepts that are dissolved by this exhaustive *tabula rasa,* we even find Merleau-Ponty's concepts of flesh and own-body. Barbaras thus criticizes Merleau-Ponty's concept of *flesh* because it is applied in two different ways to the body and to the world, thereby conveying a persistent form of dualism: "The duality of consciousness and object, which had to be overcome, is finally maintained and divides the flesh" (Barbaras 2019, 11). Barbaras also criticizes the concept of *own-body* because it bears the indirect mark of a pre-phenomenological attitude. Indeed, the very fact of using, for the experiencing own-body, the same word ("body") as for a standard experienced object, shows that Merleau-Ponty could not completely free himself from the "natural" attitude of non-phenomenologists, and from their "natural" ontology of material bodies.

So, instead of characterizing the fundamental experience of being "embodied" by broaching on half-phenomenological and half-biological *bodies,* Barbaras advocates so suspend any reference to such entities, and to undertake a more faithful and more direct description of the experience itself. As soon as the noun "body" has been dropped, the so-called experience of "embodiment" is more aptly described by verbs that express the situatedness of the experiencer (where "situatedness" means being aware both of holding the *point zero* position in the field of perception, and of being in proprioceptive continuity with what is perceived). In this context, two crucial verbs are: "to belong", and the weaker "to be embedded". "How to qualify, in a minimal way, this fundamental experience which is recklessly re-captured through the concept of the body? There is no other answer than to assert that this experience is one of *belonging*. Indeed, saying that I have (or that I am) a body is tantamount to saying that *I belong to the world*" (Barbaras 2019, 13).

Beware again and again at this point. Remember that what a phenomenologist means by "world" is no collection of intrinsically existent objects, but rather an experienced presence, a lived sense of the overwhelming "there is", a "style of styles" a "horizon of horizons". Remember also that the "world", here, can be nothing more than one of the two shadows cast by the verb "to belong" (and by the lived state of "belonging"). "We" are those who belong, and the "world" is nothing else than that to which we belong. That there can be no true phenomenological distinction between the one who belongs and that to which she belongs, that they are inextricably entangled through the very process of belonging, is clearly stated by Barbaras (2019, 15): "Thinking of *belonging* as an original fact or as the ultimate mode of being of someone, therefore amounts to affirming that *her 'other' (the world) is still her*, that she deploys what she fits into, or makes it be. In this sense, belonging must be understood as *participation*". Here again, there is no question of an ontological discontinuity between the embodied subject, her lived experience, and the world she belongs to: the world is deployed through one's lived state of belonging; and "participation" is the name given to such co-emergence of subject and world in the lived process of belonging. This is a new sense of "participation", purely phenomenological in so far as it entirely unfolds within the dynamics of lived experience. This is a sense of participation that does not imply a preliminary division of what is participating from that to which it participates, but directly expresses their radical cohesion within the unique *continuum* of lived experience. We'll see that such sense of participation is precisely what is needed to accommodate experience into the concept of "participatory realism" which is at work in QBism.

Barbaras then distinguishes between three modalities of the constitutive relation of "belonging", according to its time-orientation: present, past, and future. In French, these three modalities are stated in a few short words that are not so easy to translate: "Être *dans* le monde, être *du* monde, être *au* monde" (Barbaras 2019, 27).

"Être dans le monde" means being spatially *located* in the world, being somewhere now, occupying at present a particular place marked in a coordinate system.

"Être du monde" means *partaking of* the world, having inherited a stuff and dynamics which are the same as that of the world, having deep ontological roots in the world.

"Être au monde" means being *concerned by* the world and what it may bring in the future, i.e., being ready to face its novelties and to act on/in it.

Let's repeat tirelessly that using the noun "world" in this ambience is not a way to refer to a big "thing" awaiting us "out there"; it is just a convenience of language for expressing a lived awareness of belonging. "World" is nothing else than the threefold correlate of the threefold relation of belonging that characterizes our experienced situatedness. The world is but what we feel to be located in, to partake of, and to be concerned with.

This repeated remark has precise consequences for the interpretation of each aspect of the relation of belonging.

When a phenomenologist says we are located in the world, this does not mean that our body occupies a little volume in a pre-existing outer space; this rather means, conversely, that our sense of being situated radiates, from within it, a representation of space. The fact that the world looks "external" to us does not reduce to the fact that our small biological body is encompassed in the big body of the world; it rather means, conversely, that our lived sense of finitude, our feeling to be narrowly located somewhere, translates easily into the representation of an external world.

Similarly, when a phenomenologist says that we *partake of* the world, this does not mean that the world is an exhaustive material body from which our limited own body springs (or from which the molecular constituents of our biological body are borrowed). This rather means that we have the experience of being overwhelmed by something bigger than our individual selves, which has the capacity to support us and make us emerge at every moment. In this case, the world is not our material resource: it is the name we give to our awareness of not being self-sufficient *qua* individuals.

Finally, when a phenomenologist says that we are *concerned* by the world, this does not mean that there is a pre-existing cosmos we have reasons to beware of or to hope something from. This rather means that we experience a sense of lack, of incompleteness, of desire, that requires from us a capacity to act, and to anticipate somehow the outcome of our action. In this case, the word "world" is to be understood in two ways. "World" is the name we give to what could *a priori* satisfy our desire, and compensate in the course of time for our sense of incompleteness, without providing us with the certainty that things will turn as we would have liked. "World" is also the name of what retrospectively appears to us as a byproduct of our compensating moves "in it". "Belonging to the world means to advance in it, to go towards it, and to make it be (appear) by this very commitment; in short, to *participate* in its work of *worlding*" (Barbaras 2019, 28). Belonging to the world, participating in it, acting in/on it, is coextensive to making it appear, i.e., to "phenomenalizing" it. According to Barbaras (2019, 38), "any *belonging* to the soil of the world is a cosmophany". Living and moving as an en-worlded being lets the world-phenomenon arise. This is the principle of what Barbaras calls a "cosmological phenomenology", or in short a "cosmo-phenomenology".

However, just as the word "body" has spurious naturalist connotations, far from the strict suspension of judgment about standard objects required by phenomenology, the words "cosmos" and "world" irresistibly evoke the old concept of a circumscribed totality accessible to the gaze of some supramundane being. This is why we had to pepper this section with reminders and warnings about the non-conventional meaning of the term "world" in phenomenology. As Merleau-Ponty (2002, 385) pointed out, "belief in the thing and the world must entail the presumption of a completed synthesis": the synthesis of every actual and possible phenomenon into a holistic entity called "the world", *of which* these phenomena are supposed to be appearances. To fit with the deflationary concept of world that has been favored by the "cosmo-phenomenologist", it is then essential to play down the word "world", and to retain a minimal sense of it, similar to the "style" or "horizon of horizons" we discussed above.

Such minimal sense of "world" can be developed further by relying on the careful analysis of "ambience" offered by Bruce Bégout (2021). What Bégout pursues in this analysis is an unprejudiced attention to every aspect of lived experience: not only the recognition of salient objects but also what accompanies it, the atmosphere and emotional tonality in the midst of which objects are apprehended (and out of which they are extracted). To help him in his inquiry, Bégout borrows from Husserl an aspect of his phenomenology that is usually overlooked: the observation that "experience does not consist of a solitary face to face between subject and object, but first of all designates an original situation referring to what Husserl calls '*Umgebung*', literally 'peri-donation', or donation of a periphery" (Bégout 2021, 38). This periphery of experience does not appear in the same mode as a quasi-point-like object, of course, but it retains a form of extended and non-specific presence. What takes on a decisive importance in such experience is then no longer its directional relation to some object (Husserl's intentionality), but the non-specific, non-directional, feeling of *belonging* to a broader and somehow indefinite environment. Just as in Barbaras, a phenomenology developed along these lines "assumes the primacy of *belonging* over relation" (Bégout 2021, 40). But here, to make full sense of this primacy of belonging, what is undertaken is not a cosmo-phenomenology; it is an "*eco-phenomenology* of ambience". Unlike a cosmo-phenomenology, an eco-phenomenology does not indulge itself in a renewed discourse about the world as a whole, under the guise of an exhaustive description of experience. Instead, an eco-phenomenology tries from the outset to "respect the *immersive* mode of being of the living, and put forward the '*taking place in*' rather than the '*relationship with*'" (Bégout 2021, 39). Belonging is a mode of being; it is not a relation with "Being".

8.6 QBism as Quantum Eco-Phenomenology and Radical Participatory Empiricism

At this point, it seems we have gathered enough philosophical tools to overcome the QBist epistemological dilemma. QBists can follow the experiential thread throughout, without betraying their minimal realist intuition that phenomena are not merely fabricated by individual subjects of experience. QBists can put forward their conviction that physics has something to do with the world, while not forgetting Bohr's lesson that what physics organizes is nothing else than *human experience*. QBists can make good sense of the fact that the world has surprises in store for agents, and still accept (Pienaar 2020) that Agent and World represent *internal aspects of experience*.

But there is a condition to this peaceful coexistence. If QBists want to reconcile their interest for the world and their recognition that knowledge entirely develops on the ground of experience, they have no choice but to adopt the phenomenological, experiential, acceptation of the concept of "world" offered by Merleau-Ponty (1964) and Barbaras (2016). If they want to accommodate these two apparently unrelated sides of their thesis, they must accept, after Merleau-Ponty and Barbaras, that "world" is the name we give to an experienced sense of transcendence with respect to our individual finitude, instead of presupposing (as the dominant naturalist doctrine does) that experience must occur within an inherently transcendent, pre-given, world. In other (Jamesian) terms, QBists can harmonize the pragmatist, realist, and radical empiricist components of their philosophical outlook, but only on condition of entirely subordinating the pragmatist and realist aspects of their thesis to the radical empiricist aspect.

We will give a name to this condition. We will call it "en-experiencing". QBists should *en-experience* the body, the instrumental prostheses, and the world. But what does "en-experiencing" mean? It means putting body, prostheses, and world in ontological continuity with the only status of experimental outcomes that does not create "quantum paradoxes", namely the *experiential* status. En-experiencing the world is the correlate and the converse of en-worlding our experience, just as en-experiencing the body would be the correlate and the converse of embodying experience. En-experiencing the world and en-worlding experience are the two directions (from world to experience, and from experience to world) of a single endeavor of founding a non-dualist approach to quantum knowledge, to physical knowledge, and to knowledge as a whole.

En-worlding experience is tantamount to realizing that experience is by no means severed from what it is experience of, that it is permeated by a sense of being-in-the-world and being-with-others. En-worlding experience is tantamount to realizing that experience is haloed by a "horizon of horizons" that holds the promise of its unlimited development. Conversely, by en-experiencing the world, one recognizes that experience permeates what it is experience of, by way of (experienced) desire, (experienced) action, and (experienced) anticipation of the consequences of action. Recognizing the living body as an "own-body" in Merleau-Ponty's sense, and the experimental apparatus as a prosthesis that extends the own-body, are two steps of this process of en-experiencing. But these are not (and should not be) limiting steps, since, as we previously suggested, a proper synthesis of the two tendencies of QBism in a phenomenological framework requires to complete en-experiencing, by extending it to the allegedly "external" parts of the world, called "physical systems". A good example of this latter move was provided by Asher Peres (1995). In his book, Peres combined an en-experiencing of physical systems with an en-worldment of the physicist's experience, when he defined a "physical system" as an equivalence class of experimental preparations, and the "state" of the "physical system" as the set of *expectations* that derive from such preparations according to a physicist's prior knowledge. Indeed, "preparation" can be understood as an act performed (and experienced) by an embodied, en-prosthesized, agent; and a "physical system" arises as a horizon of expectations of this process of embodied experience, thus exemplifying the en-experiencing of the world.

This all-pervasive role of experience sounds strange, and outrageously idealist, only if one forgets its richness. Experience encompasses a sense of novelty of the perceived, a network of expectations, and a horizon of fulfillment or deception of such expectations, whose abundance has nothing to envy that of a putative "real outer world".

To better understand what is at stake, one may notice that the systematic en-experiencing of every question about the world compares with what Fine (1996) called Einstein's "en-theorizing" of questions about physical reality. Adopting a strategy of "en-theorizing" amounts to considering that the only proper way to address the issue of "physical reality" is to transform it into an investigation of the empirical success of theories that *pretend* to provide us with a "description of reality". In other terms, the strategy of en-theorizing systematically deflects the standard metaphysical concern about the "correspondence" of a physical theory to physical reality, onto the plane of the *empirical adequacy* of the theory. Here, no theoretical entity is considered to exist *a priori*, independently of both the role it plays within the theory, and the empirical credentials of the theory as a whole. Speaking as if such theoretical entities were "real", as if they existed "out there", only makes sense as a "façon de parler" justified by the inner structure of the theory that include them, and by the finding that this theory is globally corroborated by experiments.

En-theorizing is just one step short of en-experiencing, since experience includes both a capacity of anticipating events by way of theoretical structures, and a collection of the empirical findings (with their confirmation or disconfirmation of expectations) that follow the interventions of the experiencing agent. Here, speaking as if physical systems were "real", as if they existed "out there" to be manipulated by agents, only makes sense as a "façon de parler" that is justified by the inner structure of the experience of such agents, and by the experienced success of the actions they perform under the assumption that these systems "exist".

This being granted, it is time to show in some details the deep connection between the QBist understanding of quantum mechanics and the adoption of an en-worlded standpoint that manifests through our experience of "belonging". This connection is so strong that it justifies calling QBism a scientific form of "eco-phenomenology" in Bégout's sense.

In Section 8.5, we listed (though in a different order) four features that testify of our embodiment, and that can be extended to our en-worldment: holding a point zero position, being double-faced (feeling and felt), observing limitations to the perspectival presentation of what there is, and finally having an experience of free will. Let's consider these four features in turn, within the context of QBism.

To start with, the impossibility of eliminating from quantum theory the *de facto* privilege of the point-zero position, is perhaps the major discovery of QBism. QBism was born as a reaction to the fact that any attempt at reading quantum theory as a description of the world "from nowhere" has been, and still is, an inexhaustible source of conceptual intricacies and paradoxes. The lesson QBism learnt from this failure is that each quantum experiment, each feature of the quantum formalism, must be rethought as a "user's manual" for situated beings, as a guide to act "from somewhere". Thanks to this kind of "user's manual", situated beings orient themselves efficiently in an environment (*Umwelt*) of phenomena that co-emerge with their embodied actions. This ineliminable situatedness, this "belonging" of quantum agents, takes on the three forms documented by Barbaras (2019, 27): "Être dans le monde, être du monde, être au monde".

"Être *dans* le monde" points toward the spatio-temporal location of agents. In QBism (and in Rovelli's Relational Quantum Mechanics as well), the ascription of a state vector, or the probability valuation used by agents, depends crucially on their spatio-temporal location. It is precisely through the dependence of state vectors on the spatial position of agents, inside or outside the laboratory, that QBism and RQM are able to defuse the many varieties of Wigner's-friend-like paradoxes. And it is through the dependence of state vectors on the spatio-temporal position of two agents, inside or outside their respective light-cones, that QBism dispels the spurious "non-locality" usually associated with quantum entanglement. If one did not take the spatio-temporal position of agents thoroughly into account, no such dissolution of quantum paradoxes would be available.

"Être *du* monde" is the second modality of our "belonging". It refers to our sense of partaking of the world, of being rooted in its soil, of being enmeshed in it. In QBism, this is presumably what is expressed by the adjective "participatory" that qualifies "realism", but that (according to the analysis of the previous sections) should rather qualify "radical empiricism". In view of the role QBism ascribes to experience, QBism should then adopt a form of *"radical participatory empiricism"* instead of "participatory realism".

We have seen that participation is given such a strong meaning by QBism that, in its most advanced formulations, the very idea that the agent inter-acts with the world is replaced by the idea that there occurs an "intra-action" within the world, the standard split between agents and physical systems thus making flickering appearances wherever the world intra-acts. Instead of saying that experimental events are co-created in the process of the inter-action between agents and physical systems, one should then say that phenomena emerge, or are newly created, at each step of an intra-action. This formulation is still compatible with "participatory realism". But, if pushed to its ultimate consequences, it should be changed into the "radical participatory empiricist" statement that intra-action does not take place in an objectified world but *qua* phase of a quest of knowledge within the lived world. The latter process is what Barbaras (2019, 19) calls the "*phenomenalization*" of the world in the course of its becoming (a becoming that involves intra-actions). More briefly, this process is what Barbaras calls a self-produced "*cosmo-phany*".

"Être *au* monde" is the third modality of "belonging". It denotes our being *concerned* with our world, and with the way our world may undergo manifest (experienced) changes. Such concern involves both the initial vague worry that motivates our desire to know, and our focused expectations about the "cosmo-phany" to come. In QBism, this last modality of "belonging" completely determines the status that is ascribed to state vectors and probabilities. State vectors and probabilities are meant to address *our concerns*, by providing us with a bundle of expectations about future intra-active creations, namely about future experimental phenomena.

We have just seen the QBist threefold counterpart of the cosmo- and eco-phenomenology of "belonging". But what about the double-faced constitution of our "flesh"? Does anything correspond to the duality of the feeling and the felt, beyond the boundaries of our own-body? Our answer is a prudent "yes". One can indeed extend limitlessly the double-faced constitution of the flesh, provided the latter is substituted with its topological structure. Let's remind Merleau-Ponty's functional characterization of the double-face: "In so far as it sees or touches the world, my body can be neither seen nor touched" (Merleau-Ponty 2002, 105). The most essential difference between the touching and the touched, the feeling and the felt, the perceiving and the perceived, is thus not a difference of nature, but a difference of position and orientation within the attentional field. The perceiving is "wherefrom" perception arises; it lies in the background of the attentional field, and it is oriented toward a potential object of experience. Instead, the perceived is projected in front of the attentional gaze; it is "that towards which" attention is directed.

This difference of position and direction between the "wherefrom" status and the "towards which" status, was retained by Heidegger (1962) as a basis for his phenomenological distinction between tools and objects. What holds a "wherefrom" status (the tool) has the "*Zuhandenheit*" (readiness-to-hand) mode of being, whereas what holds a "towards which" status (the object) has the "*Vorhandenheit*" (presence/in-front-of-hand) mode of being. The hand itself, in some sense is "ready-to-hand" because it is permanently available to itself for directional action; it usually remains in the background of the field of attention, and it acts to manipulate or capture an object of attention. But the tools and the prostheses can acquire the same status when one becomes so accustomed to them that they are no longer objects of attention, but rather operate as means toward an end.

An example of perceptual tool and prosthesis is the "enactive torch" (Froese et al. 2012). The enactive torch is a sensory substitution implement that uses an ultrasonic sensor to measure distances to obstacles, and then translates these spatial data into a more or less intense vibration that can be felt by people holding the device in their hand. After an initial phase in which the user just perceives a vibration in her hand, a shift takes place and the user has the feeling of directly palpating walls and doors while losing the sense of vibration. The status of the enactive torch has changed from object to (perceptive) tool, from present/in-front-of-hand to ready-to-hand, from "towards which" to "wherefrom", and then, in a sense, from perceived to perceiving.

This kind of analysis was applied by Pienaar (2020) to the measuring apparatuses of laboratories of quantum physics. Indeed, such analysis is central for making sense of the QBist claim that "the apparatus is to be regarded as an essential part of the agent", "on a par with the perceptual organs of the agent". And it is therefore crucial to avoid the age-old dilemma of standard quantum mechanics as to *where* one should locate the "cut" between the quantum, object-like, part of the measurement chain, and the classical, tool-like, part of the measurement chain. But doing so, accepting that the tool status and the own-body boundaries can be shifted freely, up to a putative contact with what is to be explored, means that one is ready to extend indefinitely the domain of validity of the concept of our "flesh", thereby coming closer and closer to Merleau-Ponty's statement that "the world is flesh". In other terms, one thereby accepts that the feeling-felt double face is a basic feature of the world *qua* experience, rather than a special feature belonging only to a spatially bounded fraction of this world called "a living body". By so extending the flesh-status of our being embodied, QBism *de facto* endorses the broadest possible vision of our being "en-worlded" beyond our being merely "em-bodied".

Let's notice at this point that our being en-worlded accounts for certain well-known limitations of the manifestations of the world. It especially accounts for the lack of relevance of the perspectival model in quantum physics.

The perspectival presentations of a standard bodily object are mutually compatible, in so far as (i) the chronological order of its appearances does not affect them and (ii) nothing prevents one in principle to let them show simultaneously, in a single compound manifestation (one that uses, say, a set of mirrors).

By contrast, the presentations of what we partake of (such as our body) depart from this perspectival model. We cannot capture our own body all at once in a single manifestation, and under a single gaze. Moreover, in the case of our body, the self-perception mode of presentation is exclusive of the perceiving mode. The first one requires a phenomenological language of reflection to be described, whereas the second one is correctly apprehended by everyday object-oriented language. Interestingly, this mutual exclusivity of modes of perception and languages is a fundamental feature of the concept of "complementarity" Bohr applied to quantum observables (Heelan 1977). Mutual exclusivity is precisely the feature that makes "complementarity" of observables incompatible with perspectivism. This being granted, we can interpret Bohr's complementarity as the mark left by our en-worldment on what we (mis-)take for our knowledge of the world.

Last but not least, what about free will? We claim that having the experience of free will is a constitutive feature of our situation of "belonging", in its general form of "en-worldment". This was suspected long ago. Perhaps the first were early medieval authors (e.g., Boethius, 1999), who examined the differences between the ability to foresee from a human standpoint (within the world), and from God's standpoint (able to rise above the world). Later on, Spinoza and Kant tended to rephrase and confirm this claim, each in his own language.

Immanuel Kant is well known for having formulated the thesis that we are free according to practical reason, whereas our behavior obeys deterministic laws according to theoretical reason; or that we are noumenally free, whereas we are phenomenally determinate. Kant's dual approach of human freedom has served as the root of Arthur Schopenhauer's (2014) celebrated *World as Will and Representation*. But it has been criticized by several post-Kantian philosophers, especially by Charles Renouvier (2013) who is often quoted by Christopher Fuchs. Kant's view of human freedom has thus been rejected on the basis that, by dint of trying to remain compatible with the scientific determinism of classical mechanics, it turns out to be a weak or false defense of free will.

Yet, if properly understood, Kant's dual approach of human freedom is by no means tantamount to accepting some "real" determinism, on the basis of (classical) scientific determinism taken as a revelation of some absolute truth. On the contrary, Kant's underlying thesis is that (classical) science is nothing more than an *as if* mode of knowledge (Vaihinger 2021). According to Kant, (classical) science is a mode of knowledge that works *as if* there were objects ruled by deterministic "laws of nature", and *as if* we were passive spectators of their motion according to these laws. This *as if* ability of classical science arises from the imposition, by human understanding, of an ordering of phenomena that may hold for every rational subject. Such pre-ordering of phenomena thus generates a form of epistemic objectivity that is too easily mistaken with the discovery of some ontic stuff.

The fact that our phenomenal bodies appear to be ruled by deterministic laws is then by no means a *reductio ad absurdum* of free will. It does not reveal that we are truly, ontically, determinate, but only on the "as if" mode. It only shows that, in the framework of classical science, the epistemic approach of ourselves from a spectator-like standpoint is bound to submit our phenomenal bodies to a deterministic law. Now, according to Kant, what is more fundamental than this spectator's description of ourselves, is our status as actors of our own deeds, our status of beings who act in tacit accordance to the idea of their freedom (Beck 1963). By contrast, the so-called spectator's standpoint (from which it appears that our behaviors are determinate) is available to no concrete individual human being; it is artificially fabricated out of the coordination of a multitude of actor's standpoints sedimented in our intersubjective "pure understanding". "In view of our insuperable entanglement with what there is, the standpoint of a spectator of nature is extrapolated out of the only available standpoint, which is that of the actor" (Bitbol & Osnaghi 2016). Free will is therefore fundamental, being a necessary feature of the most fundamental standpoint we can adopt: the standpoint of an *en-worlded* actor; the standpoint of an actor thoroughly enmeshed with what she acts upon.

The reason why the experience of free will is a necessary feature of an actor's standpoint is also quite instructive. This is because, unlike an external spectator, an engaged actor cannot *in principle* capture the whole field of what is to be known all at once, in a single manifestation, and under a single gaze. A true actor cannot deploy all what is to be known under her gaze, if only because *her own body* is part of what is to be known. Such constitutive inaccessibility of part of the world to an embodied and en-worlded being is precisely what Merleau-Ponty (2002, 232) meant when he evoked the "obscurity (that) spreads to the perceived world". Irrespective of whether or not an allegedly objective world that includes our own body is ruled by deterministic laws, we, en-worlded agents, have no other option than acting under the presupposition of our free will. Indeed, the determinating factors of our actions, if any, are bound to remain hidden to ourselves in this region of obscurity which is the core of ourselves. And such hiddenness is neither accidental, nor provisional, since any attempt at unveiling the productive core of ourselves would have to be done from somewhere else that would then assume the status of another core, of another blind spot, of another "obscure" source of spontaneous actions.

The previous analysis of the organic connection between en-worldment and free will is highly relevant to quantum physics. Many connections have been established in the past between quantum physics and human free will. Some of them consist in the (highly dubious) claim that quantum indeterminism is the *natural basis* of free will (Jordan 1944). Other connections are less trivial, however. Implicitly sharing the spirit of Kant's analysis, they discard the idea that free will is the expression of some underlying natural process, and they consider instead that free will arises as a standpoint-relative presupposition and experience.

The most remarkable approach of the latter kind is likely to be Peres and Zurek's (1982). These authors first formulate three demands about physical theories: (1) strict determinism, (2) verifiability by free choice of experimental set up, and (3) descriptive universality. Then, it appears that, taken together, the three conditions are incompatible. Indeed, if a theory is descriptively universal and determinist, it is bound to deny that the decision as to which experimental set up one uses to test it, is *really* free. Viable theoretical options should therefore retain only two demands among the three previous ones. Among those viable options, quantum theory overtly satisfies demands (2) and (3), whereas it rejects demand (1). In other terms, quantum theory presents itself *prima facie* as a descriptively universal theory whose indeterminism makes it compatible with the demand of free choice of the experimental set up. But, when carefully analyzed, the quantum configuration turns out to be trickier and more interesting. To address the measurement problem, the descriptive universality of quantum theory must be qualified somehow. An arbitrary fraction of the observing must be excluded from the quantum description of the observed. Peres' and Zurek's conclusion is that, "although quantum theory is universal, it is not closed. Anything can be described by it, but something must remain unanalyzed". In other (shorter) terms, "although it can describe *anything*, a quantum description cannot include *everything*" (Peres & Zurek 1982).

Here, all the pieces of the Kantian puzzle of free will are in place; yet they are loosely adjusted. Indeed, quantum indeterminism comes together with the recognition that "something must remain unanalyzed"; but no strong relation of entailment is established between these two statements. The reason of this limitation of Peres' and Zurek's approach can easily be found: it is their repeated (but almost tacit) assumption that quantum theory is meant to afford some kind of *description* of the universe. If we now deny quantum theory any descriptive status, and rather consider that quantum symbols are meant to be purely predictive, the situation becomes much more tractable and even closer to Kant's. Indeed, in this case, the fact that some of the factors that influence what is to be predicted are in principle out of sight of the predictor (because they coincide with the predictor herself), immediately accounts for a predictive form of indeterminism (Popper 1988). (Predictive) indeterminism is not only *associated with* the *en-worldment* of the actor-predictor; *it is an unavoidable consequence of it.* More specifically, the presupposition and experience of free will (construed as self-unpredictability) is an unavoidable consequence of the actor-predictor's "obscurity" to herself due to her embodiment and en-worldment.

It would have been surprising that the most comprehensive and coherent predictive interpretation of quantum mechanics to date, namely QBism, failed to acknowledge this strong knot that binds quantum indeterminism, free will, and "participancy".

The importance of free will in the philosophical background of QBism is suggested by the 210 occurrences of the expression "free will" in Fuchs (2015). But why is free will so central in the QBist approach?

One possible reason is that QBism chooses to take the agent as its most fundamental undefined primitive notion. The agent is dealt with as a primitive notion because she does not (and should not) belong to the set of objects of the theory. The agent is not endowed with either properties or theoretical "states" (that would mostly pertain to her putative past). Instead, she is characterized by her pre-theoretical abilities to act and predict (by which she anticipates on her future). Her past is left in the blind spot of her process of knowing, since she is entirely focused on the acts by which she can trigger future events, and on the possibility of making her ready for these future events. In other terms, her past is dealt with meta-theoretically, whereas theoretical issues entirely pertain to her future.

Yet, Fuchs was not entirely convinced by Howard Barnum's insistence on his option to take the agent as a primitive: "You don't like Everett's resolution because you want to have an unanalyzed primitive around, so that it can be the locus of free will. And I say: 'it is not that'" (Fuchs 2015, 573, 1041). But then, is there another motivation for free will's being so highly praised by QBists? Here it is: "The universe has within its categories two species, one is chance, and one is free will. Free will does not rely on chance as its source. Instead, it's only through the intercourse of the two that we get a real birth" (Ibid.). Free will is said to be one of the two necessary ingredients (together with chance) of the "real birth" of radically stochastic events. Therefore, assuming "real creation" or "real birth" as the most elementary building block of the quantum realm, is a sufficient reason to posit free will.

The problem is that Fuchs' analysis, in the quoted sentence, is utterly dualist, just as much as the picture of agents intervening on "physical systems". Here, we have "free will" on the side of the subject, "chance" on the side of the object, and "real birth" as a result of the interaction of both. If we wished to settle directly in the conclusions of QBism, instead of relying on its false premises, we should take "real birth" as our unique starting point, and then wonder how such starting point can be described on the two faces of an en-worlded "intra-action". This being granted, free will and chance would reduce to the shadows casted by a "real birth" on these two co-emergent faces. And *we would thus confirm the former lesson according to which it is the en-worldment itself which co-produces free will and chance, as a consequence of leaving part of what is to be predicted in that which Merleau-Ponty called the "obscurity" of the predictor.* Here, unlike in physicalist/naturalist approaches, no reduction of free will to chance is implied.

One can hear a perfectly clear adumbration of this non-dualist thesis when Fuchs writes (though reluctantly): "Chance is what you call 'it' when viewed from the outside; free will is what you call 'it' when viewed from the inside" (Fuchs 2015, 574). If "it" identifies with the neutral, or inter-facial, "real birth", the previous sentence fits reasonably well with our previous monistic reading.

The problem is that "inside" and "outside" are themselves remnants of the dualist picture, and that they irresistibly tend to be identified, respectively, with the subject's consciousness and the object of the theory (here quantum theory). Is this standard distribution of roles not misleading? Fuchs points out, in agreement with the founding principle of QBism, that "What we call quantum theory has sadly been misidentified all these years as a 'description from the outside', when in fact it is almost completely a 'description from the inside'" (Fuchs 2015, 1172). The remark is sound, but then, the very fact of maintaining the opposition of the inside and the outside despite the collapse of the traditional locus of a description "from the outside" (namely the physical theory), becomes disputable. If physical theory, this discipline that pretends to afford a "description from the outside", has undergone such a radical mutation that it becomes akin with a "description from the inside", then the very outside-inside opposition lacks a proper foundation. Two major changes in vocabulary should be adopted in this situation: (1) since nothing is "described" by quantum symbols, since quantum symbols only purport to afford a rule to bet about event-like "real creations", one should rather speak of an "anticipative orientation from the inside" and (2) No "description from the outside" being available any longer, no "inside" is available either, since "inside" is defined only in relation and opposition to "outside". So, instead of speaking of an "anticipative orientation from the inside", one should rather use a more neutral expression inspired from the new wave of phenomenology, such as "anticipative orientation from a situated en-worlded experience".

This quantum configuration in which one *cannot even behave as if* a description from outside were available, completely escapes the Kantian scheme of determinism and free will. As we mentioned earlier, Kant (2013, 2014) ascribed the (classical) deterministic laws of *phenomena*, to physics working *as if* it were providing us with a picture from outside (from the standpoint of a spectator); as for free will, it pertained to the *noumena,* and to the standpoint of an actor (of an insider). A standard extension of this scheme would consist in ascribing the indeterministic laws of microscopic phenomena to quantum physics (mis)taken for a description from outside, and free will to an approach of the same processes from inside. But when the physical basis of any alleged "descriptions from the outside" is missing, when physics itself pertains to the standpoint of an actor, free will and chance become united in a single "creative" moment: a moment of intra-action within an en-experienced world. Free will and chance are just the reflective and intentional correlates of one and the same lived novelty. Free will and chance are just the right side and the reverse side of the "obscurity" zone of one and the same en-worlded experience of acting.

8.7 Conclusion

The birth of quantum mechanics has been an exceptional occasion for physics to turn its self-understanding upside-down. Unfortunately, this "Quantum revolution in philosophy" (Healey 2017) was postponed for almost one century, in favor of a multifarious attempt to maintain or restore the classical epistemological assumptions that had been initially challenged by Bohr (Osnaghi 2017), Heisenberg, and even Schrödinger (Bitbol 1996b). Marking the end of this interlude, the recent onrise of QBism represents a renewed attempt to finally make sense of Bohr's insight, by radicalizing it and pushing its tendencies to an unprecedented point of coherence:

- Founding a science in the first person that be compatible with its "as if" third-personal features.
- Founding a science entirely developed from the standpoint of the user of science, from the standpoint of the experiencer-agent, and yet accounting for the most highly regarded value of science, namely the objectivity of its rules and propositions.

This is a thrilling project, which has been pursued by the founders of QBism with continuity, consistency, and a sense of responsibility toward its philosophical consequences that command admiration. But, as we pointed out in Section 8.2 of the chapter, even this revolution is not entirely accomplished. The QBist attempt remains partly dependent of a pre-quantum naturalist and dualist epistemology, although, as we saw in Section 8.3, it defends an uncompromisingly phenomenological monist conception of the quantum symbols and of the experimental outcomes they tend to anticipate. This is why, in the last three sections of the chapter, every aspect of the QBist conception of knowledge and nature has been reconsidered according to phenomenological standards. In Section 8.4, we dissected the remnant "realist" component of QBism to isolate from it a demand that is both sufficient to address the accusation of solipsism, and still compatible with a purely first-person approach such as that of phenomenology. In Section 8.5, we summarized the latest developments of the phenomenology of embodiment after Merleau-Ponty, and focused on its non-standard, non-objectified, concept of "world" *qua* virtual unfolding of lived experience. This analysis culminated in the mirror-like correspondence between: (i) extending embodiment to an en-worldment of experience and (ii) bringing the world back to our experience "of it", a move that we called "en-experiencing the world". Finally, in Section 8.6 we recapitulated the characteristics of en-worlded experience as stated in recent phenomenological research. We then showed that the major characteristics of the en-worldment of experience remarkably fit with major QBist claims, and that enforcing other characteristics of en-worldment of experience would make QBism even more consistent with its own philosophical options. Having thus ascertained its feasibility, the task that awaits us is to take this path again in the opposite direction; it is to reformulate the whole of QBism and quantum physics on the basis of our fundamental situation as revealed by phenomenology: that of a present experience which opens on a world-like horizon of possible future experiences.

Notes

1 This work was supported by the Agence Nationale pour la Recherche (ANR-16-CE91-0005-01).
2 That Husserl was not oblivious of embodiment, community, and the sense of being-in-the-world can be seen in his *Ideen II* and *Krisis*. More on this later.
3 An "anhypothetical" principle is one that is not freely chosen (as an axiom would be), but is incontrovertible (see Plato, *Republic,* 6, 511b3). Some "anhypothetical" principles are deemed to be incontrovertible in virtue of the impossibility of denying them without presupposing them (see Aristotle, *Metaphysics 4,* 1005b).
4 That Husserl's mature philosophy is *narrowly* idealist is disputable, however. It is true that "at the heart of phenomenology lies a claim according to which the phenomenal stream of lived experience (*Erlebnisstrom*) derives its meaning and its being from itself, rather than from some external or underlying reality". But here, the insistence on lived experience should not be understood as a choice of the subjective sphere against the objective world. It is rather a decisive option in favor of a non-dualist ontology from which the two poles of the theory of knowledge, and the intentional directedness that unite them, are derived (Blouin 2021).

References

Aerts, D. & Sozzo, S. (2014), "Entanglement Zoo II: Examples in Physics and Cognition", In: H. Atmanspacher, E. Haven, K. Kitto, D. Raine (eds.), *Quantum Interaction 2013*, Lecture Notes in Computer Science, vol 8369, Berlin: Springer

Barad, K. (2007), *Meeting the Universe Halfway: Quantum Physics and the Entanglement of Matter and Meaning*, Durham: Duke University Press

Barbaras, R. (2016), *Le désir et le monde*, Paris: Hermann

Barbaras, R. (2019), *L'appartenance: Vers une cosmologie phénoménologique*, Leuven, Peeters

Beck, L. W. (1963), *A Commentary on Kant's Critique of Practical Reason*, Chicago, IL: The University of Chicago Press

Bégout, B. (2021), *Le concept d'ambiance*, Paris: Éditions du Seuil

Berghofer, P. & Wiltsche, H. A. (2020): "Phenomenological Approaches to Physics: Mapping the Field", in H. Wiltsche & P. Berghofer (Eds.), *Phenomenological Approaches to Physics*. Synthese Library, Berlin: Springer

Bitbol, M. (1996a), *Mécanique quantique, une introduction philosophique*, Paris: Flammarion

Bitbol, M. (1996b), *Schrödinger's Philosophy of Quantum Mechanics*, Dordrecht: Kluwer

Bitbol, M. (2010), *De l'intérieur du monde*, Paris: Flammarion

Bitbol, M. (2014), "Quantum Mechanics as Generalized Theory of Probability", *Collapse*, 8, 87–121

Bitbol, M. (2020a), "A Phenomenological Ontology for Physics: Merleau-Ponty and QBism", in: H. Wiltsche & P. Berghofer (eds.), *Phenomenological Approaches to Physics*, Berlin: Springer

Bitbol, M. (2020b), "La conscience comme origine et comme fin: une déduction teilhardienne du singulier à l'universel", *Noosphère,* Mai 2020, 48–65

Bitbol, M. (2021), "Is the Life-World Reduction Sufficient in Quantum Physics?", *Continental Philosophy Review*, 54, 563–580

Bitbol, M. (2022), "The Roles Ascribed to Consciousness in Quantum Physics: a Revelator of Dualist (or Quasi-Dualist) Prejudice", in: Shan Gao (ed.), *Quantum Mechanics and Consciousness*, Oxford: Oxford University Press

Bitbol, M. & Luisi, L. (2004), "Autopoiesis with or without Cognition: Defining Life at Its Edge", *Journal of the Royal Society (Interface)*, 1, 99–107

Bitbol, M. & Osnaghi, S. (2016), « Bohr's Complementarity and Kant's Epistemology », In: O. Darrigol, B. Duplantier, J.M. Raimond, & V. Rivasseau (eds.), *Niels Bohr 1913–2013*, Reinach: Birkhäuser

Blouin, P. (2021), *La phénoménologie comme manière de vivre*, Bucharest: Zeta Books

Boethius (1999), *The Consolation of Philosophy*, Oxford: Oxford University Press

Bohr, N. (1987), *Essays 1958–1962 on Atomic Physics and Human Knowledge*, Woodbridge: Ox Bow Press

Brukner, Č. (2020), "Facts Are Relative", *Nature Physics*, 16, 1172–1174

Carnap, R. (1967), *Logical Structure of the World*, London: Routledge and Kegan Paul

Cassirer, E. (1969), *The Problem of Knowledge. Philosophy, Science and History Since Hegel*, New Haven: Yale University Press

Caves, C., Fuchs, C., Manne, K. & Renes, J. (2004), "Gleason-Type Derivations of the Quantum Probability Rule for Generalized Measurements", *Foundations of Physics*, 34, 193–209

Crease, R. (2020), "Explaining Phenomenology to Physicists", in H. Wiltsche & P. Berghofer (Eds.), *Phenomenological Approaches to Physics*. Synthese Library, Berlin: Springer

Crease, R. & Sares, J. (2021), "Interview with Physicist Christopher Fuchs", *Continental Philosophical Review*, https://doi.org/10.1007/s11007-020-09525-6

De la Tremblaye, L. (2020), "QBism from a Phenomenological Point of View: Husserl and QBism", in H. Wiltsche & P. Berghofer (Eds.), *Phenomenological Approaches to Physics*. Synthese Library, Berlin: Springer

De la Tremblaye, L. & Bitbol, M. (2022), "Towards a Phenomenological Constitution of Quantum Physics: a QBist Approach", *Mind and Matter*, 20, 35–62

DeBrota, J., Fuchs, C. & Schack, R. (2020a), "Respecting one's Fellow: QBism's Analysis of Wigner's Friend", *Foundations of Physics*, 50, 1859–1874

DeBrota, J., Fuchs, C., Pienaar, J. & Stacey, B. (2020b), "The Born rule as Dutch-Book coherence (and only a little more)", arXiv:2012.14397v1 [quant-ph]

Destouches, J. L. (1942), *Principes fondamentaux de physique théorique*, Paris: Hermann

Destouches-Février, P. (1946), "Signification profonde du principe de décomposition spectrale", *Comptes Rendus de l'Académie des Sciences*, 222, 867–868

Destouches-Février, P. (1951), *La structure des théories physiques*, Paris: Presses Universitaires de France

Fine, A. (1996), *The Shaky Game: Einstein, Realism, and the Quantum Theory*, Chicago, IL: The University of Chicago Press

Fink, E. (1994), *Proximité et distance*, Grenoble: Jérôme Millon

French, S. (2020), "From a Lost History to a New Future: Is a Phenomenological Approach to Quantum Physics Viable?" in H. Wiltsche & P. Berghofer (Eds.), *Phenomenological Approaches to Physics*. Synthese Library, Berlin: Springer

Friedman, M. (2009), "Einstein, Kant, and the Relativized a priori" in M. Bitbol, P. Kerszberg and J. Petitot (Eds.), *Constituting Objectivity: Transcendental Perspectives in Modern Physics*, Berlin: Springer

Froese, T., McGann, M. & Bigge, W., et al. (2012), "The Enactive Torch: a New Tool for the Science of Perception", *IEEE Transactions on Haptics*, 5, 365–375

Fuchs, C. (2002), "Quantum states: what the hell are they?", http://www.physics.umb.edu/Research/QBism/WHAT.pdf

Fuchs, C. (2010), "QBism, the perimeter of Quantum Bayesianism", arXiv:1003.5209v1 [quant-ph]

Fuchs, C. (2015), My Struggles with the Block Universe, arXiv:1405.2390v2 [quant-ph]

Fuchs, C. (2017), "On Participatory Realism". In I. Durham and D. Rickles (eds.), *Information and Interaction: Eddington, Wheeler, and the Limits of Knowledge*, Berlin: Springer

Fuchs, C. (2019), "Quantifying QBism", Private communication

Fuchs, C. & Peres, A. (2000), "Quantum Theory Needs No Interpretation", *Physics Today*, March 2000, 70–71

Fuchs, C., Mermin, N. D. & Schack, R. (2014), "An Introduction to QBism with an Application to the Locality of Quantum Mechanics", *American Journal of Physics*, 82, 749–754

Fuchs, C. & Stacey, B. (2020), "QBians do not exist", arXiv:2012.14375v1 [quant-ph]

Gibson, J. (1986), *The Ecological Approach of Visual Perception*, New York: Routledge.

Harré, R. (1997), "Is There a Basic Ontology for the Physical Sciences?", *Dialectica*, 51, 17–34

Harré, R. (2006), "Resolving the Emergence-Reduction Debate", *Synthese*, 151, 499–590

Harré, R. (2013), "Affordances and Hinges: New Tools in the Philosophy of Chemistry", in: J.P. Llored (ed.), *The Philosophy of Chemistry*, Cambridge: Cambridge Scholars Publishing

Harré, R. (2014), "New Tools for the Philosophy of Chemistry", *Hyle*, 20, 77–91

Healey, R. (2017), *The Quantum Revolution in Philosophy*, Oxford: Oxford University Press

Heelan, P. (1977), "*Hermeneutics of the Experimental Science in the Context of the Life-World*", Research Resources, Fordham, (http://fordham.bepress.com/phil_research/10)
Heidegger, M. (1962), *Being and Time*, New York, NY: Harper & Row
Husserl, E. (1960), *Cartesian Meditations*, Leiden: Martinus Nijhoff
Husserl, E. (1973), *The Idea of Phenomenology*, Leiden: Martinus Nijhoff
Husserl, E. (1982), *Recherches phénoménologiques pour la constitution (Idées directrices pour une phénoménologie et une philosophie phénoménologique pures, livre second)*, Paris: Presses Universitaires de France
Husserl, E. (1983), *Ideas Pertaining to a Pure Phenomenology and to a Phenomenological Philosophy*, Leiden: Martinus Nijhoff
Husserl, E. (1989), *The Crisis of European Sciences and Transcendental Phenomenology: An Introduction to Phenomenological Philosophy*, Evanston: Northwestern University Press
Husserl, E. (2007), *De la réduction phénoménologique*, Grenoble: Jérôme Millon
James, W. (1912), *Essays in Radical Empiricism*, Cambridge, MA: Harvard University Press
Jaynes, E. T. (1990), "Probability in Quantum Theory", in W. H. Zurek (ed.), *Complexity, Entropy and the Physics of Information*, Santa Fe Institute Studies in the Sciences of Complexity 8, Boston, MA: Addison-Wesley
Jordan, P. (1944), *Physics of the Twentieth Century*, New York, NY: The Philosophical Library
Kant, I. (2013), *Kritik der reinen Vernunft,* 1781, translated as *Critique of Pure Reason*, Cambridge: Cambridge University Press
Kant, I. (2014), *Principiorum Primorum Cognitionis Metaphysicae Nova Dilucidatio,* 1755, translated as *A New Elucidation of the Principles of Metaphysical Cognition*, in: I. Kant, *Theoretical Philosophy 1755–1770*, Cambridge: Cambridge University Press
Kleinman, A. (2012), "Intra-Actions", *Mousse Magazine*, 34, 76–81
Maturana, H. & Varela, F. (1980), *Autopoiesis and Cognition*, Dordrecht: Reidel
Meling, D. (2021), "Knowing Groundlessness. An Enactive Approach to a Shift from Cognition to Non-Dual Awareness", *Frontiers in Psychology*, https://doi.org/10.3389/fpsyg.2021.697821
Merleau-Ponty, M. (1964), *Le visible et l'invisible*, Paris: Gallimard
Merleau-Ponty, M. (1995), *La nature, notes. Cours du collège de France*, Paris: Éditions du Seuil
Merleau-Ponty, M. (2002), *Phenomenology of Perception*, London: Routledge
Osnaghi, S. (2017), "Complementarity as a Route to Inferentialism", in J. Faye & H. Folse (eds.), *Niels Bohr and the Philosophy of Physics: Twenty-First-Century Perspectives*, London: Bloomsbury
Peres, A. (1978), "Unperformed Experiments Have No Results", *American Journal of Physics*, 46, 745–747
Peres, A. (1995), *Quantum Theory: Concepts and Methods*, Berlin: Springer
Peres, A. & Zurek, W. H. (1982), "Is Quantum Theory Universally Valid?", *American Journal of Physics*, 50, 807–810
Pienaar, J. (2020), "Extending the Agent in QBism", *Foundations of Physics*, 50, 1894–1920

Pienaar, J. (2021), "Unobservable entities in QBism and phenomenology", https://arxiv.org/pdf/2112.14302.pdf

Pitowsky, I. (1997), "Infinite and Finite Gleason's Theorems and the Logic of Indeterminacy", *Journal of Mathematical Physics*, 39, 218–228

Popper, K. (1988), *The Open Universe: An Argument for Indeterminism*, London: Routledge

Quine, W. V. (1969), "Epistemology Naturalized", in: W.V. Quine, *Ontological Relativity and Other Essays*, New York, NY: Columbia University Press

Quine, W. V. (1974), *The Roots of Reference*, Chicago, IL: Open Court

Renouvier, C. (2013), *Esquisse d'une classification systématique des doctrines philosophiques II (éd. 1885)*, Paris: Hachette/Bibliothèque Nationale de France

Rorty, R. (1981), *Philosophy and the Mirror of Nature*, Princeton, NJ: Princeton University Press

Ryckman, T. (2023), "QBism: Realism about what?", *This volume*

Schopenhauer, A. (2014), *The World as Will and Representation*, Cambridge: Cambridge University Press

Schrödinger, E. (1951), *Nature and the Greeks*, Cambridge: Cambridge University Press

Serban, C. (2016), *Phénoménologie de la possibilité*, Paris: Presses Universitaires de France

Smerlak, M. & Rovelli, C. (2007), "Relational EPR", *Foundations of Physics* 37, 427–445

Stacey, B. C. (2021), "Ideas Abandoned en Route to QBism," arXiv:1911.07386, 2019

Teilhard de Chardin, P. (1955), *Le phénomène humain*, Paris: Plon

Vaihinger, H. (2021), *The Philosophy of 'As If'*, London: Routledge

Varela, F., Thompson, E. & Rosch, E. (2017), *Embodied Mind: Cognitive Science and Human Experience*, Cambridge, Massachusetts: MIT Press

Von Uexküll, J. (2010), *A Foray into the Worlds of Animals and Humans*, Minneapolis, MN: University of Minnesota Press

Zwirn, H. (2021), "Is QBism a Possible Solution to the Conceptual Problems of Quantum Mechanics?", in: O. Freire (ed.), *Oxford Handbook of the History of Interpretations and Foundations of Quantum Mechanics*, Oxford: Oxford University Press

Part III
Supplementary Approaches

9 Putting Some Flesh on the Participant in Participatory Realism

Steven French

9.1 Introduction

QBism began life as 'Quantum Bayesianism' but with the distinction between Bayesian approaches to probability in general and subjectivist interpretations in particular, the name is now taken as a 'stand-alone' label (see Healey 2017; Stacey 2019).[1] The core feature of this position is that it offers 'an interpretation of quantum mechanics in which the ideas of *agent* and *experience* are fundamental' (De Brota & Stacey 2018). As a result, the wave-function should be understood solely in epistemic terms, as representing not some state of a physical system but rather that of the agent with regard to their possible future experiences.[2] It does this by encoding the agent's coherent degree of belief in each of certain alternative experiences that result from some act they perform, such as the outcomes of a measurement procedure.[3] Probability theory then acts as a normative constraint on an agent's beliefs at a given time, with additional assumptions (such as Bayes' Theorem) then invoked to capture what is taken to be the best way of updating such beliefs.

As a result, from the QBist perspective, the measurement problem simply dissolves: the observation of an outcome becomes nothing more than the acquisition of new information, leading to the reassignment of the 'state' and since that simply expresses the agent's credences, or degrees of belief, any discontinuity between the old 'state' and the new amounts to nothing more than the updating of such credences.[4] Concordance between the credences can be established via appeal to the usual devices, such as arguments that show that updating different prior probabilities in the light of new but common information will lead to convergence of the 'posterior' probabilities (although, as is well-known, there are issues with such devices; see, for example, Talbott 2016).

However those beliefs are updated, QBists adopt a subjectivist interpretation of probability and the question now is how to understand the

DOI: 10.4324/9781003259008-12

probabilities yielded by quantum mechanics, which are (apparently) objective. These are given by the Born Rule, which relates such probabilities to the square of the amplitude of the relevant wave-function.

According to Earman, QBists *could* give a straightforward answer to this question in the form of a proof that demonstrates that quantum probabilities are just 'objectified' forms of subjective probabilities (Earman 2019).[5] Unfortunately, he alleges, QBists decline to take advantage of this and, partly as a result, create 'faux difficulties' for themselves and 'fail to convey some of the strengths of their stance' (2019, p. 404). However, Earman has been accused of fundamentally misunderstanding QBism and, indeed, of begging the question (Fuchs & Stacey 2020, p. 1). In particular, appealing to features of the quantum mechanical formalism to ground the Born rule and relate subjective probabilities to the apparently objective quantum forms puts the cart before the horse, as it '... ultimately ends up re-objectifying what had been initially supposed to be subjective probabilities' (ibid., p. 3).

What Earman failed to appreciate is that according to QBism these features of the theory should not be regarded as representational, in the sense of describing how potential events are related, with the Born rule arising as a consequence; rather the latter should be understood as primitive and one of the aims of the programme is then to recover the formal structure of the theory from that basis (Fuchs & Stacey 2020, p. 3). In other words, according to Fuchs, '[t]he interpretation should come first, with the mathematical structure of the theory derivative from it' (in Crease & Sares 2020, p. 558).

As a result, it is not only the measurement problem but also other issues facing quantum mechanics that must be viewed in a new light.

9.2 Schrodinger's Cat

So, consider the example of Schrödinger's Cat: according to Earman, the QBist sidesteps the issue of whether the cat should be described as alive or dead before the box is opened because all she is concerned with is the assignment of degrees of belief to the relevant propositions about what will be found (2019, pp. 415–416). As a result, given that for the QBist the 'collapse' of the wave-function is nothing more than a change in the mathematical representation of an agent's degree of belief upon updating with the new information about the measurement outcome, she cannot *explain* why the agent experiences a definite outcome. And although this issue can be avoided as long as we think of the agent as nothing more than a disembodied probability calculator fed information by an 'oracle', it comes back to bite us once we think of ourselves as '... physically embodied observers ... whose information acquisition has to be treated quantum mechanically in terms of an interaction with the (measurement apparatus + object system)' (ibid., p. 416).

Of course, the QBist may insist that explanation is not the name of the game as far as they are concerned – all they are interested in is their own personal experiences and how they can be related via the probability calculus. However, for Earman, this suggests a form of 'solipsistic phenomenalism':

> [t]hat at least would be an interesting position. Most [QBists] deny that what they are aiming for is phenomenalism. But their subjectivist interpretation of quantum states deprives them of the resources to tackle questions about the relation of agents to a non-phenomenalistic world. Trying to make a virtue out of this seems a stretch.
>
> (Earman 2019, pp. 416–417)

In response, QBists have insisted that this characterisation fundamentally misrepresents their approach to measurement, not least because, '[t]he trinary decomposition **object + apparatus + agent** simply does not exist in QBism. QBism is all about the agent and her external world—the decomposition is a binary one' (Fuchs & Stacey 2020, p. 9). In particular, whereas other interpretations articulate the measurement process in terms of the apparatus becoming *entangled* with the system and the observer then becoming entangled with the joint system thus formed, such a 'story' contradicts the core tenets of QBism by virtue of ascribing a quantum state to the observer. Instead of thinking of the measurement apparatus as a further 'system', possessing its own state and becoming entangled with the observer, who likewise is ascribed a state, it should be regarded as merely an *extension* of the observer-as-agent, or a part of her as an individual.

Thus, Pienaar has argued that,

> QBism accommodates the idea that a sufficiently practiced scientist using an electron microscope to measure atoms might be said to literally 'sense atoms' and not merely be making inferences about them as abstract or hypothetical entities.
>
> (Pienaar 2020, p. 1918)

Furthermore, by virtue of certain measurements being lost and others being gained, with regard to the system, this 'extension' of the agent amounts to a shift in the boundary between the agent and the world: 'The World has thus shrunk by losing a System, but the Agent has grown in gaining an Apparatus' (Pienaar 2020, p. 1912). Such an extension is regarded as an act of 'free postulation', in the sense that there is no external criterion in terms of which it can be determined to be 'correct' or not. And the justification for this is embodied in von Neumann's principle: quantum mechanics does not prescribe where that boundary should be drawn, only that it must be drawn somewhere, as determined by considerations that lie out with the theory itself.[6]

This should make it clear just how wide of the mark is the suggestion that we should conceive of the engagement of the observer with the measurement apparatus in terms of entanglement, at least so far as QBism is concerned:

> QBism indeed regards agents as embodied; how could a disembodied entity take physical actions and experience consequences? The argument that because agents are embodied their interactions with the world must be treated as the generation of entangled states simply presumes its conclusion.
>
> (Fuchs & Stacey 2020, p. 9)[7]

9.3 Wigner's Friend

Such a shift in view offers an entirely new perspective on, and treatment of, the 'Wigner's Friend' thought experiment (presented as central to the development of QBism in De Brota et al 2020, p. 1860).

We recall the basic set-up: the 'friend' is in a room with a system and measurement apparatus. Wigner is outside (or simply turns his back on his friend), until the friend undertakes the measurement and then the two compare notes. Wigner then asks his friend what he saw before Wigner entered the room or turned around. It is argued that his friend will reply expressing a definite outcome, since '... the question whether he did or did not see the [dead/live cat] was already decided in his mind' (Wheeler and Zurek *op. cit.*, p. 176).[8] Since the issue as to what he saw was already decided in his friend's mind before he (Wigner) returned to the set-up and asked, Wigner concluded that the state immediately after the interaction between his friend and the system cannot be in a superposition – that is, the wave-function must have 'collapsed', in some sense – and hence consciousness must play a different role in quantum mechanics than an inanimate measuring device.

As far as QBism is concerned, however, this is a story about two agents, with the measurement apparatus treated as an extension of the friend's body (see Fuchs 2010, p. 6). A measurement is then understood in terms of the agent acting on the given system and as represented formally via a set of operators, with the action having a partially predictable consequence for the agent, namely a particular measurement outcome. Now, as already noted, the agent will entertain certain degrees of belief about such consequences, where these degrees of belief are captured formally by the quantum state, 'existing', if that could be said, in the agent's head and not in the world. Furthermore, and importantly such a consequence is a '... unique creation within the previously existing universe', albeit not subject to the agent's 'whim and fancy' (ibid., p. 6).

Once the measurement has been made, the question may be asked (by some) – what is the 'correct' quantum state that each agent should have assigned to the system? For the 'friend' it will be a definite state, either |spin up> or |spin down>, say. But what about Wigner? Regarding his friend and the box as just another isolated quantum system, and before interacting with him, Wigner would of course assign an entangled quantum state to this combined system from which the state of the system could then be extracted (using a partial trace operation). However, it would not be that assigned by his friend. Who, then, is correct?

For the QBist this question simply makes no sense, presuming as it does an agent-independent notion of 'correctness' (Fuchs 2010, p. 7). Again, as far as she is concerned, quantum states are entirely personalistic and do not represent anything 'out there'. Hence, the information gained should not be understood as 'about' some mind-independent reality but rather has to do only with the consequences of the agent's actions upon the system. However, this is not to fall into some form of extreme idealism: '... for the QBist, the real world, the one both agents are embedded in—with its objects and events—is taken for granted. What is not taken for granted is each agent's access to the parts of it he has not touched' (ibid., p. 7). As far as Wigner is concerned, with regard to his interactions with his friend, or the system, or both, he should gamble on the consequences of these interactions according to the prescription given by quantum theory. This prescription will be different as far as the friend is concerned and as long as we appreciate that difference, the QBist maintains, there is no conflict (see also De Brota et al. 2020, pp. 1866–1868).[9]

Nevertheless, one might still worry that the issue of establishing intersubjectivity has not been fully addressed and, indeed, the claim that quantum mechanics '... doesn't give one agent the ability to conceptually pierce the other agent's personal experience' (Fuchs 2010, p. 8), may raise the worry again that QBism sails a little too close to a form of solipsism (see, for example Crease & Sares 2020, pp. 545–546). In this regard it is worth noting that the QBist acknowledges that the nature of the agent is left out of this picture – indeed, to expect quantum mechanics to derive the notion of agent is akin to expecting to be able to derive the notion of the user of logic from the formalism itself, or the reader of a probability textbook from its contents – 'How could you possibly get flesh and bones out of a calculus for making wise decisions?' (Fuchs 2010, p. 8). As obvious as this may be and as sympathetic as one might be to this resistance to an unwarranted demand, one might still feel that absent some consideration of the nature of the agent, the QBist picture is incomplete (and indeed, Fuchs acknowledges in several places that there is much more to be said).[10] It is here, of course, that phenomenology may step in, particularly as the QBist insists that it is in precisely this way that quantum mechanics is different from any theory posed before, namely in being simply an addition to probability theory, understood as normative; that is, as a *theory of knowledge* (London & Bauer 1939, pp. 7–8; London 1983, p. 220).[11]

However, before we explore this resonance with phenomenology in more detail, we need to consider the issue of how we should understand the QBist's attitude towards the world.

9.4 QBism and the World

We recall the insistence that despite the position's core subjectivist elements, the real world is still 'there', in some sense. Indeed, according to Fuchs '[w]e believe in a world external to ourselves precisely because we find ourselves getting unpredictable kicks (from the world) all the time' (Fuchs 2017, p. 121). Nevertheless, the QBist cannot adopt a realist stance as standardly understood, not least because she denies that certain central features of the theory, namely quantum states and their evolution, correspond to an aspect of external reality (Glick 2021, p. 6). Having said that, the Born Rule *is* understood as objective, in the sense that any agent should use it to find their way in the world, and in that sense it is understood as correlating with something we might want to call 'real' (see Fuchs 2017). The rule expresses a relationship between probabilities associated with different sequences of measurements and together with the claim that the relevant states, their evolution and measurements are also related within the theory in a particular way (something that could only have been discovered empirically), this could be construed as supporting a form of structural realism (Glick 2021, p. 7, fn. 11; see also De Brota et al. 2020, p. 1864). As such, however, the structure that would be posited is metaphysically rather thin.[12] Nevertheless, that doesn't mean that reality, on this account, is rendered 'unspeakable' or 'ineffable', leaving us free to suggest a range of radically different ontologies.

Having said that, QBism's 'first person' approach has encouraged the thought that it should be regarded as a form of 'quasi-idealism' (Glick

2021, p. 8). Such thoughts are further supported by the emphasis on the creative aspect of quantum measurement:

> At the instigation of a quantum measurement, something new comes into the world that was not there before; and that is about as clear an instance of creation as one can imagine.
>
> (Fuchs 2010, p. 19)

One way of reading this is as suggesting '... a metaphysical picture in which we construct the world via our interactions with it' (Glick 2021, p. 10). Of course, not all such constructions are viable as the history of science demonstrates: the embeddedness of the agent in the world means that, for example adopting a classical approach to one's expectations regarding measurement outcomes would lead to disaster.[13] It is the combination of features of the world and features of us, as agents, that make it the case that measurements can be regarded as acts of creation from the perspective of the agent (ibid., p. 11).[14]

Nevertheless, there is tension here: without an appropriate description of the world, what reason do we have for following the normative constraint embodied in the Born Rule (and here we might recall Earman's concerns)?[15] We could just appeal to induction, noting that the rule has been successful in the past, so should be accepted now (Glick 2021, pp. 13–14). But then it is not clear how to cash out this notion of 'success' in QBist terms since measurement outcomes are not objective features of reality but are particular to the agent's perspective. This is rendered all the more acute by the fact that, understood as relational, the Born Rule as it stands does not make any predictions – it needs to be supplemented, either with a quantum state ascription, which is ruled out in QBist terms, or with the probability of another measurement outcome, which again is understood as subjective. Given that, it seems that QBism doesn't have the resources to ground the rule inductively in a way that would provide a compelling reason for all users of the theory.

Instead, QBists appeal to a form of coherence argument: this is a standard move in the subjectivist camp whereby the axioms of probability theory are justified on the grounds that if they're not accepted, a series of bets could be made for which the agent is guaranteed to lose money, regardless of the outcomes. Thus, QBists treat the Born Rule as the subjectivist treats the axioms of probability theory and claim that not following the rule would lead to similar incoherence. However, given that this claim holds only in worlds where quantum mechanics provides a good guide for agents in such worlds,[16] the issue returns: what is it about our world that makes the Born Rule the objectively correct constraint? One option would be to simply take the rule to be a 'brute feature of reality' (Glick 2021, p. 15) that represents, as a constraint, the limit of what we can say about the world. Fuchs himself suggests as much when he states that the rule plays some ontic role (in Crease & Sares 2020, p. 555).[17]

Alternatively, in accordance with the acknowledgment that QBism is an on-going programme, relevant empirical features could be sought that would necessitate the rule. However, given that such features would be manifested via measurement outcomes which, again, are regarded as entirely subjective, it is difficult to see what empirical resources the QBist might draw on.

That would suggest that instead of looking to the world for such support, the QBist should focus on the nature of the agent. Of course, justifying the rule by appealing to the subjective experiences of the observer would be in tension with the suggestion that it has an ontic flavour. That tension might be dissipated by stepping away from such labels as 'realist' and 'idealist' as currently understood, and adopting a phenomenological stance according to which the rule is grounded in our engagement with the 'life-world'.[18] This would then account for its ontic flavour whilst acknowledging its ultimate 'subjective-relative' nature (here we might recall Husserl 1970, p. 126). So, let's now consider how the QBist might draw on certain features of the phenomenological stance in order to philosophically underpin their position. As we'll see, the extent to which such a move can be deemed successful depends on how that stance is conceived.

9.5 QBism and Phenomenology

The suggestion that QBism might find a comfortable philosophical home in phenomenology has been the subject of considerable discussion. Bitbol, in particular, has highlighted three points of connection[19]: the first is the most obvious, perhaps, namely that just as the QBist regards quantum mechanics from a 'first-person' perspective, so phenomenology requires the adoption of the same in order to identify the contribution of consciousness to experience.[20] Thus,

> [t]he project of both phenomenology and non-interpretational approaches to quantum mechanics is to reconstruct a new, self-conscious, type of objective knowledge, starting everything afresh from the first-person standpoint of knowers and agents.
>
> (Bitbol 2020, p. 232)[21]

The second point of contact has to do with the shift in attention that we find, in both QBism and phenomenology, away from the apparently 'external' objects, whether of science or the lifeworld, and towards that contribution of consciousness. It is this shift that marks the phenomenological reduction and just as the latter is driven, methodologically, by the *epoché*, so in QBism we are urged to suspend our judgement with regard to the referential capacity of the symbols of the formalism of quantum mechanics (Bitbol 2021, p. 570). Having noted that, we might wonder if the contact is entirely smooth, given that QBism seems to go further than merely suspending judgment by adopting a stance that is closer to instrumentalism in taking these symbols to ultimately represent the probabilistic weights that agents assign to the outcomes of experiments (Bitbol 2020, p. 232).

The third similarity proceeds from the second: the QBist insistence that quantum mechanics only tells us something about the *expectations* we should have concerning the outcomes of experiments is, Bitbol has argued, similar to Husserl's understanding of perception, based as it is on his conception of 'horizontal intentionality' (see also Bitbol 2021, p. 571). Putting it simply, the idea here is that in perception only part of the perceived object is intuitively given to us but we possess an intentional awareness of the other 'profiles' or adumbrations of the object. Our anticipation of our perception of these profiles can be situated in an open manifold of such anticipations that constitutes what Husserl calls the intentional horizon.

This third point of contact is then taken up by de la Tremblaye who has used the example of our perception of a cup and suggests that,

> [t]he cup is the analogue of the microsystem, the perceptual horizon parallels the QBist quantum state, the perceptual act corresponds to the physicist's measurement and the modification of my possible horizon corresponds to the modification of the state vector after the measurement.
>
> (de la Tremblaye 2020, p. 255)

Thus, just as perception is a matter of updating the horizon of possibilities associated with our present observation of an object, such as a cup or MacBook Air, so the quantum state, on a QBist reading, expresses a 'bundle of expectations' (ibid., p. 254). Before a measurement, then, the relevant eigenstates correspond to anticipated possible profiles and '[t]he (probabilistic) estimates of subsequent measurements are … analogous to estimates of future perceptions, namely the internal perceptual horizon of an object' (ibid.). From this horizon only one possible scenario results, of course, and likewise, on a QBist reading, as we've seen, a measurement outcome is considered a personal experience:

> Observing a trace on a screen or hearing a click from an experimental apparatus is an experience that is analogous to the sensory nucleus of perception, as it is understood within Husserl's phenomenology.
>
> (de la Tremblaye 2020, p. 254)

Given these parallels between sensory perception as understood by Husserl and measurements in quantum mechanics as understood within QBism,[22] the latter can be understood as a phenomenological reconstruction of quantum mechanics (ibid., p. 255).

However, there is a concern that arises with the notion of horizontal intentionality in this context. According to de la Tremblaye, '... the bundle of expectations expressed by a quantum state [according to the QBist] can be understood as the quantum equivalent of what Husserl calls the intentional horizon' (2020, p. 254). This horizon encompasses various anticipated possible profiles; so, for example in the case of a cup about to fall off a table, we can anticipate either that it will break or will remain undamaged but never both (ibid.). Likewise, when we perform a spin measurement, via a Stern-Gerlach apparatus, say, we can anticipate either the outcome 'spin up' or 'spin down' but never both (ibid.). In the former case, our anticipations are based on our past experiences with falling cups and on our understanding of the relevant background conditions (whether the floor is carpeted or not, say) and it is on this basis that the horizon of possibilities is determined. In the case of the spin measurement, likewise, the possibilities are determined by our beliefs, at least as far as the QBist is concerned:

> In establishing the state vector, I express in a formal way my beliefs about the future of my measurements; and these beliefs arise by due consideration of my own past experience (including the experience of preparation). The (probabilistic) estimates of subsequent measurements are thus analogous to estimates of future perceptions, namely the internal perceptual horizon of an object.
>
> (de la Tremblaye 2020, p. 254)

From this horizon only one of the possible scenarios is perceived, giving priority to the role of that present perception in that, by virtue of the perceptual horizon being an integral part of our experience of the cup, say, that perception has a direct effect on the constitution of that cup, by imposing a determination on that horizon. Likewise, again, in the case of the spin measurement, only one outcome is perceived with the experience of the flash on the screen, or the click of the counter, being taken to be analogous to the 'sensory nucleus of perception' in Husserlian terms.

However, there are two worries that arise here. The first is whether we can straightforwardly draw parallels between our everyday experiences, embedded as they are in the 'lifeworld' and those that arise as what Fuchs calls 'kicks' from the world as manifested in the spin measurement. Now, the latter might be described by the *non*-QBist as 'kicks from the world as represented by quantum mechanics' but of course, that would be to beg the question! Other questions also run the risk of being begged if we were to insist on a distinction between the classical realm and the quantum, or the macro-world and the micro-, not least because whether we can draw such a distinction has been a source of contention.

Nevertheless, it is significant that the example given here, of the falling cup, is one with which many of us are reasonably familiar, to the extent that we can claim to have fairly well-formed expectations as to the possibilities in play. Of course, we don't even have to look to cases of quantum phenomena to note that those expectations are based on certain inductive inferences regarding the phenomena in question. And those inferences may well lead us astray – after all, one of the possibilities compatible with (classical) statistical mechanics is that all the air molecules in the room could suddenly be distributed to gather together beneath the cup as it falls, thereby cushioning it and even lifting it back onto the table! Granted, the probability of such an occurrence is incredibly low but now consider the (related) probabilities of different arrangements of particles over available states: calculating such probabilities as we would if considering the 'everyday' distribution of balls between boxes, say – that is, classically – would give us dramatically incorrect results when it comes to the quantum case. There is, of course, more to say here about the relationship between the 'life-world', from which Husserl himself drew his examples to help the reader understand the notion of the intentional horizon, and the idealised 'world' of physics (see Trizio 2021) but still one might well question whether that notion is sufficiently elastic in this respect (but see Alves 2021, pp. 474–475).

Furthermore, one of the differences between 'everyday' and 'quantum' cases, for want of better terms, is that whereas the cup either has to break or not break, such a disjunction may well not hold in the case of spin 'up' and 'down' – indeed, that the range of possibilities should include a superposition of such disjuncts is precisely what lies behind the Schrödinger's Cat thought experiment. Given that, it might be asked, again, how can we draw the relevant parallels here? In this case, we might recall the QBists' response already given: to assume that the cat could be in a state of alive-and-dead or the particle in a state of spin-up-and-down is, again, to beg the question as the QBist will deny the attribution of such states to the cat or particle respectively, insisting that all we have to work with are our personal experiences of a definite outcome, together with the Born Rule, however grounded. It is precisely because of that insistence that de la Tremblaye can draw the comparison that she does.[23]

Nevertheless, there are further concerns that arise from considerations of phenomenology itself. Zahavi, for example has argued that the above notion of the intentional horizon actually requires a certain kind of intersubjectivity in that it '… must be understood as the noematic correlate of the possible perception of an Other' (1997, p. 3). The idea is that it is the perceptions of another that underpin the required correlation:

> When I experience someone, I am not only experiencing another living body situated 'there', but also positing the profile which I would have perceived myself if I had been there ... Thus, my concrete experience of the Other can furnish my intentional object with an actual co-existing profile.
>
> (ibid., p. 3)

However, there is an immediate objection: surely my perception of the cup, say, cannot be dependent upon my simultaneous perception of another subject who is also actually perceiving the cup? Indeed, there would have to be a huge number of such actual subjects, given the number and variety of possible profiles.

Husserl himself was aware of this problem and suggested that this insertion of a form of inter-subjectivity leads to a certain 'openness' in that it invokes the perceptions of numerous possible others. Thus, when I perceive the cup, that object of perception is constituted by me. However, I am '... only able to perform this activity because my horizontal intentionality entails structural references to the perceptions of possible Others' (ibid., p. 4). The reciprocity involved here – in the sense that I must now accept that I am an Other with respect to one of these other perceiving egos – '... implies a dethronement of my own ego as the sole pole of constitution ... and this dethronement has far-reaching constitutive implications' (ibid., p. 5). In particular, objectivity, understood as intersubjective validity, can only be established once that reciprocity is acknowledged and the ego perceives itself to be 'one among Others'. That the world, then, has to be understood as constituted inter-subjectively was acknowledged by Husserl himself:

> ... it is an apodictic transcendental fact that my subjectivity constitutes for itself a world as intersubjective. The other self is therefore a necessary intentional 'object' of the absolutely evident structure of my awareness. Furthermore, this other self is necessarily coequal with my self. My transcendental self, by virtue of its evident structure, perceives itself as without any superiority over the other self. (I am an intentional object for him, as he is for me; he is an absolute constitutive consciousness, as I am.) This is all part of the apodictic facticity of my transcendental subjectivity. It does not depend on the fortuitous constitution of a particular object of valid *Einfühlung* in perception, but is simply an explication of the fact that I do intend a world as necessarily intersubjective. (That is what I mean by calling

> it a world. If it were not intersubjective, it would not be a world). Strictly it is an (open?) infinity of other subjects which is required by the apodictic factual structure of my transcendental consciousness, not one other subject.
>
> (Cairns 1976, pp. 82–83; reproduced in Zahavi 1997, pp. 9–10, fn. 24)

From this perspective, then, it would seem that two interacting agents cannot each consider the other as a 'system' – each has to recognise the other as an 'absolute constitutive consciousness'. There is an obvious tension with QBism here, in that once we extend the phenomenological stance to take account of inter-subjectivity, we appear to lose the 'first-person' perspective that underpins the comparison.

Indeed, Bitbol has acknowledged this when he argues that inter-subjective agreement must be based on a shared acknowledgment of the existence of the ordinary objects of the life-world (2021, p. 572). Hence,

> ... even though QBism is phenomenologically right to claim that the *de jure* basis of scientific knowledge is personal lived experience and verbal communication between subjects of experience, it should also recognize that the *de facto* basis of quantum physics is Bohr's classical-like domain of ordinary objects and instruments.
>
> (ibid.)

Indeed, he claims, the types of experience that feature in the probability assignments that QBism takes to its heart are most conveniently expressed in terms of 'classical-like' predicates regarding the relevant instruments, such as the Stern-Gerlach apparatus mentioned above (ibid.). Thus, what is typically portrayed as the basis of QBism – namely the first-person perspective – must be modified to some extent in order to maintain its alignment with phenomenology.

Further impetus to this modification is given by Bitbol's insistence that, '... when one is asked to *explain* the structure of the quantum probabilistic predictions, one must go beyond the purely subjectivistic option of QBism' (ibid., p. 573). This does not mean simply adopting one or other of the current accounts of scientific explanation, but rather we must acknowledge that the explanation of quantum phenomena lies at the 'interface' between 'outer reality' and 'inner subjectivity', involving as it does both the experimental context and the creativity of the agent (ibid.). Here Bitbol has drawn on the work of Destouches-Février (1951), whose research on the foundations of quantum mechanics is, sadly, not as well-known as it should be.[24] In particular, she had a significant influence on another phenomenologist whose thoughts have also been compared to QBism, namely

Merleau-Ponty. As Berghofer and Wiltsche note, his work goes beyond that of Husserl, not least in containing a detailed analysis of modern physics (Berghofer & Wiltsche 2020, p. 32).

9.6 Quantum 'Flesh'

Merleau-Ponty raised the fundamental question whether the picture of the world that physics presents could include the physicist *qua* observer herself (Berghofer & Wiltsche 2020, p. 33). Quantum physics, he argued, attempts to do precisely this, by placing the relationship between the subject and object in question (ibid.), and can be accommodated by shifting to a phenomenological stance, according to which the physicist is 'intermingled' with the world.[25] He argued that in physics a moment comes when it's very development calls into question the presupposition of an absolute spectator and '... "objective" and "subjective" are recognized as two orders hastily constructed within a total experience, whose context must be restored in all clarity' (1968, p. 20). Such a moment arrived with the advent of quantum mechanics which should be recognised as a physics that situates the physicist physically (!) and '... enjoin[s] a radical examination of our belongingness to the world' (Merleau-Ponty 1968, p. 27; quoted in Berghofer & Wiltsche 2020, p. 33).

In particular, measurement, for Merleau-Ponty, is an *engaged* operation and this is reminiscent of our own situation of embodiment, whereby '... any operation of our own body is an operation within the "flesh of the world"' (Bitbol 2020, p. 239).[26] As a result, he maintained, physics cannot be given the standard realist interpretation, and instead he advocated the partial or 'participationist' realism of Destouches-Février (here once again, of course, one can draw a comparison with QBism; see Berghofer & Wiltsche 2020, p. 33).

'Flesh' here should not be understood as matter, in the sense of 'corpuscles of being' that make up other beings; rather, it acts as '... a sort of incarnate principle that brings a style of being wherever there is a fragment of being' (Merleau-Ponty 1968, p. 139). Within this framework, we must abandon the old assumptions that place the body in the world and the 'seer', or observer, in the body. As Merleau-Ponty asked – and here we can recall again von Neumann's psycho-physical parallelism – '[w]here are we to put the limit between the body and the world, since the world is flesh?' (ibid., p. 138). According to Bitbol, this gives rise to '... an ontology of radical *situatedness*: an ontology in which we are not onlookers of a nature given out there, but rather intimately intermingled with nature, somewhere in the midst of it' (2020, p. 236; Pellegrini 2021, pp. 496–499).[27]

Quantum mechanics, Merleau-Ponty argued, when understood phenomenologically, incorporates such a 'radical situatedness', but, as he went on to make clear, although such a stance acknowledges that the theory represents a 'human physics', it does not reduce to a simple kind of idealism (Merleau-Ponty 2003, pp. 97–98). Instead, it transcends the opposition between object and subject and can be said to be broadly structuralist in character in setting the relations presented by the theory at its heart. Here we recall the suggestion that QBism could be accommodated within a form of structural realism (see also Berghofer & Wiltsche 2020, p. 35). However, the nature of that form depends on how one conceives of these relations. For the QBist they are, fundamentally, embodied in the Born Rule but this not how they were seen by Merleau-Ponty.

Indeed, following Destouches-Février, he asserted that these relations can claim a certain objectivity by virtue of being independent of the measurement process despite being relative to the 'species' of system being studied and refer, not to objects *per se*, but to '… certain mathematical forms that are necessary for the description of the relation of the subject to the object' (Merleau-Ponty 2003, p. 98; see again Berghofer & Wiltsche 2020, p. 33). Having said that, the fact that they are determined by the theory confers on them a form of reality going beyond the simply mathematical.

In this context, Merleau-Ponty drew heavily on the work of London and Bauer (1939 and 1983) who adopted a phenomenological approach to the measurement problem (French 2002, 2020, 2023). In particular, as Merleau-Ponty himself noted, underpinning this new picture offered by quantum physics, is the non-classical relation between measurement and the observed system. Here he emphasised the departure from the classical view of the measurement apparatus which, contra, to what the QBists assert, can no longer be regarded as an extension of our senses: 'The apparatus does not present the object to us. It realizes a sample of this phenomenon as well as a fixation' (Merleau-Ponty 2003, p. 98; see also Berghofer & Wiltsche 2020, p. 34). The very act of measurement 'fixes' the object and makes it appear as an individual existent. Here in particular Merleau-Ponty noted the crucial passage in London and Bauer's analysis where they emphasised 'the essential role played by the consciousness of the observer' in the transition from the superposition to what is taken to be the pure state, in terms of which we characterise a definite result. Looking at that situation from 'outside', as it were, they wrote:

> Objectively – that is, *for us* who consider as "object" the combined system [object, apparatus, observer] – the situation seems little changed to what we just met when we were considering only apparatus and object.
>
> (1939/1983, p. 251)

However, they continued,

> The observer has a completely different impression. For him it is only the object x and the apparatus y that belong to the external world, to what he calls "objectivity." By contrast he has with himself relations of a very special character. He possesses a characteristic and quite familiar faculty which we can call the "faculty of introspection." He can keep track from moment to moment of his own state. By virtue of this "immanent knowledge" he attributes to himself the right to create his own objectivity - that is, to cut the chain of statistical correlations ...
>
> (ibid., p. 252)

This claim as to the possession of a characteristic faculty of introspection has been much remarked upon in the history of discussions over the measurement problem but has not generally been understood correctly as indicative of a phenomenological stance (again, see French 2002, 2020, 2023). Crucially, London and Bauer went on to say that,

> ... it is not a mysterious interaction between the apparatus and the object that produces a new ψ for the system during the measurement. It is only the consciousness of an "I" who can separate himself from the former function $\Psi(x, y, z)$ and, by virtue of his observation, *set up a new objectivity* in attributing to the object henceforward a new function $\psi(x) = u_k(x)$.
>
> (*op. cit.*; their emphasis)

Thus, rather than consciousness 'causing' in some mysterious fashion, the collapse of the wave-function, the transition from a superposition to a definite state should be more suitably characterised in terms of a mutual separation of both the 'ego-pole' and the 'object-pole' through this familiar act of introspection (French 2002, 2020, 2023). It is the relational act that is central in this account, and it is of the essence of such an act and of the immanent knowledge that it yields that the ego should appear as one pole – not, crucially, something substantial, over and above or existing prior to this act. Rather it functions as a non-autonomous centre of identity or subject-pole that by virtue of the nature of the relational act, stands at one end of it, with the object under consideration as the other relatum. The latter is then 'made objective', in the sense of having a definite state attributed to it, by this objectifying act of reflection, and the 'chain of statistical correlations' is thereby cut. According to Merleau-Ponty, 'the role of the observer is not to make the object pass from the in-itself to the for-itself (as in Descartes)' (2003, p. 94) but rather, to 'make an individual existence emerge in act' (ibid.) via this cutting of the chain.

The crucial difference between London and Bauer's account and QBism is that the former, unlike the latter, explicitly takes the observer to be entangled with the apparatus and the system.[28] As far as the QBist is concerned this cannot be presumed but must be derived, on the basis of a suitable understanding of the Born Rule. Thus, at bottom, this difference comes down to the terms in which we conceive of the 'kick' of the world, as Fuchs puts it. For the QBists, with their 'first-person' agenda, this is manifested via quantum probabilities, whereas for London and Bauer, and hence also Merleau-Ponty, it is through the entanglement of the system and the observer. This difference in turn marks that between different understandings of phenomenology, namely those that emphasise the first-person point of view and those that focus on the correlational aspect (see Zahavi 2017; also French 2023).

According to Berghofer and Wiltsche,

> ... there can be no doubt that Merleau-Ponty ... accepts the perspectivity of our scientific image of reality. For Merleau-Ponty, however, this claim is not the result of a reflective analysis from outside of physics. Quite the opposite, on Merleau-Ponty's reading, quantum mechanics itself implies the strong ontological claim that the classical picture of a purely objective, observer-independent physical reality is untenable, and that every complete physical description of reality must incorporate the physicist as well as her experience. Seen from this perspective, then, quantum mechanics has the potential to live up to the ideal of a fully rationalized, critical, and ultimately *phenomenological* physics.
>
> (2020, p. 37)

It is this perspectival aspect of Merleau-Ponty's thought that encourages a positive comparison with QBism. However, as we've seen, he also drew on London and Bauer's analysis, with its explicit incorporation of a correlationist aspect, both phenomenologically and physically, as manifested via quantum entanglement. This is anathema to the QBist, of course, as is Merleau-Ponty's centring of the relations represented by the theory more generally. If, then, the QBist wants to draw on phenomenology to philosophically underpin her position, she is going to have to either modify the latter or exclude the correlationist understanding of the former.[29]

9.7 Conclusion

The compatibility of QBism with phenomenology and hence the degree to which the latter can be appealed to in order to flesh out the former, hinges on emphasising the 'first-person' perspective. This may smoothly mesh with taking the Born Rule as a primitive, as far as QBism is concerned, but runs into problems when it comes to accommodating inter-subjectivity on the phenomenological side. As we've also seen, appeals have been made to the work of Merleau-Ponty in further illustrating this compatibility. However, he explicitly drew on the earlier analysis of the measurement problem by London and Bauer that meshes with phenomenology's correlative aspect, where this is understood as consciousness and world standing in a mutually dependent context of being (Beck 1928; see Zahavi 2017). This generates an alternative form of phenomenological physics that sets that mutual dependence, as expressed by quantum entanglement, at centre stage. In either case, the role of the agent is crucial and phenomenology offers an obvious framework in which to explore and develop further aspects of that role.

Notes

1 QBists themselves insist that, '[m]any ideas voiced, and even committed to print, during earlier stages of Quantum Bayesianism turn out to be quite fallacious when seen from the vantage point of QBism' (Stacey 2019, p. 1). Fuchs, for example, sees QBism as going further, metaphysically speaking, than Quantum Bayesianism (Fuchs 2010). See also de la Tremblaye (2020, p. 246, fn. 2).
2 Antecedents can be found in the work of Bitbol, Destouches and Destouches-Février (de la Tremblaye 2020, p. 248).
3 This represents a development of the position, from taking quantum states as states of knowledge or of information to describing states of belief (Fuchs in Crease & Sares 2020, p. 556).
4 In a recent interview Fuchs indicates that he may be shifting to a more 'voluntaristic' approach whereby statements of subjective probability are not reports on one's psychological state but rather reflect certain epistemic commitments (Fuchs in Crease & Sares 2020, p. 556).
5 The demonstration hinges on Gleason's theorem, understood as the fundamental representation theorem for quantum probabilities (Earman 2019, pp. 407–408).
6 Having said that, continuity through such an extension can be established by demonstrating that possible measurements post-extension can be obtained from those before the incorporation of the apparatus (Pienaar 2020, p. 1918).
7 This does not exhaust the debate. Thus, Earman has argued that for QBists there can be no Schrödinger evolution between events at which their credence function is updated on new information, because, by definition, such evolution would correspond to belief change that is uninformed by new information (Earman 2019, p. 414). Hence, in order to account for the statistics of measurement outcomes, typically explained by the Schrödinger evolution of states yielding the relevant probabilities, the QBist must invoke the Heisenberg perspective, according to which such evolution is characterised via the observables rather than the states. However, this leads to a disquieting dualism: '... there is something like a realist/objectivist commitment to the structure of quantum observables and their temporal evolution but an instrumentalist/subjectivist attitude towards quantum states' (ibid., pp. 414–415). By this point the QBist

response should come as no surprise: observables and the corresponding time-evolution operators are treated exactly the same as states, namely as subjective information or doxastic quantities and hence there is no dualism (Fuchs & Stacey 2020, p. 8). Again, to think of Schrödinger evolution as giving rise to the possibility of uninformative belief change is to misunderstand and beg the question against the QBist picture: Schrödinger evolution is '... simply a deformed counterpart of classical stochastic evolution' (ibid., p. 8).

8 It is here, of course, that Wigner gave, in support of this latter claim, London and Bauer's assertion of the observer's 'characteristic and quite familiar faculty' of introspection; we shall come back to London and Bauer shortly.

9 When it comes to Wigner's concern that his ascribing a superposition state to the arrangement that includes his friend, the apparatus and the original system implies that his friend must be regarded as in a 'state of suspended animation', the QBist's response is, bluntly, that Wigner's state ascription simply has no bearing on the state of consciousness of his friend (De Brota et al. 2020, p. 1868).

10 'QBism knows that its story cannot end as a story of gambling agents—that is only where it starts' (2010, p. 27).

11 The thought experiment has been extended to yield a 'No-Go Theorem', which puts pressure on the claim that the predictions of different agents will not be contradictory (Frauchiger & Renner 2016, 2018); for a response see (De Brota et al. 2020).

12 QBism may be understood as a kind of 'normative structural realism': it is *structural realist* in the sense that quantum theory is taken to describe a fundamental relation that is agent-independent but normative in that this relation is understood as fundamentally captured by the Born rule (Crease & Sares 2020, p. 553).

13 Pienaar states that according to QBism 'reality is inherently subjective' (2020, p. 1898) with objectivity secured via the 'holistic structural features of the theory that apply equally to all Agents' (ibid.).

14 This is something that QBism shares with Healey's pragmatist approach and Fuchs himself repeatedly draws attention to the connections with pragmatist philosophers, especially William James, stating that '[a]long with James again, the key thing about QBism's understanding of quantum theory is that it indicates "reality is not ready-made and complete"' (in Crease & Sares 2020, p. 548; Healey 2017).

15 According to Bitbol and de la Tremblaye, such tensions within QBism arise because of its 'dual image' of an agent 'really' acting on a 'real' system (Bitbol & de la Tremblaye 2024, p. XX). As we'll see, they propose a thorough 'phenomenologisation' of QBism, according to which 'neither the nature of the world nor the nature of its objects is fundamentally different from the nature of experience' (ibid., p. XX).

16 Pienaar distinguishes this from standard 'Dutch book' coherence and calls it 'World-coherence' (2020, p. 1900).

17 Fuchs also acknowledges that another 'ontic element' is associated with the Bell and Kochen-Specker theorems, for example, which impose certain structural constraints typically understood in terms of non-locality and contextuality (Crease & Sares 2020, p. 555).

18 According to Bitbol and de la Tremblaye, the rule was derived by Destouches-Février from the non-contextuality of the probabilities (that is, their coherence) and the contextuality of the phenomena, understood as 'the fact that the world

cannot be neatly separated into an observing system and an observed system' (Bitbol & de la Tremblaye 2024, p. XX).

19 QBism is the 'most consistent phenomenological approach' towards quantum mechanics (Bitbol 2021, p. 570).

20 Having said that, Pfänder and fellow members of the Munich school urged the inclusion of inter-subjective relationships.

21 Fuchs himself has indicated some resistance to a wholly first-person standpoint (in Crease & Sares 2020, p. 546).

22 In both cases the processes of knowledge acquisition and decision making is dynamic, crucially involving an active role on the part of the agent (de la Tremblaye 2020, p. 256).

23 That our expectations must be governed by the Born Rule might also be alluded to in these considerations but that point doesn't impact on the parallels drawn with regard to the notion of the intentional horizon at least.

24 Paulette Destouches-Février presented a 'principle of subjectivity' according to which measurement results should not be considered as pertaining to intrinsic properties of the system but as properties of the 'system-apparatus' complex (Bitbol 2001; Pellegrini 2021, pp. 487–496). This represents one 'level' of reality but there also exists a further 'level of structures' which is composed of '... the most essential properties of the equations expressing the fundamental properties of physical systems: these are reflected in the evolutionary equations by the algebraic structures associated with these equations' (trans. by Pellegrini 2021, p. 490).

25 Thus, as Bitbol notes, according to Merleau-Ponty, '... no one can truly understand quantum mechanics without accepting a deep transformation of our conception of knowledge' (Bitbol 2020, p. 239).

26 Bitbol goes further and asserts that the situation in quantum mechanics is an extension of our situation of embodiment, so that '[a]t the end of the day, quantum physics testifies that the world behaves as a big flesh, of which our flesh is a sample' (2020, p. 241). As he then acknowledges, it is thought that cuts the measurement chain, thereby yielding a definite outcome.

27 As a result, Bitbol has argued, the role of constituting objectivity is extended to anything that expresses this principle of incarnation – in effect, then, the world as flesh becomes self-objectifying (2020, p. 236). The worry, of course, is that we lose any sense of 'objectivity' in such a move.

28 Indeed, it affirms what Fuchs denies, namely that consciousness can enter into a superposition.

29 Bitbol has argued that the claim that quantum mechanics describes the correlations expressed by the notion of entanglement, for example, '... can only arise from a descriptive, and therefore "realist," construal of quantum states; and therefore, deriving the "reality" of correlations from this argument is a *petitio principii*' (2021, p. 578). However, as what has been said above indicates, we can accept that the theory presents such correlations, *qua* theory of knowledge, without giving it or the correlations themselves a realist construal.

References

Alves, P. (2021). Fritz London and the Measurement Problem: A Phenomenological Approach. *Continental Philosophy Review* 54, 453–81.

Beck, M. (1928). Die Neue Problemlage der Erkenntnistheorie. *Deutsche Vierteljahrsschrift für Literaturwissenschaft und Geistesgeschichte* 6, 611–39.

Berghofer, P. & Wiltsche, H. (2020). Phenomenological Approaches to Physics: Mapping the Field. In H. Wiltsche & P. Berghofer (Eds.), *Phenomenological Approaches to Physics* (pp. 1–47). Cham: Springer.

Bitbol, M. (2001). Jean-Louis Destouches: théories de la prévision et individualité. *Philosophia Scientiae 5*, 1–30.

Bitbol, M. (2020). A Phenomenological Ontology for Physics: Merleau-Ponty and QBism. In P. Berghofer & H. Wiltsche (Eds), *Phenomenological Approaches to Physics* (pp. 227–242). Cham: Springer.

Bitbol, M. (2021). Is the Life-World Reduction Sufficient in Quantum Physics? *Continental Philosophy Review 54*, 563–580.

Bitbol, M. & de la Tremblaye, L. (2024). QBism: An Eco-Phenomenology of Quantum Physics. This volume.

Cairns, D. (1976). *Conversations with Husserl and Fink*. The Hague: Martinus Nijhoff.

Crease, R. P. & Sares, J. (2020). Interview with Physicist Christopher Fuchs. *Continental Philosophy Review 54*, 541–561.

De Brota, J. B. & Stacey, B. C. (2018). FAQBism. arXiv:1810.13401.

De Brota, J. B., Fuchs, C. A. & Schack, R. (2020). Respecting One's Fellow: QBism's Analysis of Wigner's Friend. *Foundations of Physics* 50, 1859–1874.

de la Tremblaye, L. (2020). QBism from a Phenomenological Point of View: Husserl and QBism. In P. Berghofer & H. Wiltsche (Eds), *Phenomenological Approaches to Physics* (pp. 243–260). Cham: Springer.

Destouches-Février, P. (1951). *La structure des théories physiques*. Paris: Presses Universitaires de France.

Earman, J. (2019). Quantum Bayesianism Assessed. *The Monist 102*, 403–423.

Frauchiger, D. & Renner, R., (2016). Single-world interpretations of quantum theory cannot be self-consistent. arXiv:1604.07422.

Frauchiger, D. & Renner, R., (2018). Quantum Theory Cannot Consistently Describe the Use of Itself. *Nature Communications* 9, 3711. https://doi.org/10.1038/s41467-018-05739-8

French, S. (2002). A Phenomenological Approach to the Measurement Problem: Husserl and the Foundations of Quantum Mechanics. *Studies in History and Philosophy of Modern Physics 33*, 467–491.

French, S. (2020). From a Lost History to a New Future: Is a Phenomenological Approach to Quantum Physics Viable? In H. Wiltsche and P. Berghofer (Eds.), *Phenomenological Approaches to Physics* (pp. 205–226). Cham: Springer.

French, S. (2023). *A Phenomenological Approach to Quantum Mechanics: Cutting the Chain of Correlations.*

Fuchs, C. (2010). QBism, the perimeter of Quantum Bayesianism. arXiv:1003.5209.

Fuchs, C. (2016). On Participatory Realism. arXiv:1601.04360.

Fuchs, C. (2017). On Participatory Realism. In I.T. Durham and D. Rickles (Eds), *Information and Interaction: Eddington, Wheeler and the Limits of Knowledge* (pp. 11–134). Cham: Springer.

Fuchs, C. & Stacey, B. (2020). QBians Do Not Exist. arXiv: 2012.14375.

Glick, D. (2021). QBism and the Limits of Scientific Realism. *European Journal for the Philosophy of Science* 11. https://doi.org/10.1007/s13194-021-00366-5

Healey, R. (2017). Quantum-Bayesian and Pragmatist Views of Quantum Theory. In E. N. Zalata (Ed.), *The Stanford Encyclopedia of Philosophy* (Spring 2017 Edition). https://plato.stanford.edu/archives/spr2017/entries/quantum-bayesian/

Husserl, E. (1970). *The Crisis of European Sciences and Transcendental Phenomenology*, trans. D. Carr (1954). Evanston: Northwestern University Press.

London, F. & Bauer, E. (1939). *La Théorie de L Observation en Mécanique Quantique*. Paris: Hermann.

London, F. & Bauer, E. (1983). The Theory of Observation in Quantum Mechanics. In J. A. Wheeler & W. H. Zurek (Eds.), *Quantum Theory and Measurement* (pp. 217–259). Princeton, NJ: Princeton University Press.

Merleau-Ponty, M. (1968). *The Visible and the Invisible*. Evanston: Northwestern University Press.

Merleau-Ponty, M. (2003). *Nature. Course Notes from the Collège de France*. Evanston: North- western University Press.

Pellegrini, P. (2021). Merleau-Ponty's Phenomenological Perspective on Quantum Mechanics. *Continental Philosophy Review 54*, 483–502.

Pienaar, J. (2020). Extending the Agent in QBism. *Foundations of Physics 50*, 1894–1920.

Stacey, B. C. (2019). Ideas Abandoned en Route to QBism. arXiv:1911.07386.

Talbott, W. (2016). Bayesian Epistemology. In E. Zalta (Ed.), *The Stanford Encyclopedia of Philosophy* (Winter 2016 Edition). https://plato.stanford.edu/archives/win2016/entries/epistemology-bayesian/

Trizio, E. (2021). *Philosophy s Nature: Husserl s Phenomenology, Natural Science, and Metaphysics*. Abingdon: Routledge.

Zahavi, D. (1997). Horizontal Intentionality and Transcendental Intersubjectivity. *Tijdschrift voor Filosofie 59*, 304–321.

Zahavi, D. (2017). *Husserl's Legacy*. Oxford, UK: Oxford University Press.

10 Back to Kant! QBism, Phenomenology, and Reality from Invariants

Florian J. Boge

10.1 Introduction: Routes to QBism and the Road Ahead

QBism is all about personal matters, so let me begin with a bit from my own personal story. When I first became seriously interested in Quantum Theory (QT), I was close to finishing a master's degree in philosophy, so I already had a fair bit of philosophy under my belt. Given this critical training, I had a hard time taking everything that physics textbooks were suggesting quite seriously. For instance, why would everyone make such a fuzz about the double slit experiment? Couldn't the physical setup simply alter the behaviour of tiny bits of matter in such ways that they would distribute as observed? Why should we assume them to follow straight-line trajectories anyways?

Looking for answers, I first stumbled upon Landé's early attempts to provide QT with new foundations, and later became attracted to the de Broglie-Bohm theory. As a long-time fan of the popular science fiction show *Star Trek: The Next Generation*, I also became fascinated with the Everett interpretation, which was the basis for the episode 'Parallels'. However, I should soon discover that each of these had serious problems, like being in conflict with relativity, or being unable to recover the Born rule in any sensible way – not to mention the difficulties associated with objective collapse views.

My story actually *begins* a little differently though. As an undergraduate, I had started out as a kind of sceptical epistemologist, interested in the limits of what can be known. However, I found no compelling reason to not take metaphysics at least seriously and so became interested in trope theory, the metaphysical theory that ultimately everything breaks down into particular properties (this red, that hardness ...). This process culminated in my supervisor, a trained chemist, telling me that my ideas about objects being nothing but particular properties 'meeting up' wouldn't work: the Pauli principle, he told me, implies certain properties for a whole system of electrons that are not reducible to the electrons'

DOI: 10.4324/9781003259008-13

individual properties. This was my first serious encounter with QT and there was something quite remarkable here: *entanglement*.

But a function of the form $\alpha\psi_a \otimes \psi_b + \beta\psi_b \otimes \psi_a$ didn't look very 'ontic' to me. Rather, it seemed to say something like: 'Either electron 1 is in state *a* and electron 2 is in state *b*, or vice versa'.[1] I was hence relieved to find that Spekkens (2007) had created a toy model which seemed to allow one to view the quantum state as (broadly) 'epistemic', i.e., something characterizing the experimenter's knowledge, information, convictions, etc. I was then very much disappointed to see that this model could not reproduce violations of Bell-type inequalities. Apparently, no (serious) epistemic model could!

So Schrödinger (1935, 555; orig.emph.) seemed to have it right: Entanglement, unlike superposition, non-commutativity, or uncertainty, was not just *some* feature of QT; it was '*the* characteristic trait of quantum mechanics, the one that enforces its entire departure from classical lines of thought'. But if one's interpretation was to be (broadly) epistemic, meaning that the quantum state was *not* a representation of the goings on in a radically mind-independent reality, and there was apparently also no *other* way to refer in an empirically adequate way to what goes on between two misaligned Stern-Gerlach magnets – then how should one think about reality, the quantum state, and the relation between the two *at all*?

It was through the popular-level writings of N. David Mermin, who reminded me that 'there is [...] a split [...] between the world in which an agent lives and her experience of that world' (Mermin, 2012, 8), that I realized it was time to reverse my metaphysical turn. I also realized that *QBism*, next to positions like those of Healey (2017) and Friederich (2015), was 'by far the most interesting game in town' (Mermin, 2012, 9). So this is my personal 'route to QBism', i.e., the sequence of steps that led to me becoming interested in it.

Famously, QBism has itself followed an interesting route to its present development (cf. Stacey, 2019). When Chris Fuchs and Rüdiger Schack first proved Dutch Book theorems for quantum states (Caves et al., 2002a) together with Carlton Caves (Fuchs's PhD supervisor), as well as a de Finetti-style representation theorem (Caves et al., 2002b), the whole project was still executed under the name 'Quantum Bayesianism'. Fuchs and Schack then, however, took the whole thing into a philosophically more radical direction, in turn changing the name to 'Quantum Brunoism' (after Bruno de Finetti's subjectivist interpretation if probability), or 'Quantum Bettabilitarianism' (as the world, at least, allows us to bet on it), or simply 'QBism' (not an acronym for anything). This philosophical radicality certainly wasn't lessened when N. D. Mermin joined the QBist ranks and Fuchs et al. (2014) discarded the 'intuition that correlations in the experiences of agents in widely separated regions ought to find their explanation in correlations in conditions prevailing in those regions'. They claimed:

> The variable λ [encountered in derivations of Bell-type inequalities – FJB] is nothing more than a version of the discredited EPR elements of reality. For a QBist the nonexistence of such objective facts-on-the-ground as λ no more implies nonlocality than does the nonexistence of elements of reality in the original EPR argument.

Today, QBists' main focus is on what they call the *Urgleichung*. This is a particular representation of the Born rule, which reads

$$Q(j) = (d+1)\sum_{i=1}^{d^2} P(i)R(j \mid i) - 1, \tag{10.1}$$

where we recognize the usual law of total probability if we replace the $d+1$ by 1 and remove the -1. Most notably, there is no mention of state vectors or operators here at all, and so the notorious measurement problem – that we do not know any Lorentz-invariant interpretation of the transition $\alpha_1 \mid a_1\rangle + \alpha_2 \mid a_2\rangle + \cdots \mapsto \mid a_j\rangle$, especially when the state is entangled, or any coherent way to circumvent the assumption of such a step – *vanishes*: If quantum 'states' are basically probability assignments, then a transition of the aforementioned form is no more mysterious than a probability update.

This version of the Born rule follows if one is in possession of a 'SIC': a symmetric informationally complete positive operator valued measure (POVM); something known to exist for a large number of Hilbert space dimensions d, but presently not for arbitrary dimensionality (DeBrota et al., 2020). More precisely, a 'MIC' (minimal informationally complete POVM) is a POVM $\{E_i\}_{1 \le i \le d^2}$, where E_i are linearly independent, and form a basis of the space $\mathcal{L}(\mathcal{H}_d)$ of bounded operators on Hilbert space $\mathcal{H}_d$ of dimension d. Furthermore, any density operator (positive trace-one operator) ρ on $\mathcal{H}_d$ can be expressed as a linear combination of E_i, which allows to characterize ρ completely in terms of the probabilities it generates via the Hilbert-Schmidt inner product. A MIC whose elements are defined by $E_i = \Pi_i / d$, with Π_i projections satisfying $\mathrm{tr}[\Pi_i \Pi_j] = (d\delta_{ij} + 1)/(d+1)$ is called a SIC.

But assuming SICs exist in all dimensions, what is the meaning of the Urgleichung? First of all, note that already the expressibility of ρ in terms of probabilities generated by inner products with MICs has some interesting implications:

> the mapping $\rho \mapsto (p(1), \ldots, p(d^2))$, although injective, cannot be surjective; only some probability distributions in the simplex are valid

> for representing quantum states [...]. If quantum states are nothing more than probability distributions, a significant part of understanding quantum mechanics is understanding what restrictions there are on the set of valid distributions.
>
> (Fuchs and Schack, 2013, 1698; notation adapted)

In other words: QT generally *constrains* the credences we can entertain. Furthermore, according to Fuchs and Schack (2013), the difference between the law of total probability and the Urgleichung is contained in the fact that the measurement with outcomes i remains *counterfactual*, i.e., that the bet in which i is an outcome has been called off, whereas this is not the case in Dutch book arguments for the law of total probability.

So QBism tells us that we only need to figure out the right ways to look at these probabilities, and then everything will fall into place. Problems solved, at least tentatively, right? Alas, if only this were true!

10.2 Challenges for QBism

10.2.1 Prelude: What's the Explanandum?

Let's assume that the Urgleichung is indeed best construed as representing a situation where the bet for the events conditioned on has been called off. It may then be true that it would be 'irrational in some situations' to bet in accordance with the law of total probability (Fuchs and Schack, 2013, 1697). But this does not even touch on the question as to why it would be *rational* to bet *in accordance with* the Urgleichung.

It seems that the discrepancy between the law of total probability and the Urgleichung points us to something 'out there'; something which 'makes it so' that we have to constrain our credences in a different way when faced with the sort of counterfactual dependency explored by Fuchs and Schack (2013). And it must be this 'something out there' which provides the reason why we should bet differently. QBists agree:

> Now, if you accept that the Born rule is an extra normative rule, you might ask me, 'Why that rule; why not some other way of relating the probabilities?' When you ask me that question, I answer, 'Because that's the way the world is.' There is something about the world that has led us all to adopt this as the best adapted method for living in our world.
>
> (C. Fuchs, as cited in Crease and Sares, 2021, 14)

However, there is something important that, to my mind, QBism *neglects*: That this divergence from the law of total probability is ultimately forced upon us by calibrating credences on observed *frequencies*.

Consider, e.g., the form of the quantum de Finetti theorem (Caves et al., 2002b):

$$\rho^{(N)} = \int \mathrm{d}\rho\, \varrho(\rho) \rho^{\otimes N}, \tag{10.2}$$

where $\rho^{\otimes N}$ is an N-fold tensor product of the same state, $\rho^{(N)}$ is 'exchangeable' in the sense that it is permutation symmetric and extendable as $\rho^{(N)} = \mathrm{Tr}_M \rho^{(N+M)}$, and $\mathrm{d}\rho\, \varrho(\rho)$ defines a probability measure over density operators. The gist of this theorem is that, when different agents with non-extreme priors update this state Bayesian-style on a growing number of measurements on the N measured systems, they will converge on a common state, regardless otherwise of the shape of their priors. Hence, the quantum-tomographical notion of an 'unknown state', which is then found out, can be *replaced* by that of a state *agreed upon* after several measurements. This is an important result, mirroring de Finetti's semantic replacement of 'unknown' probabilities. However (capital 'H'), *what* quantum state these agents will agree upon *will be determined by the distribution of outcomes they observe.*

Furthermore, while some singular observations may strike us as profoundly significant, we are more likely to discard them as illusions, misconceptions, or coincidences than unusual frequencies of certain types of events. It is those which we consider the 'scientific phenomena' to be accounted for (Massimi, 2007; van Fraassen, 1991). Experimental results which inevitably appealed to observed frequencies indeed stand at the very *inception* of QT: It was the fact that spectral lines always occurred, as predicted, 'Breit-Wigner distributed' within a very narrow interval around fixed relative distances to one another that convinced physicists of the formalism's utility back in the day, and it is the agreement to eight significant digits between the experimental *average* for the anomalous dipole moment and the Q(F)T prediction that convinces us of it today. Hence, relative frequencies are something quite important: They guide us both in shaping our credences and in collecting scientific evidence.

Here is what I take to be the most important set of relative frequencies not well accounted for by QBism. Recall from the introduction that I followed Schrödinger in considering quantum entanglement and the correlations it implies to be *the* characteristic feature of QT. Superpositions, uncertainty, etc. could all be interpreted as expressions of ignorance (Bartlett et al., 2012; Spekkens, 2007), to be surpassed by a future theory that makes better descriptions available. That this would be the case also for entanglement was certainly the hope of Einstein et al. (1935). But things did not turn out in this way – entanglement reflects the 'true quantumness' (Jennings and Leifer, 2016). Yet the remarkable correlations it implies tend to be downplayed in QBism:

> Correlations are just a special case of more general probability assignments. To explain a correlation is therefore no different than to explain a probability assignment. [...] [This] remain[s] unchanged in the case that the correlations $p(x,y \mid a,b)$ implied by the prior state and measurement operators violate a Bell inequality.
>
> (Fuchs and Schack, 2014, 5)

Let's consider the QBist rendering of the usual Alice-Bob story in a little more detail. Alice and Bob sit at space-like distance to one another, rotating, at agreed upon times, their Stern-Gerlach magnets at will to one of two arbitrary positions, resulting in three possible angles between them. In this way, they elicit experiences they each call 'spin up' or 'spin down', meaning visual impressions of dots on the upper or lower half of a screen (relative to the orientation of the magnet), respectively. They write down a table which codifies their experiences and then get together and compare. To their surprise, they find a remarkable number of coincidences between up and down in their respective tables, especially whenever there was no misalignment between both magnets. After some error-correction, they even find this correlation to be *perfect*.

Smart as they are, Alice and Bob realize that it would be very difficult to supplant a causal model for this correlation: The settings were chosen at such points in time that whatever locally caused the dots on the screen (say, invisible 'particles' transmitted from 'the source'), together with their own interventions, could not have interacted causally, at least not at (sub) luminal speeds.

Now, admittedly on QBism,

> quantum mechanics explains why the agent should expect the measured frequencies to lie in a certain range, but does not provide an explanation for the particular numbers the agent obtains in a given realization of the data table.
>
> (Fuchs and Schack, 2014, 5)

But we were never *interested* in explaining what frequencies agents should *expect* in the first place, were we? The puzzles associated with QT are the dots successively building up fringes on the screen in the double-slit experiment, the changing count rates in particle detectors in delayed choice-experiments, or the surprising relative frequencies with which Alice and Bob find coinciding values. If one hasn't explained *these*, one has arguably not explained *anything*. This concern I share with John Earman:[2]

> The [QBist] story explains why both [Bob] and Alice *expect*, with degree of belief one, to find anticorrelated spins, but [...] does not explain why the measured spins *are in fact* anticorrelated.
>
> (Earman, 2019, 418; orig.emph.)

Replace 'spins' by 'values on the two lists' and 'are in fact' by 'are experienced to be', and this objection should get even the most immutable QBist nervous: If Alice, a firm QBist, seeks an explanation for the correlated entries on the two lists, but wants to avoid invoking the 'discredited EPR elements of reality', she should actually consider her interaction with *Bob* to be the cause of the observed correlations between table entries. Worse yet, Alice might justifiably take the whole situation to be a hoax concocted, and never resolved, by Bob and the team of scientists setting up the experiment – something akin to a conspiracy theory. Holding Bob causally responsible like this might seem ludicrously incompatible with other conceptions Alice entertains about him, based on her past experience. So how is Alice to regard this situation, if she does not simply want to shut up and bet on it?[3]

10.2.2 *With Apologies to Chris Fuchs: QBism and Solipsism*

Let's step back a moment and recall that, according to QBism, QT is 'a single user theory' (Fuchs and Schack, 2014, 3). Like Norsen (2016), I believe that this makes for a connection to *solipsism*, though that connection is more subtle than Norsen claims.

Unlike Norsen, I am not claiming that QBism *is* a kind of 'FAPP' solipsism, i.e., that it is practically indistinguishable from solipsism. It is true that QBism discards the representational character of our currently most successful theory while also dismissing other 'objective facts-on-the-ground', as could be represented by additional variables λ. And it is also true that, by the standards of most philosophers, a position like this would count as *anti-realist*.[4] However, Fuchs in essence urges that this should *not* be mistaken as a kind of *metaphysical* anti-realism, and so not as (substantive) solipsism:

> We do [...] hold evidence for an independent world [...] external to ourselves [...] because we find ourselves getting *unpredictable kicks* (from the world) all the time.
>
> (Fuchs, 2002, 11; emph. Added)

This reason for not becoming an idealist or solipsist of sorts is very similar to d'Espagnat's:

> We sometimes build up quite beautifully rational theories that experiments falsify. Something says no. This something cannot be 'us.' There must be something else than just 'us.'
>
> (d'Espagnat, 1995, 314)

In my own words (with a nod to Quine, 1951): the *recalcitrance* of experience provides a wonderful reason for postulating (or abducing) the existence of other agents and material entities.

Clearly, one must be careful not to mistake this reason for *postulating* the *existence* of 'others' and a 'mind-independent reality' with definite *knowledge* of their respective *constitutions*. And this is an element certainly also present in Fuchs's *participatory* realism (inherited from Wheeler), according to which 'reality is *more* than any third-person perspective can capture' (Fuchs, 2017, 113; original emphasis). Nevertheless, it should thus be clear that the mere rejection of additional variables λ and the representational status of QT does not imply solipsism, so long as 'solipsism' is understood either as the *metaphysical* position that reality is a figment of one single mind, or the *epistemological* position that reality beyond that one single mind is unknowable.

What I do find to be correct, however, is that QBism is properly described by the term '*methodological solipsism*', for it 'amounts [...] to an application of the form and method of solipsism' even if 'not to an acknowledgment of its central thesis' (Carnap, 2003, 102; orig.emph.). What I do find to be correct as well is that the combination of QBism's focus on (guided) subjective credences, its appeal to recalcitrance as the ground for abducing an external reality, and the *neglect* of the 'stronger than classical'-correlations as a relevant datum makes it hard to see how QBism has any *advantage* over 'actual' solipsism. For our strongest reasons for *not* being solipsists lie exactly in our success with stipulating further variables (like λ) in other circumstances – say, *your* consciousness as creating *my* impressions of talking to some*one*, or the lawn 'out there' as creating my present impression of green. Hence, rejecting the hidden variables λ too easily out of hand yields a *slippery slope* towards solipsism.

To see this more clearly, consider a *single* scientist, using what Jarrett (2009) coined the 'Mermin contraption' (after Mermin, 1981), and writing down a protocol as in Table 10.1. Here, *A* and *B* refer to the 'measuring devices' placed on two diametrically opposite sides of C (the 'source'). They each have two settings, 1, 21,2, and a red (r) and green (g) light on top of them. At certain times, lights flash on each of the two devices simultaneously, and the nob on each of *A* and *B* may or may not switch

Table 10.1 Hypothetical Protocol

A	B	A	B	$\cdots$
$1, r$	$1, r$	$1, g$	$2, r$	$\cdots$
$2, g$	$1, g$	$2, g$	$2, g$	$\cdots$
$\vdots$	$\vdots$	$\vdots$	$\vdots$	$\vdots$

automatically just shortly before the lights flash, with no discernible correlation between both nobs (Figure 10.1).[5]

After watching a very long sequence of some thousands of flashes and writing down settings and lights flashed, our scientist notices that the frequency of joint occurrences of r and g on A and B, respectively, is somewhere near $(2+\sqrt{2})/8$ among the runs in which A was set to 1 and B was set to either 1 or 2, and equally when A was set to 2 and B to 1, but somewhere near $(2-\sqrt{2})/8$ when both are set to 2. However, for each setting, the occurrences of either r or g on either A or B individually settle down around 1/2. Consequently, she notices the following empirical correlation between the frequencies, f, of flashing lights for given joint settings on both devices:

$$f(A=g \wedge B=r \mid a \wedge b) \approx f(A=r \wedge B=g \mid a \wedge b) < f(A=g \mid a) f(B=r \mid b) \approx f(A=r \mid a) f(B=g \mid b) \quad \text{for } a=b=2, \tag{10.3}$$

$$f(A=g \wedge B=r \mid a \wedge b) \approx f(A=r \wedge B=g \mid a \wedge b) > f(A=g \mid a) f(B=r \mid b) \approx f(A=r \mid a) f(B=g \mid b) \quad \text{else}, \tag{10.4}$$

where a denotes A's setting and b denotes B's.

The scientist of course realizes immediately that if these visual experiences would be caused by some state of an external reality, λ, – maybe featuring also the states of two invisible particles being emitted from C – this

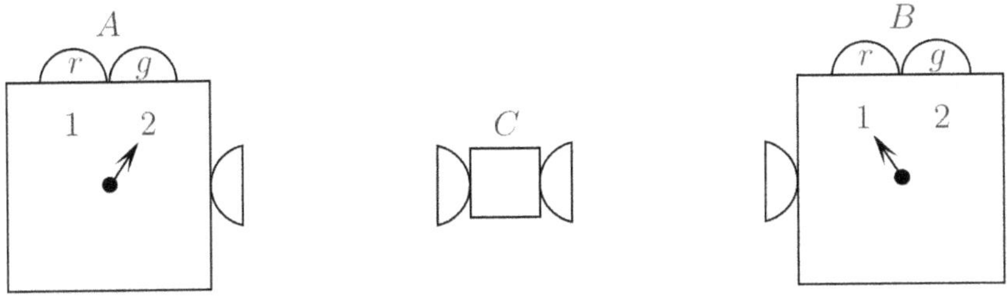

Figure 10.1 Mermin Contraption.

would *predict* an observable correlation between the flashes. So without crunching many numbers, she may assume that a probabilistic model of the following form could help explain the correlation:

$$P(A \wedge B \mid a \wedge b \wedge \lambda) = P(A \mid a \wedge \lambda)P(B \mid b \wedge \lambda) \tag{10.5}$$

Our scientist cannot be a convinced QBist: QBism not only denies the representational status of QT, but also the suitability of further representational variables λ. But, as I will show, rejecting the possibility of such a causal model tout court also makes for a subtle connection to solipsism.

Compare the above situation to that contemplated by Reichenbach (1938), which he utilized to provide a *refutation* of solipsism. Reichenbach offered a philosophical thought experiment, reminiscent of Plato's cave, that delivered 'a beautiful analogy for the problem of the external world' (Sober, 2011, 20). In this thought experiment, mankind is confined to a cube on whose outside walls shadows appear. The shadows displayed on two adjacent walls of the cube exhibit astonishing correlations, and the story's hero, 'Copernicus', comes up with the following explanation: There are objects outside the cube, and two matching shadows are just images caused as common effects by some mechanism involving one single object. That's why they are so remarkably correlated! And indeed, Copernicus is right: Outside the cube, there are birds flying around and a single bird's shadow is simultaneously projected onto two adjacent walls by a cleverly contrived system of lights and mirrors.

The appeal of the story as an argument against solipsism is that the cube represents an individual's confinement to her own immediate experience. But in contrast to any conceivable solipsistic hypothesis about this experience, an outside world-hypothesis could *predict* the correlations met with in experience, and so has the higher evidential support:[6] The correlations exhibited between, say, our auditory and visual experiences can be predicted on the assumption that they are both simultaneously caused by the states of an external reality, but typically not from mental states alone. In Sober's words: 'It's the external world that is doing the work, stupid' (Sober, 2011, 18).

More concretely (cf. Sober, 2011, §7, §8), closing my eyes and making the visual impression, W, of waves crashing on the beach disappear does not make my auditory impression, S, of the crashing-sound go away. Similarly, shutting my ears eliminates W but not S. Hence, it doesn't seem that W and S are related as cause and effect.

This empirical correlation between W and S is extraordinarily robust, and so doesn't seem 'spurious'. And it also doesn't seem 'analytical', i.e., directly given by the *meanings* of W and S. Hence, this is a situation in which we might expect Reichenbach's *principle of the common cause*

(PCC; Reichenbach, 1965) to apply;[7] that if X and Y are (robustly) correlated but logico-analytically independent, then either X causes Y, Y causes X, or X and Y are joint effects of a common cause λ that renders X and Y conditionally probabilistically independent ('screens them off').

Now the most obvious candidate common cause from within one's own mental world is certainly the *intention* to go to the beach. But this intention doesn't screen off W and S: I will experience their joint occurrence after having this intention *more often* than their individual occurrences in that very case. Hence,

> the solipsist must concede that [...] *something* is going on in the production of my wavy visual and auditory sensations besides my intending to go to the beach. [...] Suppose, for example, that right after I form the intention to go to the beach that I am rendered unconscious; the next thing I know, I am either experiencing wavy visual and auditory experiences, or I am experiencing neither. When I introspect, I find no further experiences that I can cite to explain this uncanny correlation of [W] and [S].
>
> (Sober, 2011, 18)

The inability to manipulate, say, S by intervening on W could still be the result of 'ham fisted' operations within one's mind. I.e., there might not be any proper ('surgical') *intervention* available, and so also no way to *exhibit* the causal dependency properly: Shutting my eyes might simultaneously eliminate and, *via* a different causal path, *create* S anew. But that is an obviously conspiratorial story, and so is not an attractive option for the solipsist: It features a causal connection that does not show up in the statistics, and so corresponds to what causal modelers call an 'unfaithful' causal model.

But now comes the saucy part. As pointed out above, our scientist abduces the existence of λ by appeal to the PCC, or even just from the fact that postulating λ predicts the correlation and so has higher evidential support than a solipsistic hypothesis. However, as is well known, she soon runs into trouble. If she also assumes that λ is causally unrelated to the setting (i.e., $P(\lambda \mid a \wedge b) = P(\lambda)$) and uses a numerical variable that maps g to +1 and r to −1 to compute expectation values, she easily derives the infamous CHSH inequality (Clauser et al., 1969). But plugging the values approximated by her noted frequencies into this inequality, she finds that this yields $2\sqrt{2} \leq 2$.

The scientist might now ponder whether the two flashes are not related as cause and effect. But similarly to the waves/sound case, she will quickly find that option to be undermined, by the observation that there are no conceivable interventions she could possibly undertake to alter the statistics in A by manipulating B, or *vice versa*; at bottom because the situation at *A* and *B* is perfectly symmetric (Friederich, 2015, 132). She could also come up with weirder causal stories, such as the latent variable λ being influenced by, or influencing, the settings (an option known as 'superdeterminism'), or maybe even that there are causal influences going back in time.

But being a skilled causal modeler, she will soon figure out that all these options are possible only on pains of *stipulating causal relations that have no manifestation in the frequency data* (Wood and Spekkens, 2015). Hence, while in the *W*/*S* case, the realist *can* point to some state of an external reality as facilitating the common cause and the solipsist is forced to embrace unfaithful causal models, for the single-scientist Mermin contraption *neither* can offer a causal story without invoking conspirational, unfaithful models. So the solipsist might counter: 'Maybe it's *not* the external world that's doing the work after all, stupid!'

Now, someone who, like a QBist, uses QT as a mere calculus for adjusting expectations about future experiences should thus raise *neither* question: that as to why some correlations are screenable and some aren't, nor that as to why, more generally, *some experiences seem like 'unpredictable kicks' whereas others seem 'fairly regular'*. For, if *anything* counts as 'recalcitrant' in the relevant sense, it is the correlations exhibited by the Mermin contraption: They were totally unexpected, whence the infamous EPR paper 'came down [...] as a bolt from the blue' on Bohr and Rosenfeld (1983 [1967]). And they are radically out of our control – remember: no possibility to intervene locally on the remote outcome which is strongly correlated but, like the local one, apparently random. Hence, if we stick to recalcitrant experiences as evidence of a mind-independent world, these correlations should serve as such evidence if anything does.

In sum, if QBism *selectively* uses recalcitrant experiences as evidence of an external world it has no advantage over actual solipsism, since it offers no coherent basis for abducing reality from experience. I will call this the *epistemic problem* (EP) of QBism.

10.2.3 *Wallace's Challenge for Non-Representationalism*

There is a second, maybe more profound problem lingering in the constructivist elements of QBism; much as the late Wittgenstein's (1968) *meaning* scepticism was judged to be more profound than knowledge scepticism by Kripke (1982, 60). For how, exactly, is 'everything any of us knows about the world [...] constructed out of his or her individual private experience' (Fuchs et al., 2014, 753), if this involves, say, talk of '"particles" [...] that come to Alice and Bob from a common source' (ibid., 752)?

The problem I am hinting at is, of course, a version of a problem well-known to philosophers: The problem, associated with all constructivist or reductionist empiricisms, of first identifying certain 'purely observational' concepts and then showing how all other concepts that do not directly refer to experience can be *defined* in terms of these. *Prima facie*, that is a necessary step in showing how the world is 'constructed out of experience'; but as is well known, it is a formidable task. I will call this the *semantic problem* (SP) of QBism.

I am obviously not the first one to raise this concern. The following objection (or, let's say, observation) has recently been voiced by Wallace (2020, 91–92; orig.emph.):

> the central idea in the logical-positivist and logical-empiricist pictures of science [was to] make a principled distinction between the 'observation language' in which our observations are described, and the 'theory language' in which the non-observational parts of our scientific theories are stated. [...] Non-representationalist strategies at least seem to be committed to making the same division, whether the analogue of the 'observation language' is Copenhagen's use of classical mechanics, or pragmatism's 'non-quantum' description, or QBism's appeal to direct experience. The problem with these approaches [...] is not that making such a distinction is *unreasonable* or *illegitimate*, but that—at least at present—we do not know how to do it.

Wallace's criticism is not specific to QBism; a lengthy passage is devoted to Healey's pragmatism. However, Wallace *is* more specific as to where, precisely, he sees the troubles arise for *all* these approaches. Consider the following fictional dialogue concerning the meaning of 'quark-gluon plasma', presented by Wallace (2020, 20) as a means for showing the semantic difficulties faced by non-representationalist approaches:[8]

Q_1: What's the quark-gluon plasma?

A_1: It's the state of a quantum-chromodynamics (QCD) system above a certain temperature, at which a phase transition occurs to a state where the fermionic elementary excitations are associated with the quark field rather than to colour neutral products of that field.

Q_2: Slow down. What's 'temperature' in QCD?

A_2: A quantum system, including a field-theoretic system, is at (canonical) thermal equilibrium when its quantum state is

$$\rho(\beta) \propto \exp(-\beta\hat{H}), \tag{10.6}$$

where $\hat{H}$ is the Hamiltonian and β is a real number. For a system at thermal equilibrium – or that is reasonably close to thermal equilibrium – its temperature T is given by $\beta = 1/k_B T$.

Q_3: And what's an 'elementary excitation'?

A_3: Generally in quantum field theory, we can analyse systems in states reasonably close to the thermal equilibrium state as gases of weakly interacting particles. Those weakly interacting particles are the elementary excitations.

Q_4: 'Particles' as in classical point particles?

A_4: Not really. 'Particles' as in subsystems whose Hilbert space bears an irreducible representation of the Poincaré group, at least in the interaction-free limit.

Q_5: So the quark-gluon plasma is associated with one sort of particle, colder systems with another. Shouldn't I be able to say what the 'particles' are once-and-for-all?

A_5: Not in quantum field theory: the optimal choice of particle depends on the state of the system. Hot systems are described most naturally in terms of quarks, colder systems, in terms of protons and neutrons.

Q_6: Can't I just think of a proton or neutron as an agglomeration of three quarks?

A_6: Only heuristically. The more precise way to explain the relation between the protons and quarks is at the field level: the proton is associated with a certain triple product of the quark field.

Q_7: How is a particle supposed to be associated with a field?

A_7: If a quantum system is in thermal-equilibrium state $\rho(\beta)$, the 'two-point function' of that system with respect to field $\hat{\phi}(x)$ and that state is

$$G_2(x-y;\phi,\beta) = \mathrm{Tr}(\rho(\beta)\hat{\phi}(x)\hat{\phi}(y)) \tag{10.7}$$

If the Fourier transform of that state has a pole, there's a particle associated with it.

Q_8: That's a weird postulate.

A_8: It's not a postulate; it's something you derive, by looking at the dynamics of states obtained by excitations of the thermal-equilibrium state. Where there's a pole, there's a subspace of states which can be interpreted as superpositions of singly localized excitations and which is preserved under the dynamics.

The point Wallace is trying to make here is that he has

> *not the faintest idea* how to make sense of any of this without taking the quantum state of the QCD system, and its dynamical evolution under the Schrödinger equation, as representational.
> (Wallace, 2020, 90; orig.emph.)

After all, '[e]ven the claim that the system has temperature T is a claim about its state'. The point, then, is that a *high-level* concept, such as 'quantum state' is arguably fundamental for the semantic content of the entire dialog, and any reductionist or even constructivist approach seems to be doomed to failure. Let's call this *Wallace's challenge*.

What could a non-representationalist about the state respond? I believe a dialogue between the representationalist, providing the answers, and a more sceptical inquirer would go a little differently:

A_2: A quantum system, including a field-theoretic system, is at (canonical) thermal equilibrium when its quantum state is

$$\rho(\beta) \propto \exp(-\beta\hat{H}), \tag{10.8}$$

where $\hat{H}$ is the Hamiltonian and β is a real number. For a system at thermal equilibrium – or that is reasonably close to thermal equilibrium – its temperature T is given by $\beta = 1/k_B T$.

Q_{2a}: Remind me what a 'quantum state' is?

A_{2a}: It' s a positive trace-one operator on a suitable Hilbert space.

Q_{2b}: Yes, I recall. But that's an abstract mathematical definition. What does it represent, physically?

A_{2b}: It represents the state of the system.

Q_{2c}: O…k, but then why would it *have* these mathematical properties? Or maybe more importantly: Why should it take on the form you just showed me when the system is in an equilibrium? Frankly, what's a 'thermal equilibrium' anyways?

A_{2c}: It's complicated. Quantum theoretically, thermal equilibrium pertains to a system which has just this sort of state. This is how you define it. A little more illuminatingly, you could maybe say that the defining property is that the energy remains constant within the system.

Q_{2d}: And whence the form?

A_{2d}: Well originally, the thermal equilibrium of a system was defined by the property that its temperature is homogeneous across the system and constant in time. Maxwell and Boltzmann then came up with negative exponential distributions for systems in thermal equilibrium, and …

Q_{2e}: Wait, besides the fact that you just used 'temperature' in your explanation, it seems that you are telling me now that the form of the state ultimately gets its justification from a correspondence with a probability distribution. Is that correct?

A_{2e}: No. These were merely historical remarks. I mean, it's true that von Neumann did call it an 'analog' for Boltzmann's distribution, when he introduced his famous entropy formula. But you can derive this form of the density operator by requiring unit normalization and that the expectation value of the Hamiltonian be constant.

Q_{2f}: But didn't I just hear 'normalization to unity' and 'expectation value' in your formal explication? And aren't these still hallmarks of a probability distribution? … I'm starting to get terribly confused here. Let's set this issue back for the moment. (*continues with* Q_3)

The point of this exercise is not to discredit Wallace's challenge altogether. In a way, I believe it to be quite serious. However, the first part of this extended dialogue is useful for rebutting what I would like to call the *first level* challenge: That the quantum *state* has to be taken as representational.

As the extended considerations tickled out by the sceptical inquirer show, it is far from clear that we have to take the state as representational just because it figures importantly in the explication of other concepts. By the same token, we would otherwise have to take a classical probability density as (directly) representational. But while the choice and definition of a probability density may elucidate the concepts one entertains about the system in question and *why* one apportions one's credences in a certain way (say: maximizing entropy), this doesn't mean that the density itself *represents* the state of the system. Same thing with density *operators*: that something is *important* for understanding something *else* does not mean that the first thing is representational.

However, what I would like to call the *second level* challenge arguably persists: That in order for the whole dialogue to make sense, *something* about the formalism has to be taken as at least *tentatively* representational, and not in any obviously reducible or deconstructable way. In other words, it would mean throwing out the baby with the bathwater to conclude, from the observation that the quantum state has much in common with a probability distribution or density, that *no* concept connected to the quantum formalism, and not in any obvious way to direct sense experience, is at least *intended* as a representation *sui generis* of the goings on in a mind-independent reality.

I believe that a defence of *state*-non-representationalism – one that takes to heart many of the messages of QBism while avoiding EP and SP – is possible on the grounds of this distinction. I will offer my own one in Section 10.4. For now, however, let me first establish a connection to *phenomenology* – which is, after all, the defining philosophy of this volume.

10.3 Help from Phenomenology?

I suppose I should have qualified my personal story even a little further: When I became interested in QT, I had a fair amount of *analytic* philosophy under my belt. In contrast, my knowledge of continental brands, such as phenomenology, was – and still is – rather limited. Hence, what follows is almost certainly a caricature of actual phenomenology. Following, however, Gallagher and Zahavi (2008, 28; orig.emph.), we may take the heart of phenomenology to be a certain methodology that can be specified in terms of four basic steps:

1 The *epoché* or suspension of the natural attitude.
2 The *phenomenological reduction*, which attends to the correlation between the object of experience and the experience itself.
3 The *eidetic variation*, which keys in on the essential or invariant aspects of this correlation.
4 *Intersubjective corroboration*, which is concerned with replication and the degree to which the discovered structures are universal or at least sharable.

The 'natural attitude' here means that 'Reality is assumed to be out there, waiting to be discovered and investigated. And [that] the aim of science is to acquire a strict and objectively valid knowledge about this given realm' (ibid., 22). But practicing the epoché does not mean engaging in radical scepticism or metaphysical anti-realism:

> the epoché entails a change of attitude towards reality, and not an exclusion of reality. The only thing that is excluded as a result of the epoché is [...] the naïvety of simply taking the world for granted, thereby ignoring the contribution of consciousness.
>
> (ibid., 23)

The aim of the phenomenological reduction, on the other hand,

> is to analyse the correlational interdependence between specific structures of subjectivity and specific modes of appearance or givenness. [O]nce we adopt the phenomenological attitude, we are no longer primarily interested in what things are [...] but rather in how they appear, and thus as correlates of our experience.
>
> (ibid., 25)

Hence, these first two steps are, in a sense, preparatory: They set the stage of investigating 'things' as 'things-for-us', rather than 'things-out-there'. What is arguably the most important step, then, is the eidetic variation. It is the key to tickling out

> what Plato called the *eidos* or *essence* of things. [...] If the object that I am examining happens to be a book, what features of it can I imaginatively vary without destroying the fact that it is a book. I can change the colour and design of the cover; I can imaginatively subtract from the number of pages, or add to them; I can change the size and weight of the book; I can vary the binding. [...] [T]he core set of properties that resist change [...] constitute the essence, the 'what makes a book a book'.
>
> (ibid., 27; orig.emph.)

Finally,

> another tool at the phenomenologist's disposal [...] is simply the fact that phenomenologists do not have to do their phenomenological analyses alone. Descriptions allow for intersubjective corroboration. And again, the quest for invariant, essential structures of experience is not narrowly tied to the peculiarities of my own experience.
>
> (ibid., 28)

In sum, we find out the essence of something by first retreating from our inclination of thinking in terms of things being simply 'found out' by science, by then focusing on what things are *to us*, by abstracting away as much as possible, and by then comparing what's left (the purported 'eidos') with what others may have found.

Admittedly, I see some fundamental problems associated with this method. Before I turn to a critique, however, let me first point out in what ways it could be of help to QBism and also express my sympathies for parts of it.

Recall that I had claimed the two main problems of QBism to be a want of a coherent basis for abducing reality from experience (EP) and the want of a coherent basis for constructing complex concepts, as appealed to by QBists in explicating their own position, from their acclaimed foundation in experience (SP). Now if phenomenology contains the central realization that experience comes *pre-structured*, i.e., that some properties of the objects we encounter are *essentially* attached to them and cannot even be 'stripped away in thought', then this might allow one to circumvent the pertinent problems. For instance, Nagel (2000, 346), and Ladyman (2000, 2010) following her, argue that:

> To make the kind of epistemic use of experience that empiricism demands, we need at least the capacity to sort out its deliverances from other products of the mind [...] and this sorting task is [...] a rational enterprise [...] that demands substantive a priori knowledge for its execution.

Hence, insofar as the eidetic variation is such a source of substantive a priori knowledge, it might help circumvent these problems, and so help QBism solve the SP. Here is Merleau-Ponty (2012 [1945], xxx):

> if I am able to speak about 'dreams' and 'reality,' to wonder about the distinction between the imaginary and the real, and to throw the 'real' into doubt, this is because I have in fact drawn this distinction prior to the analysis, because I have an experience of the real as well as one of the imaginary.

Furthermore, phenomenology could give rise to what might be seen as a *dis*solution, rather than a solution, of the EP:

> To believe in [...] a pure third-person perspective is to succumb to an *objectivist illusion*. [...] It is a view that *we* can adopt of the world. It is a perspective founded upon a first-person perspective, or to be more precise, it emerges out of the encounter between at least two first-person perspectives; that is, it involves intersubjectivity.
>
> (Gallagher and Zahavi, 2008, 40; orig.emph.)

Hence, phenomenologist methodology seems to offer a way to bypass any abductive, inferential step: 'The world' is, in a sense, directly given, in the way it appears to 'us'. And by reflecting on experience in the ways suggested by phenomenology, and corroborating results by intersubjective exchange, we can get a clear view of *that* world we so experience. Claims to any 'world beyond', however, are thus effectively rendered moot.

Now I am deeply sympathetic to taking a step back from 'normal' or 'natural' modes of thinking – including, say, the Husserlian rejection of a hypostatization of mathematical concepts as directly indicative of 'the real' (see Gurwitsch, 1974, 44 ff.) – to reflecting deeply on one's own consciousness, and even to using variational methods in sorting out what may count as 'essential' or maybe 'objective'. However (capital 'H'), I see various problems associated with the ways phenomenologists suggest to proceed from these initial steps, and what they believe these can establish.

For instance, take the idea of intersubjective corroboration. The first kind of problem I see here is that *you*, dear reader, are an object to me. Don't take that personally: It's just meant as an epistemological claim. If you are in physical pain or suffering a terrible loss, I may feel compassion for you however strongly. But that does not mean that I actually feel what you feel. These are just the feelings I experience in relation to my auditive, visual, olfactory, and maybe even haptic experiences I have of you otherwise. These feelings may incline me to think of you as a very special object; one that has an 'inner world', very much like my own. *But that doesn't change the fact that, for me, you are among the objects constructed out of experience.*

The question thus arises why I should prefer my experiences of *that* sort of object over my experiences of *other* objects. At the level of fundamental epistemology I see no compelling reason – especially when taking into account how I can feel compassion *even for a car*. In general, it seems to me that the status of *objects* in phenomenology is all but clear. Here is how Berghofer and Wiltsche (2020, 14; orig.emph.) phrase the relevant point:

> The main question [...] concerns the relationship between consciousness and the external world: Does transcendental phenomenology only imply that the *meaning* or *sense* of the intended objects is constituted by consciousness? Or does transcendental phenomenology advance the more radical claim that the objects themselves are constituted by consciousness and that, consequently, there is no reality beyond the phenomena?

Frankly, I have no clue as to what the most natural answer to be gathered from the writings of Husserl or Merleau-Ponty would be. But I find it telling that different Husserl scholars – and presumably different phenomenologists in general – have come up with remarkably different answers:

> First, there are those who understand Husserl's [...] as a purely methodological endeavor that is consistent with both metaphysical realism and metaphysical idealism [...]. Second, there are those who

> argue that [it] inevitably culminates in a form of metaphysical idealism [...]. Third and finally, some commentators argue that transcendental phenomenology [...] can be considered [...] a rejection of metaphysical realism [...] without thereby collapsing into some sort of metaphysical idealism[.].
>
> (ibid.)

Let us zoom in very briefly on the third alternative, as defended, e.g., by Zahavi (2017). Zahavi (2017, 186) argues that Husserl's phenomenology

> is an explication of the sense that the world has for all of us 'prior to any philosophizing' [...]. Indeed, there is nothing wrong with the natural attitude and with our natural realism; what Husserl takes exception to is the philosophical absolutizing of the world that we find in metaphysical realism

However, contrast this with the following two observations:

> Husserl is adamant in rejecting the notion of an inaccessible and ungraspable *Ding an sich* as unintelligible and nonsensical [...]. To posit a hidden world that systematically eludes experiential access and justification, and to designate that world as the really real reality, would for Husserl involve an abuse of the term 'reality'[.]
>
> (ibid., 69)

> The phenomenological credo 'To the things themselves' calls for us to let our experience guide our theories. We should pay attention to the way in which reality is experientially manifest.
>
> (ibid., 151)

The problem I see associated with this rejection of the (Kantian) 'Ding an sich' and the claim to the 'things themselves' as being experientially manifest is this: If there is no thing independent of my experience, whose mere *thought* is the Kantian notion of a 'Ding an sich' (Allison, 2004, 3.I), then either phenomenology *does* collapse into an idealism of the Berkeley-variety, or it verges on something incomprehensible: Claiming that the (philosopher's beloved) chair-as-experienced is the chair itself while simultaneously denying that it also has an 'an sich'-ness, i.e., an existence completely independent of my (modes of) experiencing it, must mean that the chair *pops in and out of existence whenever I turn towards/away from it* – worse than Einstein's bed![9]

This leaves the other two options (pure methodologism or Berkeley-anism), the former of which might seem promising for *science*. However,

I honestly doubt that 'letting experience guide our theories' does justice to actual scientific practice, when the latter's success is measured by (use-) novel predictions and the production of new phenomena and technologies. In particular, I doubt that the phenomenological method could have brought us to QT: 'Nobody has ever understood what the hell Heisenberg was [...] smoking [...] when he invented matrix mechanics' (Susskind, 2008, 15:19–15:31).[10]

So apparently, there is something other than attention to intersubjectively communicable experiences going on in theory construction, especially in modern physics.[11] But it is equally unclear to me whether phenomenology can help ground talk of scientific *entities* such as '"particles" [...] that come to Alice and Bob from a common source'.[12]

Take a quark. Exercising the epoché, I should maybe not take the QCD Lagrangian, the collective evidence from colliders, or even the fact that I can somehow associate particles with isolated mass shells as providing a picture of the quark being 'discovered out there'. I am more than happy with that. The next step, the phenomenological reduction, might mean attending to the displays of apparent 'tracks' in computer-generated images of detectors, and how they can be traced back to a certain interaction point and matched up with certain types of interaction. But I'm beginning to feel that this is hardly what phenomenologists have in mind. Furthermore, when it comes to intersubjective corroboration, I am completely lost: Why would it help to compare the result of the foregoing process with the results of similar processes as undergone by others, in an effort to sort out what a quark, essentially, is?[13]

Quite certainly, a lot more could be said here and there will be more charitable ways of reading phenomenology. Ryckman (2007, Ch. 5–6), for instance,[14] carefully argues for a definite impact of phenomenology on Weyl's thinking and demonstrates the influence of phenomenological ideas such as 'Wesensschau' on Weyl's development of his geometry. Nevertheless, even in this case, several things remain unclear to me.

First, Weyl's approach is displayed by Ryckman (2007, 144) as an exercise in 'regional ontology', which presupposes the (local) acceptance of the natural attitude. It remains unclear to me in how far this approach really trades on basic phenomenological ideas and cannot be largely detached from them. Second, it is not clear to me in how far Weyl's approach even succeeds, given his somewhat *ad hoc* response to the Einstein-Pauli 'prehistory' objection (discussed at some length in Ryckman, 2007, § §4.2.4 ff.), of an effective, dynamical washing out of history-induced effects on atomic spectra. Einstein's elevating the apparent approximate invariance of atomic spectra to something motivating a general physical principle (see Giovanelli, 2014, 27) might be seen as a superior move, and, I believe, is somewhat consistent with the epistemological position I advertise below. Third, it is not clear just how close Weyl's attachment to phenomenology really is, as pointed out, e.g., by Bernard and Lobo (2019) or Sieroka (2019), and underscored by Weyl's self-admitted interest in thinkers such as Fichte, Plato, Hume, and others.[15] Finally, even if geometry was a field in which the phenomenological method could be put to good use, it remains unclear to me whether the same is true of quantum physics – wherein 'Evidenz' and 'Anschaulichkeit' (Ryckman, 2007, §5.4.1; §6.3.1) clearly become touchy subjects (however, see French, 2020, in this connection).

Hence, I believe the concerns raised in this section do point to some serious challenges for phenomenology, wherefore I now turn to a philosophical stance that, to me, seems more clearly capable of circumventing these or similar concerns.

10.4 Back to Kant! (... and Then a Big Leap Forward)

When the 19th-century 'neo-Kantians', such as Otto Liebmann, took issue with German idealism's reception of Kant, they coined a notion that was later paraphrased as *Back to Kant!* (cf. Ollig, 2017, 9 ff.). It is interesting to realize, in this context, that 'despite all kinds of [...] differences' the basic approach and methodology of phenomenology is also 'firmly situated within a certain Kantian or post-Kantian framework'. For it takes to heart

> the realization that our cognitive apprehension of reality is more than a mere mirroring of a pre-existing world. Rather, a philosophical analysis of reality, a reflection on what conditions something must satisfy in order to count as 'real', should not ignore the contribution of consciousness.
>
> (Gallagher and Zahavi, 2008, 23–24)

As I have argued above, there are several respects in which phenomenology employs this Kantian heritage in a way that I find objectionable. I will hence follow the neo-Kantians' call and take inspirations more directly from Kant (though maybe also not too stringently).

Now Kant [CPR, A158/B197] was famously concerned with 'conditions of the possibility of experience in general' that would 'at the same time' be 'conditions of the possibility of the objects of experience themselves, and thus possess objective validity in a synthetical judgment *a priori*'. In the course of sorting these out, he declared space and time 'pure forms of our sensibility', and objects to be 'representations [...] which [...] are connected and determinable [...] in space and time [...] according to laws of the unity of experience' (A494/B522). Hence, in a Kantian view, objects are (involuntarily) *constructed*, or constituted, out of experience by the mind according to a fixed scheme.

However, thus declaring space and time pure forms of sensibility also misled Kant into endowing the principles of *Euclidean* geometry (the only geometry he knew of)[16] with the status of a synthetic a priori (A47/B64), and this move became untenable with the rise of non-Euclidean geometries in the 19th century (Friedman, 1999, 6). However, Reichenbach (1920, 46) pointed out that 'the notion of an a priori has two distinct meanings in Kant. First, it means something like "apodictically valid", "valid for all times", and second it means "constitutive for the concept of an object"' (my translation – FJB).

Several authors (d'Espagnat, 2011; Friedman, 1999, 2001; Mittelstaedt, 2009; Reichenbach, 1920) have hence suggested to dispose of the first meaning while keeping the second intact. One might then sort out what Reichenbach called the 'axioms of coordination' of a given theory Θ, which contrast with 'axioms of connection'. The former ones 'must be laid down antecedently to ensure [...] empirical well-definedness in the first place' (Friedman, 1999, 61), and so provide 'structurally and functionally [...] that without which the rest of a theory would lack content' (Howard, 2010, 337) – or in yet other words, that which is merely *constitutively* a priori. The latter ones are 'empirical laws in the usual sense involving terms and concepts that are already sufficiently well defined' (Friedman, 1999, 61).

In the version endorsed by Friedman (1999, 66), what is constitutively a priori may be sorted out by determining invariants of a given theory under a relevant group of transformations. Such invariants are also sometimes called *symmetries* (though sometimes this name is also given rather to the transformation-group), where 'the symmetry of a 'something' (a figure, an equation, ...) is defined in terms of its invariance with respect to a specified transformation group, its symmetry group' (Castellani, 2003, 322).[17]

The attentive reader will have already noted a vague similarity between the general description of symmetries and the eidetic variation. In the eidetic variation, the task is to sort out the essence of a given something, and this is done by removing as many contingencies as possible. More abstractly, all possible conditions under which a given something can be viewed are considered, and that which is unchanging under this variation of perspective or circumstance is then considered the eidos. But on the same level of abstraction, nothing else really happens in theoretical considerations of symmetries in modern physics.

To see how this more clearly, let's once more resume the dialog that establishes Wallace's challenge and extend it even a little further:

Q_7: How is a particle supposed to be associated with a field?

A_7: If a quantum system is in thermal-equilibrium state $\rho(\beta)$, the 'two-point function' of that system with respect to field $\hat{\phi}(x)$ and that state is

$$G_2(x-y;\phi,\beta) = \mathrm{Tr}(\rho(\beta)\hat{\phi}(x)\hat{\phi}(y)) \tag{10.9}$$

If the Fourier transform of that state has a pole, there's a particle associated with it.

Q_8: That's a weird postulate.

A_8: It's not a postulate; it's something you derive, by looking at the dynamics of states obtained by excitations of the thermal-equilibrium state. Where there's a pole, there's a subspace of states which can be interpreted as superpositions of singly localized excitations and which is preserved under the dynamics.

Q_{8a}: Wait, but didn't you say the particle was a subsystems whose Hilbert space bears, in the interaction free limit, an irreducible representation of the Poincaré group?

A_{8a}: Yes, this subspace bears the irreducible representation.

Q_{8b}: But isn't *that* a postulate: that the irreducible representation you retrieve somehow identifies the particle?

A_{8b}: Well, no. Wigner actually *showed* that there are two invariants under the Poincaré group, which, for a massive particle, are mass and spin.

Q_{8c}: But that doesn't 'show' that a particle *has* the properties corresponding to these invariants, does it? For otherwise we would appear to be moving in circles. So I take it that this is the postulate then: That a particle can be identified through that whichever remains invariant under the transformations specified by the Poincaré group?

A_{8c}: *mumbles* In a way … I guess, but … *mumbles*

To unpack this dialog a little, recall some details.[18] As is well known, Wigner (1931) first showed that symmetry transformations in QT are represented by unitary or antiunitary operators. Focusing on the proper orthochronous (inhomogeneous) Lorentz group ($|\Lambda| = 1$ and $\Lambda^0{}_0 \geq 1$), which is a connected Lie group and can be combined with parity and time inversion to reproduce any element of the whole Lorentz group, one can use a continuity argument (the connection to the unit element) to argue that any operator representing an element of that group must be unitary rather than antiunitary.

Focusing, further, on infinitesimal transformations $\Lambda^{\mu}{}_{\nu} = \delta^{\mu}{}_{\nu} + \omega^{\mu}{}_{\nu}$ with additional infinitesimal translation ε^{μ}, it is then possible to show that the group of unitaries representing these has generators $J^{\rho\sigma}$ and P^{ρ} whose commutation relations exhibit the relevant Lie algebra structure. Furthermore, defining all states of definite momentum p^{μ} in terms of states of a fixed reference momentum k^{ν} (e.g., that defining the system's rest-frame, or, for massless systems, one where its three-momentum lies along the z -axis), it is possible to induce a representation of the whole group from what is known as the 'little group', the subgroup of transformations $W^{\mu}{}_{\nu}$ that leave k^{ν} unchanged.

After sorting out the physically interpretable cases ($p^2 \leq 0$ and $p^0 > 0$), it becomes possible to classify massive representations according to their (continuous) mass value and their spin-quantum number j and massless representations by the remaining three-momentum component and the helicity, *because these are invariants of the little group*.

This doesn't tell us anything about QFT and poles yet. To make the connection, first note that the trace functional $\mathrm{Tr}(\rho(\beta)\hat{\phi}(x)\hat{\phi}(y))$ computes the expectation value of $\hat{\phi}(x)\hat{\phi}(y)$ w.r.t. $\rho(\beta)$. In what is called the 'thermofield dynamics'-formalism (e.g., Khanna et al., 2009), one can seek out a 'thermal vacuum state' $|0(\beta)\rangle$ such that this can be recast in the more homely notation $\langle 0(\beta)|\hat{\phi}(x)\hat{\phi}(y)|0(\beta)\rangle$, where we can vividly see the 'excitation' of the equilibrium by means of the operators. This induces the need to introduce additional 'thermal operators' and changes the Lie algebra generating the unitary representation of the ('thermal') Poincaré group, but we can ignore this here.

Assuming, for simplicity, also that $\hat{\phi}(x)$ is a scalar operator and using $|\Omega\rangle$ as the vacuum state of a generic interacting theory, we recall that the time-ordered version of $\langle\Omega|\hat{\phi}(x)\hat{\phi}(y)|\Omega\rangle$ can be rewritten as

$$\langle\Omega|T[\hat{\phi}(x)\hat{\phi}(y)]|\Omega\rangle = \sum_{\lambda}\int \frac{d^4p}{(2\pi)^4}\frac{i}{p^2+m_{\lambda}^2-i\varepsilon}e^{-ip(x-y)}\left|\langle\Omega|\hat{\phi}(0)|\lambda_0\rangle\right|^2, \quad (10.10)$$

where $|\lambda_0\rangle$ is an energy eigenstate in the rest-frame ($\boldsymbol{p}=0$).

The Fourier-transform of the first term, which, up to normalization and choice of vacuum, corresponds to the interaction-free limit, is thus proportional to $1/(p^2+m^2-i\varepsilon)$. For $\varepsilon \to 0$, it has a pole at $p^2 = -m^2$, which coincides with the particle being 'on shell' and satisfying the relativistic energy-momentum relation. Hence, we exactly retrieve something which has a space of states associated with it that can be expanded in terms of momentum and labelled by mass and spin.[19] Furthermore, we can immediately also see in what sense that space is 'preserved under the dynamics' by realizing that the infinitesimal generators of the unitary representation of the Poincaré group commute with the energy operator P^0, which generates the dynamics. Hence, it makes no difference as to whether we 'move around' in the space first and then let the dynamics run or *vice versa*; the reachable space remains the same.

Two things are crucial here: (a) The group itself has two invariants, which are spin and mass (or three-momentum and helicity) and (b) the dynamics has an invariant, which is the entire (sub)space bearing the group's representation. But this really gives us quite a lot: a particle of a given *type* can, up to considerations of charge, be identified by the unchanging values of mass and spin, and the particular particle itself can be identified as that whichever is described by the ('non-Boolean', contextual set of) assertions allowed by (projections onto) the dynamically unchanging subspace. Hence, invariants can be used, exactly, for sorting out what a particle *is*.

I'm neither the first to deliver such an analysis nor the first to associate it with Kant (e.g., Auyang, 1995; Falkenburg, 2007; Mittelstaedt, 1978). What I would hence like to do here is (i) draw the red line from Kant to Friedman, with a few added details (cf. Boge, 2021a); (ii) draw the line between Kant and phenomenology more clearly, with an eye on the constitutive use of *theoretical* symmetries; and (iii) draw conclusions as to how a Kantian line of reasoning might help QBism defend itself against EP and SP.

10.4.1 *From Kant to Friedman (and Beyond)*

It is interesting to first expound on how the use of invariants in constituting objects is, indeed, essentially Kantian. This fact is mentioned almost in passing by several Kant scholars (Allison, 2015, 340; Rosenberg, 2005, 250; Schrader, 1951, 520), but given the weight I will put on it here, it may be helpful to look more deeply into the way in which Kant himself deploys invariance-based arguments

Let's begin with the transcendental aesthetic. Recall that Kant called 'all cognition *transcendental* that is occupied not so much with objects but rather with our *a priori* concepts of objects in general' (CPR, A11–2/B25; orig.emph.). The task of the transcendental aesthetic, construed as 'a science of all principles of *a priori* sensibility' (A21/B35), was to sort out the 'mere form of sensibility in the mind' (ibid.), where by 'form' Kant means 'that which allows the manifold of appearance to be intuited as ordered in certain relations' (A20/B34). In the end,

Kant famously comes down with two such principles, namely space and time; but what really interests us here is the *way* in which he arrives at them:

> In the transcendental aesthetic we will […] first isolate sensibility by separating off everything that the understanding thinks through its concepts, so that nothing but empirical intuition remains. Second, we will then detach from the latter everything that belongs to sensation, so that nothing remains except pure intuition and the mere form of appearances, which is the only thing that sensibility can make available *a priori*.

Hence, in order to isolate what constitutes the pure (i.e., content-free) form of sensibility, it is necessary to detach it from any particular context of understanding as well as from any particular sensation. However, thus detaching it from all particular sensations, one realizes that spatiality itself remains throughout all possible sensations:

> in order for certain sensations to be related to something outside me (i.e., to something in another place in space from that in which I find myself), thus in order for me to represent them as outside one another, thus not merely as different but as in different places, the representation of space must already be their ground. […] One can never represent that there is no space, although one can very well think that there are no objects to be encountered in it.
>
> (A23–A24/B38–B39)

In other words: regardless of its content and regardless of how this content is ordered, we cannot 'imagine space away', even if we can at least *think* an endpoint to this variation in content wherein it is entirely empty (cf. Mohanty, 1991, 262). Hence, space (or spatiality) is an invariant of this variational process, and so must be part of what constitutes an object of sensation.

A similar reasoning chain is encountered in the transcendental analytic, when Kant offers his 'deduction' of transcendental consciousness:

> it is this one consciousness that unifies the manifold that has been successively intuited, and then also reproduced, into one representation. This consciousness may often only be weak, so that we connect it with the generation of the representation only in the effect, but not in the act itself, i.e., immediately; but regardless of these differences one consciousness must always be found; even if

> it lacks conspicuous clarity, and without that concepts, and with them cognition of objects, would be entirely impossible.
>
> (A104)

Here Kant tells us that, no matter how much we transmute the *content* of experience, the fact that this experience is *conscious* cannot be altered. That *one* consciousness to which all these experiences are attached is the invariant of all experiences.

There are several things to note here, the first being that many a philosopher of physics will certainly object to the connection between Kant and the use of symmetries in modern physics I am trying to promote here because, say, abstract symmetries like local gauge invariance do not fit the bill. I am not convinced that this is true, and I consider myself in good company with this (e.g., Falkenburg, 2007; Janas et al., 2022; Lyre, 2009; Mittelstaedt, 1978, 2009). To my mind, the crucial observation is Kant's apparent, illegitimate slide from constituting a *manifest* image to considering the foundations of that image as also necessarily being the foundations of any possible *scientific* image.

With 'synthetic' replaced by 'constitutive', we can detach the 'apodictic validity' from the constitutive elements of the manifest image so constituted, and see how a specifically scientific image – which is preferable for some but not *all* purposes – can be constituted by means of much more abstract and less intuitive symmetries.

That being said, the second thing to note is that both the *generation* and constitutive *utilization* of these more abstract symmetries can be guided by decidedly pragmatic elements, as I have argued at some length in Boge (2021a). To illustrate this very briefly, refer once more to the modified Wallace-dialog considered in this section. I had the sceptical inquirer claim, in Q_{8c}, that it is a postulate that a particle can be identified through that whichever remains invariant under the Poincaré group. But did we not see mass and spin *follow*, by a symmetry-based argument, as two 'natural' identifying properties qua Poincaré invariants in relativistic QFT?

In a way, this is correct, but why should one even *use* the Poincaré group in the first place? Why a *unitary* representation thereof? Ultimately, this is justified by empirical success, and the limitations of this very success may ultimately justify the move to some working quantization of general covariance. But neither Poincaré invariance nor unitarity were *directly suggested* to us by evidence.

I already mentioned above how '[n]obody has ever understood what the hell Heisenberg was [...] smoking [...] when he invented matrix mechanics' (Susskind, 2008, 15:19–15:31). But of course the use of Hermitian matrices, first introduced as tables of numbers by Heisenberg, is ultimately the root of the unitary evolution in QT. These weren't *forced* upon Heisenberg by experience, but he somehow managed to get his head around the idea by appealing to a mix of values, preferences, purposes, etc. Similarly, Einstein wasn't forced to embrace the principles of special relativity by sheer evidence: Lorentz's ether program, in which the Poincaré transformations or their divergence from the Galilei transformations would have an *empirical* meaning, was still relatively respectable at the time. And Einstein was at best vaguely aware of the Michelson-Morley experiment: He based his ideas on a desire for a theory that *uniformly applies* to both mechanics and electrodynamics in a coherent fashion (Zahar, 1989, chapter 3).

The point is that any 'Kantianism' able to cope with this sort of development in modern physics must be a kind of *pragmaticized* Kantianism. Such a position has been attributed to Bohr by (Folse, 1994, 121–122), and described as follows:

> Pragmatized Kantians defend their claims to knowledge through appeal to the pragmatic virtues of the categories under which the content of experience is subsumed. [...] Bohr's work in philosophy is in effect simply this: a campaign to revise the limits of application of key concepts in the physicist's synthesis of the experiences which form the empirical basis of our knowledge of the atomic domain.

However, above I pointed out that there is a further sense in which pragmatic elements enter into constitutive efforts; namely in *comprehending* certain symmetries *as* being constitutive. Repeating what has become my favourite example, the *scaling invariance* of certain cross sections can be used to *define* what is meant by the 'pointlikeness' of elementary particles (Drell and Zachariasen, 1961; Falkenburg, 2007).

Quite often, particles are introduced simply as 'bumps in the field', based on the fact that the free propagator corresponds to something which (almost) satisfies the relativistic energy-momentum relation, and that its Fourier transform can be formally read as the probability of 'finding something at x', given that there 'was something at y'. Hence, when $|\Omega\rangle$ is naïvely considered the 'state of a field', the structure of these results looks suspiciously as if QFT was telling us to expect 'a disturbance in the field to propagate from [y to x]' (Zee, 2010, 24). However, not only do several results (Halvorson and Clifton, 2002; Malament, 1996) imply tight limitations on the localizability of this 'disturbance'; any alternative in terms of 'properties of the field', wherein $|\Omega\rangle$ is literally thought of as a 'state', obviously runs into the notorious measurement problem.

A slightly less problematic phrasing has particles be 'pointlike' in the sense that 'we construct interaction Hamiltonians by multiplying the relevant fields at exactly the same spacetime point' (Duncan, 2012, 164).

However, that phrasing still encourages the problematic interpretation of particles as properties of fields. The only *coherent* way to make sense of this I know of is that 'pointlike' really means 'structureless', and that in the sense of 'not being breakable into finer pieces' (Falkenburg, 2007). However, the very definition of pointlikeness is thus an experimental condition: that of a scattering cross section being (approximately) *scaling invariant* under conditions where the scattered particles can be considered 'essentially free'. For, this tells us that what it means for entities to be 'pointlike' is to scatter off of one another in the same ways, no matter how hard they are smashed together. This 'hardness' is quantified by the interaction scale Q^2, and for protons scattered at a Q^2 where they can be construed as 'collections of almost free quarks and gluons', this is supported by evidence.

Now scaling invariance is certainly nothing like Poincaré invariance: the former is usually construed as a *fundamental* symmetry of the theory, the latter somehow as an emergent property exhibited under constrained conditions.[20] In Boge (2021a), however, I have argued that scaling invariance can be seen as the symmetry of a *sub-theory* of the Standard Model, namely, its *scattering theory*. In brief, the idea is as follows.

Many of the steps necessary for deriving a certain cross section are in no way sufficiently constrained by the fundamental assumptions defining the Standard Model; so it makes sense to take the 'theory' in 'scattering theory' seriously. Nevertheless, the shape a cross section can take on is constrained also by the principles underpinning the Standard Model, whence it makes sense to call its scattering theory a *sub*-theory (Boge, 2021a, for more details). In this way, however, scaling invariance can have a non-accidental, even though also in a sense non-fundamental, status; and being non-accidental, it can thus also function as a constitutive principle induced by the *application* of a more encompassing theory to a certain problem-set (scattering scenarios).

A second crucial observation here is that scaling invariance is always only *approximate*. For partons, this could be expected even with little knowledge of QFT, as presumably anybody has heard about 'quark confinement'. And if quarks and gluons must be considered constituents of hadrons, any 'probe' will have to interact with the whole hadron (thus indicating substructures through scaling-non-invariant scattering behaviour). However, even an electron is best thought of as only ever *asymptotically* free; for otherwise one's theory is necessarily interaction-free (cf. Bain, 2000).

Thus, if we think that scaling invariance tells us to constitute quarks as structureless, an approximate, non-fundamental symmetry can have a constitutive function. Furthermore, this symmetry would become exact when the theory would be literally free. It is interesting to note, in that respect, that for most practical purposes, any cross section involving hadrons can usually be expressed as a weighted sum of cross sections on the parton level, i.e., as defined in terms of definite patron momenta ('free parton wavefunctions'). This has only been 'proven' (in the physicist's sense of the word) for a number of selected cases, but the resulting 'factorization' of cross sections involving hadrons into parton-level cross sections is applied ubiquitously for high energetic scattering events with hadronic scatterers, on account of heuristic arguments for generalizability (e.g., Schwartz, 2014, section 32.5).

When provable, this factorization bears the hallmarks of a *decoherence theorem* (also Schwartz, 2014, 674): The final form nicely separates the non-elementary cross section into a probability-weighted sum over elementary cross sections (for leptons and partons), and so suggests a basis of partonic rather than hadronic (momentum) states. Since cross sections include a matrix element squared, this is formally analogous to computing with a mixed state over different parton momenta and flavours. Furthermore, since the interaction scale will be determined by the four-momenta of the scatterers (the electron and the proton) and thus increases as both or one of them is accelerated in the lab-frame, the result becomes more and more valid as Q^2 increases.

What is interesting about this in the present context is that we can see another *approximate symmetry* at work here, namely the dynamical invariance of the preferred basis (the pure parton states) under increasing acceleration. Quite generally, decoherence can be described by the condition $UA \approx AU$, where A is a (preferred) system-observable, and U the interaction between system and environment. Multiplying both sides from the left with $U^\dagger$, we see that this corresponds to $A \approx U^\dagger AU$; so A remains approximately unchanged when the time is evolved according to U.

Decoherence usually has the consequence that 'classical' observables, such as position and momentum, are dynamically singled out – and both simultaneously, to a respectable degree. Hence, decoherence is often equated with an 'emergence of classicality' (Joos et al., 2003; Schlosshauer, 2007). However, as a general assessment, this seems inappropriate for two reasons: (a) In the scattering case considered here, the preferred states which are used in the 'dynamically emergent mixture' are plane-wave states (momentum eigenstates), and so not very classical (perfectly indefinite in position). Hence, decoherence at best *often* sanctions a classical treatment. And (b), a state-non-representationalist obviously cannot interpret the transition to an approximately diagonal density matrix as a dynamical process in which the interference between terms in a superposition become suppressed, and a classical trajectory (or maybe a multitude of neatly separated 'worlds') literally 'emerges' in consequence.

I suggest a more general view of decoherence as a *bridging principle*: It sometimes specifies under what conditions we can treat physical systems the way we always have (as buzzing around in space); sometimes under what conditions we can treat protons as mere collections of quarks and gluons with well-defined speeds; and sometimes maybe yet other things. But unless we buy into a many-worlds interpretation – and there are pretty good reasons for abstaining from this (Boge, 2016, 2018) – this ability to think in terms of trajectories or fixed momenta for elementary particles is *only* given if the content of the quantum state is considered inherently *probabilistic*: These theorems do not deliver a 'selection step' (Fuchs and Schack, 2012, 246).

In at least some agreement with Healey (2012), I suggest that decoherence theorems urge us to distribute our credences in a certain way, namely across the different values of the magnitudes that are approximate dynamical invariants under the evolution considered. Certainly, decoherence theorems do not exist for all contexts in which we may want to assign credences in accordance with one density matrix or another. However, when they do exist, they tell us that, relative to the model of the dynamics we have chosen, certain quantities come out as those across whose values we *should* distribute our credences, and hence as in a certain sense *objective*.

To sum up: In an extension of the Friedmanian program, I suggest to pay attention not only to fundamental and exact symmetries, but also to symmetries that become important in certain contexts of application, as well as certain approximate symmetries. The former ones can be seen as fundamental, or at least constitutive, symmetries of *sub-theories* which arise in the context of applying a theory to a problem set and have a life of their own; the latter ones as bridging principles between two different theories, such as hadron and parton-level, or even quantum and 'classical' theories. This I see as the coherent execution of at least part of the Kantian program, wherein invariant elements occupy a crucial role in determining what is *objectively fixed*. However, in contrast to (what I take to be) Kant's original doctrine, I concede that *scientific* objectivity is determined far more *opportunistically* – relative to a sufficiently large but not boundaryless *problem set* – and so not in the same sense transcendental.

10.4.2 *Kant vs. Phenomenology*

As already indicated above, there is an obvious connection between Kant's method of reasoning and the eidetic variation: Clarke (2014, 268) describes Kant's method in the transcendental aesthetic as 'a precursor to Husserl's method of eidetic variation', and similar parallels are drawn by Wiesing (2014, 64) or Mohanty (1991). The question thus arises what the defining differences are.

I have highlighted the difficulties I see associated with phenomenology, and I believe that the most important difference – for my purposes at least – lies in the status of synthetic a priori judgments. For instance, Gallagher (1972) argues that Kant employs the synthetic a priori to refer to necessary *forms* of experience only; and thus to *structural* knowledge about the way things are bound to appear to us (also Ladyman, 2020, section 3.1). The particular necessary structures he thought he could derive, among other things, from the shape of the existing mathematics and mathematical physics of his time (Gallagher, 1972, 342; Friedman, 2001, 10). But as I have argued above, (a) this makes for an important *continuity* to the project I am undertaking here and (b) the *necessity* can be removed on the pains of a loss of certainty alone. By contrast,

> Husserl is holding for [...] a necessity which is based upon insight into essential connections between the *content* of subject and predicate. In this sense, the insight into necessity, far from being a formal condition for the experience of objects, is rendered possible *through* the experience of certain objects.
>
> (Gallagher, 1972, 343; orig.emph.)

So whereas Neo-Kantians in the Reichenbach-Friedmanian tradition essentially suggest to significantly *weaken* the synthetic a priori, while simultaneously retaining its connection to scientific and mathematical theories, Husserl, and presumably most phenomenologists following him, urge to *strengthen* it into an all-pervading guide to impermutable conceptual structures. But it was the extraordinary genius of certain scientists to *let go of* certain apparently impermutable conceptual structures which made possible the scientific revolutions that gave us modern science and technology.

Hence, unlike phenomenology, I suggest to let go of certainty and embrace progress. Nevertheless, insight into conceptual structures that we *create* on the way in this need-driven, flexible, opportunistic process can be generated in ways that are rooted in Kant and compatible with at least part of the phenomenological project as I understand it.

An important objection might cross one's mind at this point: Did I not say that one major problem of phenomenology was the ineffable connection between consciousness and reality therein, and isn't every form of Kantianism haunted by that same problem? Frankly, didn't Husserl reject the Ding and sich due to its very ineffability, as evidenced by the notorious 'problem of affection' in the Kant-literature? For 'affection' appears to be a causal notion and we are thus, apparently, (i) either only affected by appearances (so Kant's position collapses into Berkeley's), or (ii) causation fulfils a dual role and can reach out into a 'noumenal world' (in defiance of Kant's view of causation as a form of experience), or (iii) that there are two distinct kinds of causation ('noumenal' and 'phenomenal' causation); a kind of silly fix. However, I believe that the thorough Kant exegesis of Henry Allison (2004) has led to a satisfying solution to this problem:

> [T]he Kantian theory of sensibility not only requires that something 'affect' or be 'given to' the mind; it also maintains that this something becomes part of the content of human cognition [...] only as the result of *being subjected* to the apriori forms of human sensibility (space and time). [...] The point is [...] that, insofar as [the spatiotemporal objects of human experience] are to function in a transcendental account as material *conditions* of human cognition, they cannot, without contradiction, *be taken under* their empirical description.
>
> (Allison, 2004, 67–68; emph. added)

The key to (dis)solving the problem of affection is hence to read Kant as an epistemologist, not a metaphysician in the modern sense of the word: All talk of 'things in themselves' and 'affection' is intended as an analysis of what it means for sensibility to be receptive and for cognition to be discursive (to require concepts and sensible intuitions). Beyond that, Kant can simply *remain silent* about the status of 'super-empirical entities' or their relation to us (Allison, 2004, 73). It is unclear to me whether such a move is available to phenomenology, with its focus on the 'things themselves'.

10.4.3 *What's to Gain for QBism (or State Non-Representationalism More Generally)*

Let us recall that I had identified two basic problems that I see associated with QBism:

(EP) By *selectively* using recalcitrant experiences as grounds for postulating a 'world out there', QBism undermines its own basis for doing so.
(SP) By putting 'naked' experiences at centre stage, QBism undermines its own basis for employing 'higher level' concepts.

As for SP, recall also that I argued that phenomenology offers a source of 'substantive a priori knowledge' (Nagel, 2000, 346), and hence could help QBism circumvent SP. However, the same, I believe, is true of the Neo-Kantianism I am advocating here, even if the amount of a priori knowledge is significantly sparser and more detached from everyday-life modes of thinking. All we need to circumvent the SP is a creative contribution of the mind in terms of not directly experience-related concepts. Hence, the invariants contained in theories of modern physics may do this job just as well – even if our vision of what we take the world to be when viewed through the lens of the given theory may thereby end up being equally sparse.

To see this more clearly, refer once more to Wallace's second level challenge; that a dialog on such things as the quark-gluon plasma becomes incomprehensible if we do not consider concepts connected to the quantum formalism as at least intended as representations. We had seen above how a symmetry-based argument can, in the interaction-free limit, identify a particle as that whichever is characterized by spin and mass (or three-momentum). However, more generally speaking, we can also always associate a preferred basis to a given system, relative to a given situation, whose projections represent the properties it has in that situation, with the basis either singled out by decoherence or in some other way (e.g., by considerations of preparation and measurement). Since unitaries only leave the whole 'non-Boolean lattice' invariant, however, the system is that whichever carries the whole collection of properties; not something characterized by a select set of continuously evolving properties (Mittelstaedt, 2009; see similarly Janas et al., 2022).

Now QBists may object that this view is too narrow: POVMs as a generalization of PVMs (projector valued measures) have established themselves as representations of properties in a quantum context. While true, I believe this is not an objection: Obviously, POVMs too have symmetries, and these determine the possible situations the system identified by the collection of properties represented by the POVM can be in cf. Decker and Grassl (2007, for an example). Furthermore, singling out the symmetries of a *collection* of POVMs, with elements from different POVMs *not* jointly resolving the unity, one may similarly generalize the treatment in terms of non-Boolean *lattices* sketched above. So I see no principled differences between PVMs and (general) POVMs.

So much for the SP. The treatment of the EP I consider far more interesting. For first of all, how is *abduction* at all related to the constitutive Kantian project I have bought into here? Usually, abduction is used in the context of stronger realist positions; say in the form of no miracles-style arguments: This and that theory has unrivalled success; if that success was due to a correspondence between (crucial aspects of) the theory and a mind-independent reality, it would be expected; therefore, it probably is due to that.

However, that is certainly not what the Kantian can have in mind (if and) when she talks of an 'abduction base' for constructing reality. A very first point to note are the scare quotes I have employed when I have considered Kant's 'deductions' above. For surely, the arguments Kant deploys in the Critique are not stringent deductive arguments in the modern sense of 'deduction':

> Jurists, when they speak of entitlements and claims, distinguish in a legal matter between the questions about what is lawful (*quid juris*) and that which concerns the fact (*quid facti*), and since they demand proof of both, they call the first, that which is to establish the entitlement or the legal claim, the *deduction*.
>
> (Kant, 1998, A84/B116)

This establishment of what is lawful in court is clearly similarly uncertain (and thus non-deductive, in the modern sense) as is the establishment of facts. For otherwise, there could not be precedents which determine the future understanding of what is lawful according to a set of laws. What jurists seek out in 'deductions' is the best *interpretation* of the existing laws. However, that is clearly not what Kant is after.

Lyre (2009, 493), indeed, analyses Kant's reasoning as *abductive*: 'the transcendental argument structure [...] should perhaps [...] be reconstructed as an inference to the best explanation: the existence and validity of [preconditions of experience] PE is the most plausible explanation for [experience] E'. Hence, transcendental arguments deliver a preferred *explanation* for the structure of particular *experiences*. However: 'it is Kant's special claim that synthetic judgments a priori are accompanied by necessity and generality. But it is exactly this demand, which should better be weakened in view of a modern revised use of transcendental arguments[.]' (ibid.)

Now, with the a priori already constitutive, handling the EP by abducing certain preconditions of experience doesn't seem sufficient; for in the original Kantian line of argument, these were always maximally *general* preconditions. Hence, at face value, the sort of abductive argument given by Kant, when viewed under the terms of a relativized, constitutive a priori, has to do rather with the discovery of *theoretical frameworks*.

I suggest that two further realizations are important for making sense of how to deploy 'transcendental-style' reasoning in order to arrive at a solution to EP. First, several scholars identify Neo-Kantianism as an anti-realist position, because it seems to 'reject [...] the metaphysical dimension of realism' (Chakravartty, 2017, section 4.1). According to the epistemological reading I have sided with here, that is not right in the sense that Neo-Kantianism implies a rejection of the metaphysical thesis that there exists a mind-independent reality. That would be idealism of the Berkeley-variety. Rather, the claim must be that questions as to the mind-independent reality of x are answerable only by reference to science's presupposing x's reality. And science is, in fact, not pursued as a mere exploration of the mind.

So in other words, Neo-Kantianism does *not* commit to the metaphysical thesis of the *non-existence* of 'noumenal' entities; it rejects the very *question* as to their 'really, really real' existence as unanswerable.

This has clarified the sense in which Neo-Kantianism is 'anti-realist'. Second, however, it is crucial to realize that 'to requests for explanation [...] realists typically attach an objective validity which anti-realists cannot grant' (van Fraassen, 1980, 13). Hence, when we have settled for a certain theory (say, the Standard Model) with certain constitutive principles, and then infer the existence of *particular* entities (say, Higgs bosons) from observations that accord with its principles, *then this cannot be understood by the Neo-Kantian as an inference to the really real reality of those entities.* I am suggesting, in other words, that the Neo-Kantian can happily indulge in abductive practices, all the while being aware that she thereby engages, in fallible, revisable ways, in the *construction* of an empirical reality in accord with the principles of an accepted theory.

The solution of the EP should now become obvious: The basis for *what* to abduce, in this constructive sense, is delivered by the subsumption of experience under the fundamental concepts of the theory accepted at the time of the inference; or, if necessary, in a process in which the fundamental conceptual structures underlying that theory are revised. Thus, the pragmaticized Neo-Kantian can embrace the existence of 'pointlike' entities with spin and mass that carry experimentally measured properties without committing either to their 'super-empirical' existence, their picturability as spinning solid spheres, or the fact that in the future she will still embrace them.

Let me here finally reassess the problem raised by the correlations encountered by Alice and Bob, when they compare their protocols of an experiment on the Mermin contraption. In particular, consider the spin singlet $|\chi\rangle = 2^{-1/2}(|\uparrow\rangle|\downarrow\rangle - |\downarrow\rangle|\uparrow\rangle)$, recently used for acclaimed loophole-free violations of Bell-inequalities (Hensen et al., 2015): Any unitary which models the two particles involved here as traveling from a common source towards two detection devices without affecting the spin-part will, in virtue of the invariance of the norm under unitaries ($|U|\chi\rangle|^2 = ||\chi\rangle|^2$), leave the perfect anti-correlation $p(\uparrow_{\text{one side}}|\downarrow_{\text{other side}}) = 1$ intact. Hence, we can identify it as an objective property of the two-particle system on the Neo-Kantian analysis.

In addition, the state is *rotation invariant*, so if both spins are measured along the same axis, the (anti-)correlation is perfect *regardless of the axis of measurement.* This fact may be spoiled in a theory that models the whole situation in a spacetime curved under the influence of gravity, because the agreed upon axis of measurement is not invariant under parallel transport (von Borzeszkowski and Mensky, 2000). But this just means that, with the change of conceptual system, the symmetry involved in rotation invariance is watered down to a merely approximate one. Hence, while the *correlation* is still constitutive of the two-particle system, its *perfectness* is only so in a limiting case.

Given the problems I had outlined with causal models of the situation above, what is the status of this relation between both particles? A fruitful view has been offered by Salmon (1984) and Gebharter and Retzlaff (2020), as that of an extra-causal, *nomological* relation. Similarly Mermin (1990, 184) observes that 'some physicists today might regard (the Aspect experiments) as no more than an extremely complicated confirmation of Malus's classical law'; or more generally, of *angular momentum conservation.*[21]

Caveat emptor: I first do not think that we thus retrieve a satisfying *explanation* of the correlations. At best, one recovers a kind of deductive-nomological explanation which is clearly in conflict with our intuitive requirements on 'explanation'. Second, one must not mistake this proposal as providing a metaphysical story in the sense that the 'law intervenes and makes it so' that the two particles correlate. Rather, so long as we commit to representing our credences by a singlet (or relevantly like) state, we commit ourselves to an *image* in which the 'pair of particles' always comes out with opposing values when measured along the same axis. This may be dissatisfying for a metaphysician, and certainly flies in the face our desire to know and understand. But it can be acceptable to a Kantian *epistemologist* as much as 'affection' and 'things in themselves' can be.

10.5 Conclusions

I have argued here that QBism, or state-non-representationalism more generally, could profit from a Neo-Kantian philosophy of science. The reason is that this allows for a solid, comprehensible abduction-basis and a solid framework for a non-reductionist semantics, thus doing justice to actual physical practice. The suggestion should actually not come as a big surprise, since a connection has been made before (e.g., Chalmers, 2014), and since not only phenomenology, the philosophy currently 'flirted with' by QBism, has its roots in Kant, but also QBism's 'old love' *pragmatism* (cf. CP, 5.452).

Furthermore, with theoretical symmetries occupying a central role, it is possible to not only hold fast to the beautiful non-representationalist solution to the measurement problem, but to also to reserve a righteous place for '*the* characteristic trait of quantum mechanics, the one that enforces its entire departure from classical lines of thought'. On the other hand, the scientific image that results from this move is a highly deprecated desert-landscape in which a lot of things we would like to ask and say must be relegated to the questions that do not have a possible answer, or to the claims that do not refer to any state of affairs, respectively. As a corollary of my treatment, QT hence most forcefully reminds us of the difficult and notorious 'split [...] between the world in which an agent lives and her experience of that world' (Mermin, 2012, 8).

Acknowledgements

I have profited from discussions with other contributors to this volume, including (in alphabetical order by last name) Philipp Berghofer, Michel Bitbol, Steven French, Jacques Pienaar, Thomas Ryckman, Harald Wiltsche, and Hervé Zwirn, as well as from comments by members of the Wuppertal Philosophy of Physics reading group, including (same order) Radin Dardashti, Oliver Passon, Erhard Scholz, and Marij van Strien. While writing this chapter, I was employed with the DFG research unit *The Epistemology of the LHC* (grant FOR 2063), so I acknowledge the funding.

Notes

1 Notably, the tensor-product/conjunction correspondence *is* presupposed in recent arguments such Frauchiger and Renner (2018); I make this explicit in Boge (2019).

2 In all fairness, I should note that Earman's paper otherwise misrepresents QBism in several ways (cf. Fuchs and Stacey, 2020), culminating in the fact that he claims QBists call themselves 'QBians'.

3 Like Mermin (1989, 10), who almost certainly coined the original slogan (cf. Mermin, 2004), I hence urge to 'rather celebrate the strangeness of quantum theory than deny it'. However, I don't find it '*foolish* [...] to demand an explanation for the correlations beyond that offered by the quantum theory' (ibid., 11; my emph.), for I'm hesitant to say that QT *delivers* an explanation (cf. Boge, 2022).

4 See Lewis (2019); see also Glick (2021) for an assessment of the relation between QBism and realism.

5 We can imagine this to be brought about by further devices hooked up to A and B respectively.

6 This is actually Sober's version of Reichenbach's argument, which avoids problems associated with Reichenbach's original account, connected to frequentism and positivism.

7 Sober (2001) actually holds the PCC to be restricted in its scope, offering an example of a seemingly robust correlation between Venetian sea-levels and British bread prices. I am not convinced that this (and similar) correlation(s) cannot be explained in terms of further causal structure with a screener, or that interesting 'classical' correlations that cannot be explained causally are ever robust (Boge, 2021b).

8 I have numbered questions and answers for reasons to become clear below.

9 Recall that Einstein explained his discomfort with the Dirac-von Neumann interpretation of QT to Putnam (2005, 624) as follows: 'Look, I don't believe that when I am not in my bedroom my bed spreads out all over the room, and whenever I open the door and come in it jumps into the corner'.

10 Cf. Heisenberg's autobiography or Rovelli (2021) for the details.

11 My scepticism here is nurtured also by certain passages from the relevant phenomenological literature, such as Gurwitsch's (1974, 59) remark that his own thorough analysis of Husserl's approach to physics results in 'no more than sketchy hints for a phenomenological theory of the natural sciences', or Wiltsche's (2021, 468) concession that 'Husserl's most noteworthy engagement with physics is a rather general analysis of the early modern mechanics of Galileo Galilei'. An oft-cited counterexample is Weyl's development of a purported unified field theory, to which I will turn briefly below.
12 See French, this volume, for similar doubts about phenomenology's scope.
13 I have here bracketed the eidetic variation for reasons to become clear below.
14 See also Wiltsche (2021) or, more generally, the other contributions to this volume.
15 I owe thanks to Erhard Scholz for pointing me to this literature.
16 This is but almost right; cf. Cuffaro (2012).
17 Nothing depends on using only *group theory* for inspecting symmetries though (Guay and Hepburn, 2009, Dardashti, 2019, e.g.). It rather requires some well-defined, structured set of transformations.
18 E.g., Peskin and Schroeder (1995); Weinberg, (1995). I assume a metric with signature $(-+++)$.
19 'Superpositions of singly localized excitations' essentially means that the representation is not co-diagonal with a spacetime one.
20 I have been confronted with this objection for the first time by Dean Rickles at the Stellenbosch Institute for Advanced Study, and I should thank David Glick for helping me sort out the right response to it.
21 This is ironic, insofar as Kant attributed a synthetic a priori status to causal closure (e.g., CPR, B134) However, if we follow (Cassirer, 1956) in equating causation with *law-likeness* – a move objectionable on certain grounds, but presumably consistent with Kant's thinking – then, given the above considerations on 'pointlikeness' and merely contextual attributability of properties, the main message of QT could indeed be taken to be that the *res* it treats of does not 'possess a substantial thinglike being, a being immediately describable by analogy with ordinary perceivable objects' (ibid., 150).

References

Allison, H. E. (2004). *Kant's Transcendental Idealism*. New Haven, London: Yale University Press, revised and enlarged edition.

Allison, H. E. (2015). *Kant's Transcendental Deduction: An Analytical-Historical Commentary*. Oxford, New York, NY: Oxford University Press.

Auyang, S. Y. (1995). *How Is Quantum Field Theory Possible?* Oxford, New York, NY: Oxford University Press.

Bain, J. (2000). Against particle/field duality: Asymptotic particle states and interpolating fields in interacting QFT (or: Who's afraid of Haag's theorem?). *Erkenntnis*, 53(3):375–406.

Bartlett, S. D., Rudolph, T., and Spekkens, R. W. (2012). Reconstruction of Gaussian quantum mechanics from Liouville mechanics with an epistemic restriction. *Physical Review A*, 86(1):012103 (25 pp).

Berghofer, P., and Wiltsche, H. A. (2020). Phenomenological approaches to physics: Mapping the field. In Berghofer, P. and Wiltsche, H. A., editors, *Phenomenological Approaches to Physics*. Cham: Springer Nature.

Bernard, J., and Lobo, C. (2019). Introduction: Structure and philosophical foundations of Hermann Weyl's work on space. In Bernard, J. and Lobo, C., editors, *Weyl and the Problem of Space*, pages v–xxiv. Cham: Springer Nature.

Boge, F. (2016). *On probabilities in the many worlds interpretation of quantum mechanics. Universität zu Köln*. BSc thesis.

Boge, F. J. (2018). *Quantum Mechanics Between Ontology and Epistemology*. Cham: Springer International Publishing.

Boge, F. J. (2019). Quantum information versus epistemic logic: An analysis of the Frauchiger–Renner theorem. *Foundations of Physics*, 49(10):1143–1165.

Boge, F. J. (2021a). Quantum reality: a pragmaticized neo-Kantian approach. *Studies in History and Philosophy of Modern Physics*. https://protect-us.mimecast.com/s/M8EQCDkY05ij1xq3ASW-YdV?domain=doi.org.

Boge, F. J. (2021b). Realism without interphenomena: Reichenbach's cube, Sober's evidential realism, and quantum solipsism. *International Studies in the Philosophy of Science*. https://protect-us.mimecast.com/s/R-LeCERZP5f1mqLpxIwiLku?domain=doi.org.

Boge, F. J. (2022). The positive argument against scientific realism. *Journal for General Philosophy of Science*. https://protect-us.mimecast.com/s/pZOtCG6Y9jfWm8kA4sp2bac?domain=doi.org.

Carnap, R. (2003). *The Logical Structure of the World and Pseudo-Problems in Philosophy*. Open Court.

Cassirer, E. (1956). *Determinism and Indeterminism in Modern Physics*. Yale University Press.

Castellani, E. (2003). On the meaning of symmetry breaking. In Brading, K. and Castellani, E., editors, *Symmetries in Physics: Philosophical Reflections*, pages 321–334. Cambridge, New York, NY: Cambridge University Press.

Caves, C. M., Fuchs, C. A., and Schack, R. (2002a). Quantum probabilities as Bayesian probabilities. *Physical Review A*, 65(2):022305.

Caves, C. M., Fuchs, C. A., and Schack, R. (2002b). Unknown quantum states: The quantum de Finetti representation. *Journal of Mathematical Physics*, 43(9):4537–4559.

Chakravartty, A. (2017). Scientific realism. In Zalta, E. N., editor, *The Stanford Encyclopedia of Philosophy*. Metaphysics Research Lab, Stanford University, Summer 2017 edition.

Chalmers, M. (2014). Quantum weirdness: All in the mind? *New Scientist*, 222(2968):32–35.

Clarke, E. (2014). Kant, Husserl, and analyticity. In Fabianelli, F. and Luft, S., editors, *Husserl and Classical German Philosophy*, pages 265–279. Springer.

Clauser, J. F., Horne, M. A., Shimony, A., and Holt, R. A. (1969). Proposed experiment to test local hidden-variable theories. *Physical Review Letters*, 23(15):880–884.

Crease, R. P., and Sares, J. (2021). Interview with physicist Christopher Fuchs. *Continental Philosophy Review*, 54:541–561.

Cuffaro, M. E. (2012). Kant's views on non-Euclidean geometry. *Proceedings of the Canadian Society for History and Philosophy of Mathematics*, 25:42–54.

d'Espagnat, B. (1995). *Veiled Reality. An Analysis of Present Day Quantum Mechanical Concepts*. Reading, MA: Addison-Wesley Publishing Co.

d'Espagnat, B. (2011). Quantum physics and reality. *Foundations of Physics*, 41(11):1703–1716.

Dardashti, R. (2019). Physics without experiments? In Dardashti, R., Dawid, R., and Thébault, K., editors, *Why Trust a Theory?*, pages 154–172. Cambridge University Press.

DeBrota, J. B., Fuchs, C. A., Pienaar, J. L., and Stacey, B. C. (2020). The born rule as Dutch-book coherence (and only a little more). *arXiv*, quant-ph(2012.14397).

Decker, T., and Grassl, M. (2007). Implementation of generalized measurements with minimal disturbance on a quantum computer. In Schleich, W. P. and Walther, H., editors, *Elements of Quantum Information*, pages 399–424. Wiley.

Drell, S., and Zachariasen, F. (1961). *Electromagnetic Structure of Nucleons*. London, Glasgow: Oxford University Press.

Duncan, A. (2012). *The Conceptual Framework of Quantum Field Theory*. Oxford University Press.

Earman, J. (2019). Quantum Bayesianism assessed. *The Monist*, 102(4):403–423.

Einstein, A., Podolsky, B., and Rosen, N. (1935). Can quantum-mechanical description of physical reality be considered complete? *Physical Review*, 47:777–780.

Falkenburg, B. (2007). *Particle Metaphysics. A Critical Account of Subatomic Reality*. Berlin, Heidelberg: Springer.

Folse, H. J. (1994). Bohr's framework of complementarity and the realism debate. In Faye, J. and Folse, H. J., editors, *Niels Bohr and Contemporary Philosophy*, pages 119–139. Dordrecht: Springer Science + Business Media.

Frauchiger, D., and Renner, R. (2018). Quantum theory cannot consistently describe the use of itself. *Nature Communications*, 9(1):3711.

French, S. (2020). From a lost history to a new future: Is a phenomenological approach to quantum physics viable? In Berghofer, P. and Wiltsche, H. A., editors, *Phenomenological Approaches to Physics*, pages 205–225. Springer.

Friederich, S. (2015). *Interpreting Quantum Theory: A Therapeutic Approach*. Basingstoke: Palgrave Macmillan.

Friedman, M. (1999). *Reconsidering Logical Positivism*. Cambridge, New York, NY: Cambridge University Press.

Friedman, M. (2001). *Dynamics of Reason*. Stanford: CSLI Publications.

Fuchs, C. A. (2002). The anti-Vaxjo interpretation of quantum mechanics. *arXiv preprint quant-ph/0204146*.

Fuchs, C. A. (2017). On participatory realism. In Durham, I. T. and Rickles, D., editors, *Information and Interaction*, pages 113–134. Springer.

Fuchs, C. A., Mermin, N. D., and Schack, R. (2014). An introduction to QBism with an application to the locality of quantum mechanics. *American Journal of Physics*, 82(8):749–754.

Fuchs, C. A., and Schack, R. (2012). Bayesian conditioning, the reflection principle, and quantum decoherence. In Ben-Menahem, Y. and Hemmo, M., editors, *Probability in Physics*, pages 233–248. Springer.

Fuchs, C. A., and Schack, R. (2013). Quantum-Bayesian coherence. *Reviews of Modern Physics*, 85(4):1693.

Fuchs, C. A., and Schack, R. (2014). QBism and the Greeks: Why a quantum state does not represent an element of physical reality. *Physica Scripta*, 90(1):015104.

Fuchs, C. A., and Stacey, B. C. (2020). QBians do not exist. *arXiv*, [quant-ph] (1911.07386).

Gallagher, K. T. (1972). Kant and Husserl on the synthetic a priori. *Kant-Studien*, 63(3):341.

Gallagher, S., and Zahavi, D. (2008). *The Phenomenological Mind*. London, New York, NY: Routledge.

Gebharter, A., and Retzlaff, N. (2020). A new proposal how to handle counter-examples to Markov causation à la cartwright, or: Fixing the chemical factory. *Synthese*, 197(4):1467–1486.

Giovanelli, M. (2014). 'But one must not legalize the mentioned sin': Phenomenological vs. dynamical treatments of rods and clocks in Einstein's thought. *Studies in History and Philosophy of Science Part B: Studies in History and Philosophy of Modern Physics*, 48:20–44.

Glick, D. (2021). QBism and the limits of scientific realism. *European Journal for Philosophy of Science*, 11(2):1–19.

Guay, A., and Hepburn, B. (2009). Symmetry and its formalisms: Mathematical aspects. *Philosophy of Science*, 76(2):160–178.

Gurwitsch, A. (1974). *Phenomenology and the Theory of Science*. Northwestern University Press.

Halvorson, H., and Clifton, R. (2002). No place for particles in relativistic quantum theories? In Kuhlmann, M., Lyre, H., and Wayne, A., editors, *Ontological Aspects of Quantum Field Theory*, pages 181–214. New Jersey, London: World Scientific Publishing.

Healey, R. (2012). Quantum theory: A pragmatist approach. *The British Journal for the Philosophy of Science*, 63(4):729–771.

Healey, R. (2017). *The Quantum Revolution in Philosophy*. Oxford University Press.

Hensen, B., Bernien, H., Dréau, A., Reiserer, A., Kalb, N., Blok, M., Ruitenberg, J., Vermeulen, R., Schouten, R., and Abellán, C., et al. (2015). Loophole-free Bell inequality violation using electron spins separated by 1.3 kilometres. *Nature*, 526(7575):682–686.

Howard, D. (2010). 'Let me briefly indicate why I do not find this standpoint natural.' Einstein, General Relativity, and the contingent a priori. In Domski, M. and Dickson, M., editors, *Discourse on a New Method*, pages 333–356. Chicago, La Salle, IL: Open Court.

Janas, M., Cuffaro, M. E., and Janssen, M. (2022). *Understanding Quantum Raffles*. Springer.

Jarrett, J. P. (2009). On the separability of physical systems. In *Quantum Reality, Relativistic Causality, and Closing the Epistemic Circle*, pages 105–124. Springer.

Jennings, D., and Leifer, M. (2016). No return to classical reality. *Contemporary Physics*, 57(1):60–82.

Joos, E., Zeh, H., Kiefer, C., Giulini, D., Kupsch, J., and Stamatescu, I.-O. (2003). *Decoherence and the Appearance of a Classical World in Quantum Theory*. Berlin, Heidelberg: Springer, second edition.

Kant, I. (1998). *Critique of Pure Reason*. London: MacMillan and Co. Translated by Paul Guyer and Allen W. Wood.

Khanna, F., Malbouisson, A. P. C., Malbouisson, J. M. C., and Santana, A. E. (2009). *Thermal Quantum Field Theory: Algebraic Aspects and Applications*. New Jersey, London: World Scientific.

Kripke, S. (1982). *Wittgenstein on Rules and Private Language: An Elementary Exposition*. Cambridge, MA: Harvard University Press.

Ladyman, J. (2000). What's really wrong with constructive empiricism? Van Fraassen and the metaphysics of modality. *The British Journal for the Philosophy of Science*, 51(4):837–856.

Ladyman, J. (2010). The epistemology of constructive empiricism. In Dicken, P., editor, *Constructive Empiricism: Epistemology and the Philosophy of Science*, pages 46–61. Springer.

Ladyman, J. (2020). Structural realism. In Zalta, E. N., editor, *The Stanford Encyclopedia of Philosophy*. Metaphysics Research Lab, Stanford University, Winter 2020 edition.

Lewis, P. J. (2019). Bell's theorem, realism, and locality. In Cordero, A., editor, *Philosophers Look at Quantum Mechanics*, pages 33–43. Cham: Springer Nature.

Lyre, H. (2009). Structural realism and abductive-transcendental arguments. In Bitbol, M., Kerszberg, P., and Petitot, J., editors, *Constituting Objectivity*, pages 491–501. Springer.

Malament, D. (1996). In defense of dogma: Why there cannot be a relativistic quantum mechanics of (localizable) particles. In Clifton, R., editor, *Perspectives on Quantum Reality*, pages 1–10. Dordrecht: Kluwer.

Massimi, M. (2007). Saving unobservable phenomena. *The British Journal for the Philosophy of Science*, 58(2):235–262.

Merleau-Ponty, M. (2012 [1945]). *Phenomenology of Perception*. London, New York, NY: Routledge. Translated by Donald A. Landes.

Mermin, D. N. (1989). What's wrong with this pillow? *Physics Today*, 42(4): 9–11.

Mermin, N. D. (1981). Quantum mysteries for anyone. *The Journal of Philosophy*, 78(7):397–408.

Mermin, N. D. (1990). A bolt from the blue: The Einstein-Podolsky-Rosen paradox. In Mermin, N. D., editor, *Boojums All the Way Through: Communicating Science in a Prosaic Age.*, pages 177–185. Cambridge University Press. Reprinted from *Niels Bohr, A Centenary Volume*, A. P. French and P. J. Kennedy (eds.), Harvard, 1985, p. 141.

Mermin, N. D. (2004). Could Feynman have said this. *Physics Today*, 57(5):10.

Mermin, N. D. (2012). Commentary: Quantum mechanics: Fixing the shifty split. *Physics Today*, 65(7):8–10.

Mittelstaedt, P. (1978). *Quantum Logic*. Dordrecht, Boston, MA: Reidel.

Mittelstaedt, P. (2009). Cognition versus constitution of objects: From Kant to modern physics. *Foundations of Physics*, 39(7):847–859.

Mohanty, J. N. (1991). Method of imaginative variation in phenomenology. In Horowitz, T. and Massey, G., editors, *Thought Experiments in Science and Philosophy*. Rowman & Littlefield.

Nagel, J. (2000). The empiricist conception of experience. *Philosophy*, 75(293): 345–376.

Norsen, T. (2016). Quantum solipsism and nonlocality. In Bell, M. and Gao, S., editors, *Quantum Nonlocality and Reality: 50 Years of Bell's Theorem*, pages 204–237. Cambridge: Cambridge University Press.

Ollig, H. (2017). *Der Neukantianismus*. J.B. Metzler.

Peirce, C. S. (CP) (1974). *Collected Papers of Charles Sanders Peirce Volumes V and VII*. Cambridge MA: Harvard University Press. Edited by C. Hartshorne, P. Weiss, and A. Burks.

Peskin, M. E., and Schroeder, D. V. (1995). *An Introduction to Quantum Field Theory*. Reading, MA: Perseus Books.

Putnam, H. (2005). A philosopher looks at quantum mechanics (again). *British Journal for the Philosophy of Science*, 56(4):615–634.

Quine, W. V. (1951). Main trends in recent philosophy: Two dogmas of empiricism. *The Philosophical Review*, pages 20–43.

Reichenbach, H. (1920). *Relativitätstheorie und Erkenntnis Apriori*. Berlin: Springer.

Reichenbach, H. (1938). *Experience and Prediction: An Analysis of the Foundations and the Structure of Knowledge*. University of Chicago press.

Reichenbach, H. (1965). *The Direction of Time*. Mineola, New York, NY: Dover Publications, Inc. Edited by Maria Reichenbach.

Rosenberg, J. F. (2005). *Accessing Kant*. Oxford, New York, NY: Oxford University Press.

Rosenfeld, L. (1983 [1967]). Bohr's reply. In Wheeler, J. A. and Zurek, W. H., editors, *Quantum Theory and Measurement*. Princeton, NJ: Princeton University Press.

Rovelli, C. (2021). *Helgoland*. Penguin Books Limited.

Ryckman, T. (2007). *The Reign of Relativity: Philosophy in Physics 1915–1925*. Oxford University Press.

Salmon, W. (1984). *Scientific Explanation and the Causal Structure of the World*. Princeton, NJ: Princeton University Press.

Schlosshauer, M. (2007). *Decoherence and the Quantum to Classical Transition*. Berlin, Heidelberg: Springer, second edition.

Schrader, G. (1951). The transcendental ideality and empirical reality of Kant's space and time. *The Review of Metaphysics*, 4(4):507–536.

Schrödinger, E. (1935). Discussion of probability relations between separated systems. *Mathematical Proceedings of the Cambridge Philosophical Society*, 31(4):555–563.

Schwartz, M. (2014). *Quantum Field Theory and the Standard Model*. Cambridge, New York, NY: Cambridge University Press.

Sieroka, N. (2019). Neighbourhoods and intersubjectivity. In Bernard, J. and Lobo, C., editors, *Weyl and the Problem of Space*, pages 99–122. Springer.

Sober, E. (2001). Venetian sea levels, British bread prices, and the principle of the common cause. *The British Journal for the Philosophy of Science*, 52(2):331–346.

Sober, E. (2011). Reichenbach's cubical universe and the problem of the external world. *Synthese*, 181(1):3–21.

Spekkens, R. W. (2007). Evidence for the epistemic view of quantum states: A toy theory. *Physical Review A*, 75:032110.

Stacey, B. C. (2019). Ideas abandoned en route to QBism. *arXiv*, [quant-ph] (1911.07386).

Susskind, L. (2008). Quantum Entanglements, Part 1 | Lecture 4. Stanford University. https://protect-us.mimecast.com/s/FxNwCJ6YPmfpRnZKwFLwdP1?domain=doi.org (accessed 07 April 2021).

van Fraassen, B. C. (1980). *The Scientific Image*. Oxford: Clarendon Press.

van Fraassen, B. C. (1991). *Quantum Mechanics: An Empiricist View*. Oxford, New York, NY: Oxford University Press.

von Borzeszkowski, H., and Mensky, M. B. (2000). EPR effect in gravitational field: Nature of non-locality. *Physics Letters A*, 269(4):197–203.

Wallace, D. (2020). On the plurality of quantum theories: Quantum theory as a framework, and its implications for the quantum measurement problem. In French, S. and Saatsi, J., editors, *Scientific Realism and the Quantum*, pages 78–102. Oxford University Press.

Weinberg, S. (1995). *The Quantum Theory of Fields, Vol. I*. Cambridge University Press.

Wiesing, L. (2014). *The Philosophy of Perception*. Bloomsbury Publishing.

Wigner, E. (1931). *Gruppentheorie*. Braunschweig: Vieweg und Sohn.

Wiltsche, H. (2021). Physics with a human face: Husserl and Weyl on realism, idealism, and the nature of the coordinate system. In Jacobs, H., editor, *The Husserlian Mind*, pages 468–478. Routledge.

Wittgenstein, L. (1968). *Philosophical Investigations*, page 302. Oxford: Blackwell. Translated by G.E.M. Anscombe.

Wood, C. J., and Spekkens, R. W. (2015). The lesson of causal discovery algorithms for quantum correlations: Causal explanations of bell-inequality violations require fine-tuning. *New Journal of Physics*, 17(3):033002.

Zahar, E. (1989). *Einstein's Revolution: A Study in Heuristic*. Chicago, La Salle, IL: Open Court.

Zahavi, D. (2017). *Husserl's Legacy: Phenomenology, Metaphysics, and Transcendental Philosophy*. Oxford University Press.

Zee, A. (2010). *Quantum Field Theory in a Nutshell: Second Edition*. Princeton University Press.

11 The Role of Reconstruction in the Elucidation of Quantum Theory*

Philip Goyal

11.1 Introduction

A physical theory must balance two very different demands. On the one hand, it must allow us to better *grasp* some aspect of the workings of the physical world; or—as it would have been common to say in a bygone era—to better understand the mind of God. On the other hand, it must actually *work*—it must provide a conceptual and mathematical framework of some generality within which one can describe actual laboratory experiments and can make precise, novel predictions that conform to the brute facts of experience[1].

In the *developmental* phase of a theory, if push comes to shove, the demand for workability usually wins out. Consequently, a freshly developed physical theory is inevitably a *compromise*, one manifestation of which is that certain mathematical features of the theory's formalism may lack clear physical motivation or meaning. Once the theory has been tested and the physics community is sufficiently convinced that its formalism captures some basic regularities in nature's workings, there usually follows a *reflective* phase in which efforts are made to elucidate these physically obscure features.

Reconstruction is a methodology for elucidating physically obscure features of a theory's mathematical formalism by deriving these features from a set of physical principles and auxiliary assumptions. An ideal reconstruction is one that traces these features either to extant broadly accepted fundamental physical principles or desiderata, or to newly formulated physical principles of a widely accepted type (such as symmetry, compositional, or extremal principles). The target of reconstruction varies according to whether one wishes to elucidate the

* Based on a talk given at the 'Phenomenological approaches to Physics' conference, Linköping, Sweden, June 2022. A video of the talk is available on the author's website (https://www.philipgoyal.org/).

DOI: 10.4324/9781003259008-14

physical basis of a specific feature of a theory's formalism, or the formalism as a whole.

The process of reconstruction—especially wholesale reconstruction of a theory's formalism—can be viewed as the construction of a *metatheory* that yields an existing theoretical formalism (or part thereof) as an output. From this perspective, the existing theory's formalism is *data*, a brute fact that one seeks to understand through the principles of the metatheory. That is, reconstruction *iterates* the theory-building process: the original theory explains patterns in the brute facts of sensory experience; the metatheory in turn elucidates the physical meaning of the mathematical features and structures in the thus-devised theory[2].

The reconstruction of a theory (or a part thereof) tends to require concepts, mathematical tools, and sometimes ways of thinking about the physical phenomena of interest, that are quite different from those that were employed in the theory's development. Accordingly, successful reconstruction of a theory must usually await the development of the appropriate concepts, mathematical tools, or new ways of thinking, and may not be achieved until many decades after the theory's formulation.

The degree to which a reconstruction enhances a theory's intelligibility depends on the extent to which the theory in question was shaped by general physical principles. In the case of Newtonian mechanics, which was substantially shaped by general principles (such as the principle of inertia and Galileo's principle of relativity), reconstructive work has tended to clarify interconnections between parts of the theory[3] without shaking its deeper conceptual foundations. In contrast, in the case of Faraday–Maxwell electromagnetism, which was largely shaped by Faraday's imaginative and detailed engagement with electromagnetic phenomena (rather than by new general principles comparable in scope to those that underpin classical mechanics), reconstruction had a correspondingly greater impact: Einstein's reconstruction of the Lorentz transformations—a mathematical structure that was abstracted from Maxwell's equations only decades after their formulation—led to a profound reconceptualization of the nature of space and time, and to the addition of light alongside matter in the inventory of fundamental physical entities.

Since its formulation almost a century ago, quantum theory has stubbornly resisted elucidation. It is broadly—if not universally—accepted that the theory violates numerous basic convictions about the constitution of the physical world and its relation to observers, convictions that sustained the development of classical physics for three centuries. Ideally, one would like to know what aspects of the classical conception of physical reality can be retained, what aspects need to be modified, and which abandoned; and to be in possession of an overarching conception of physical reality (analogous to the mechanico-geometric conception which underpins classical physics; Berghofer et al., 2021) which renders these changes intelligible. However, although traditional elucidative methods (such as no-go theorems, reformulations, and interpretations) have provided valuable insights, we still lack a comprehensive, compelling account of just what kind of physical reality is so extraordinarily elegantly encoded in the mathematical formalism of the theory.

In this chapter, I argue that, in order to make further decisive progress, a new elucidative strategy is called for, one based on reconstruction of the quantum formalism. In particular, I propose a *two-step reconstruction-based strategy*[4]:

1. *Reconstruct the quantum formalism.* First, reconstruct the quantum formalism, with the specific goal of distilling the full physical content of the formalism into physical principles and assumptions that can be expressed in natural language and that are amenable to philosophical reflection.
2. *Interpret the reconstruction.* Second, reflect on the principles and assumptions of the reconstruction, bringing to bear whatever philosophical traditions may be appropriate.

Ideally, the second, reflective step will yield a set of physical principles and assumptions that can be laid alongside those that comprise the classical conceptual framework, enabling a point-by-point comparison which makes clear what aspects of the classical framework have been retained, modified, or abandoned. Ideally, it will also yield an overarching conception of physical reality which broadly motivates this new set of physical principles and assumptions.

This reconstruction-based interpretive strategy has many advantages over most traditional elucidative approaches. In particular, the reconstructive step potentially makes the full content of the formalism available for philosophical reflection. Hence, in the reflective step, it is possible to simultaneously take into account a larger number of the nonclassical features of quantum theory. In contrast, traditional elucidative approaches take most or all of the quantum formalism as a given and typically only seek to offer explanation of specific aspects of the theory. As a result, they each harness only a small fraction of the physical content of the formalism, and generate fragmentary insights which are difficult to unify into a coherent conception of reality.

A reconstruction-based interpretative strategy is particularly timely: the *quantum reconstruction program* has galvanized the efforts of many in the quantum foundations community over the past twenty or so years, during which period several detailed reconstructions of key parts of the quantum formalism have been developed. Philosophical reflection on certain reconstructions has already been carried out, and some intriguing insights into long-standing puzzles have already recently been obtained. One of the broader aims of this chapter is to stimulate the kind of collaborative work that will likely be needed to fully harvest the fruits of the quantum reconstruction program.

The chapter is organized as follows. I begin in Section 11.2 with a discussion of classical physics, with the aim of clearly identifying what underlies the widespread view that classical physics is intelligible. I argue that the intelligibility does not consist primarily in classical physics' comportment with everyday intuitions about the physical world (as is often supposed or presumed), but rather in classical physics possessing a *coherent tripartite structure* that connects a clearly articulated project, a conceptual framework, and specific physical theories. Accordingly, such a tripartite structure provides a template and benchmark for the elucidation of quantum theory.

In Section 11.3, I discuss the lack of intelligibility of quantum theory, describing how this is a legacy of its complex and convoluted historical development. I outline the main traditional approaches for elucidating quantum theory, and analyse their limitations. I then turn to the methodology of reconstruction, and describe how it can help us to overcome the key interpretative bottleneck of the traditional approaches by potentially making the full content of the quantum formalism available to philosophical reflection.

In Section 11.4, I discuss the methodology of reconstruction *per se*, showing that it is part of the natural life-cycle of physical theories. I then summarize how reconstruction has been used in classical physics to elucidate the physical meaning of key mathematical features of both classical mechanics and electromagnetism, with the latter theory providing an excellent illustration of the potential of the reconstructive methodology for the elucidation of quantum theory.

In Section 11.5, I return to the reconstruction of quantum theory, describing the operational and informational perspectives which lie behind so much of the recent success of the reconstruction program, and then summarize some recent reconstructive work.

In Section 11.6, I describe some of the key interpretational insights that can be gleaned through interpretation of reconstructions of the quantum formalism, both of the abstract quantum formalism (aQF) and of the quantum symmetrization algorithm (QSA) (which is needed to handle systems of identical particles).

I conclude in Section 11.7 with a summary and a few remarks on future outlook.

11.2 Intelligibility of Classical Physics

The overarching goal of physics is to explore the physical world through precise, controlled experimentation; to develop a precise conception of physical reality; and on these twin bases to develop mathematically precise, predictive theories of nature's workings. As such, the project of physics seeks to integrate three distinct facets of human activity—experimental, natural philosophic, and mathematical/theoretical. The creative interplay between these facets is most clearly visible in the development of classical physics, where they at once challenge and inspire, guide and constrain each other.

Alongside this goal is a *knowability* assumption (or article of faith) that we, no matter our limitations (in spatial and temporal reach, cognitive abilities, and experimental capacities), can—with the aid of a methodology based on the synthesis of the above three facets—aspire to a substantial knowledge of the physical world. This assumption expresses the belief or faith that reality—or a significant part thereof—is so constituted as to be amenable to discovery through a specific method.

As elaborated below, this overarching goal and knowability assumption are deeply intertwined with the fundamental assumptions that lie at the heart of classical physics and that comprise its conceptual framework. Specific theories of classical physics—such as Newtonian mechanics—are then built within that conceptual framework by proposing additional assumptions and principles appropriate to the specific group of phenomena of interest.

Thus conceived, classical physics has a tightly woven tripartite structure:

1. *A clearly articulated project* in the form of an overarching goal and knowability assumption.
2. *A conceptual framework* which articulates a specific conception of reality, which is consonant with the overarching goal and knowability assumption.
3. *Specific theories,* with their domain-specific assumptions and mathematical predictive formalisms, which are built within the conceptual framework.

As elaborated below, I contend that this coherent tripartite structure is primary origin of our shared sense that classical physics is *intelligible*, more so than any particular way in which classical physics appears to comport with certain of our pre-scientific intuitions about the nature of reality.

11.2.1 *The Tripartite Structure of Classical Physics*

11.2.1.1 *Overarching Goal of Physics*

The above-stated overarching goal of physics combines two distinct poles—the experimental and the theoretical. The experimental ideal contains two key ideas:

E1. *Isolability.* One can *isolate* a physical system of interest from its past and future history, and to some extent shield it from uncontrollable contemporaneous events occurring elsewhere.
E2. *Observability.* An experimenter can precisely observe a system's behaviour when placed in a spatiotemporal physical context under experimental control.

Satisfaction of the isolability requirement ensures that the system of interest is insulated from that which an experimenter cannot possibly entirely know or control. Observability ensures that all aspects of the behaviour of a system are open to view, and moreover can be *precisely* measured; and that the aspects of physical context relevant to this behaviour can be controlled.

According to the theoretical ideal, an ideal physical theory has the following characteristics:

T1. *Generality.* The theory can be applied to all things, no matter where or when they be.
T2. *Mathematical precision.* The theory is cast in exact, mathematical terms.
T3. *Testability.* The theory generates empirically testable predictions of sufficient precision as to permit its falsification.

The desideratum of generality focusses attention on the most deep lying regularities in events rather than the particular events themselves, and in particular seeks to eliminate any distinctions between types of physical object (for example, between the living and nonliving), their spatial locations (for example, between terrestrial and heavenly bodies) or the universal epoch during which they exist. The ideal that a theory be cast in mathematical terms and generates precise predictions establishes a tight linkage between sense and thought, while testability ensures that theoretical errors or limitations can be exposed by empirical data.

11.2.1.2 *Classical Conceptual Framework*

The Newtonian conceptual framework of classical physics posits that what exists is matter that moves on a stage of space in step with a universal time, with all properties of that matter open to the gaze of the ideal observer.

In short, it posits a *mechano-geometric* conception of physical reality. A summary of the key assumptions underlying the Newtonian conception is given in Figure 11.1, organized by those assumptions that describe (i) the space-time canvas, (ii) the matter that exists on this canvas, (iii) the motion of matter, and (iv) the observation of matter. As summarized in Section 11.2.1.3, most of the assumptions can be traced back to experimental and theoretical ideals that are incorporated in the project of physics.

Canvas

C1. *Canvas quantifiability*. Space and time are quantifiable.
C2. *Spatial homogeneity*. Space is homogeneous.
C3. *Spatial isotropy*. Space is isotropic.
C4. *Temporal homogeneity*. Time is homogeneous.

Composition of bodies

P1. *Simple mereology*. Bodies are simply composed of particles.
P2. *Particle eternality*. Particles are indefinitely persistent.
P3. *Property universality*. Particle property-types are universal.
P4. *Property sparsity*. Particle property-types are few.
P5. *Property quantifiability*. Particle properties are quantifiable.

Motion

M1. *Determinism and Reversibility*. Motion is (a) deterministic and (b) reversible.
M2. *Continuity*. Motion is continuous.
M3. *Isolated motion*. An isolated body moves uniformly and rectilinearly.
M4. *Conservation*. Collective motion conserves certain quantities of motion.
M5. *Differential composability*. Collective motion is differentially composable.
M6. *Influence fall-off*. Interparticle influences monotonically diminish in strength with interparticle separation.

Observation

O1. *Sight-like observation*. Ideal observation passively registers particle properties.
O2. *Exact observability*. All properties of bodies are exactly observable.
O3. *Position categoricity*. Particle position is the only directly accessible particle property.
O4. *Principle of relativity*. All inertial frames are physically equivalent.

Figure 11.1 Conceptual framework of classical physics. A list of the fundamental assumptions underlying classical physics in the Newtonian era, up to (but not including) the development of electromagnetism.

As classical physics developed over the course of some three hundred years, the classical conceptual framework underwent significant change—the introduction of the energetic framework, the development of the field concepts of electromagnetism, and the theories of special and general relativity, all challenged the framework's fundamental assumptions, and it either stretched to accommodate or underwent modification. However, to keep the following discussion sufficiently focussed, these changes are not systematically considered here, although the challenge posed—and changes wrought—by electromagnetism, being of special interest in connection with the potency of the reconstructive methodology, are discussed in Section 11.4.2.

11.2.1.2.1 CANVAS

Assumptions C1–C4 license a mathematical description of space and time, so that what one experiences as an instant of time and as a location in space can be described as a region—in the limit, a *point*—in a *mathematical* manifold. Due to the assumed homogeneity and isotropy of space, the spatial portion of the manifold is ascribed particular topological and metrical structure—it is assumed to be simply connected and Euclidean—and is assumed to be three dimensional. Similarly, due to the homogeneity of time, the temporal part of the manifold is ascribed a uniform metric. Thus, the underlying canvas is highly symmetric, bereft of any structure corresponding, for example, to the spatial and temporal asymmetries—*up* vs. *down* or *past* vs. *present* vs. *future*—so characteristic of lived experience, although spatiotemporal continuity and three-dimensionality, for instance, seem to be reasonable extrapolations of our lived experience.

11.2.1.2.2 COMPOSITION OF BODIES

Assumptions P1–P5 reflect the atomistic conception of physical bodies, namely that every body can be reduced to an arrangement of particles. Bodies are assumed to be *composed* of these particles in a manner that precludes strong emergence (i.e. the 'whole' is nothing more than the arrangement of its 'parts') (P1). The particles themselves are assumed to be indefinitely persistent (they are not created or destroyed) (P2), which can be viewed as arising from the desideratum of maximizing symmetry (the existence of particles of finite lifetime would introduce an inherent distinction between different moments in time). Moreover, all particles are assumed to be characterized by the values of just a few property-types, such as mass and charge (P3–P5), with position being a particle's only *directly* observable property (O3) (all its other properties are dispositional, manifest to an observer only indirectly through their effect on position in certain experimental contexts).

11.2.1.2.3 MOTION

Assumptions M1–M6 *concern* the motion of matter, and range from the general to the specific. The twin requirements of determinism and reversibility (M1), together with continuity (M2), ensure that the persistent particles which compose all bodies are *reidentifiable*—that is, each particle has its own trajectory throughout time, along which it can (due to O1, O2, and O3) be precisely tracked. If reversibility (M1b) were not to hold, it would be possible for two particles' trajectories to merge, which would prevent their reidentification were they to be identical in their time-independent properties. Similarly, were continuity (M2) not to hold, so that particles could 'jump' from place to place, then reidentification would be rendered at best probabilistic.

The assumption that an isolated particle moves uniformly in a straight line (M3) establishes a direct connection between the posited mathematical structure of space-time and the behaviour of particles immersed therein—isolated particles move equal spatial distances in equal times (as judged by the spatial and temporal metrics previously posited).

The formulation of the specific equations of motion of classical mechanics depends centrally on continuity (M2), conservation (M4), and differential composability (M5). Conservation posits that certain physical quantities remain unchanged during such fundamental processes as the collision of two particles, and provides the basis for the later introduction of specific *quantities of motion* (such as kinetic energy and momentum). Continuity enables one to specify motion via *differential* equations, which specify how a system will change in an infinitesimal of time. And differential composability posits that, in an infinitesimal of time, the change in motion of a given particle in a system of particles can be understood as a composition of the changes due to the two-particle interactions between that particle and each of the other particles in the system. This assumption enables the 'scaling up' of the equations of motion for a two-particle system to a system of any number of particles (in Newtonian mechanics, the assumption leads to a rule for the composition of forces) (see Section 11.2.1.4).

11.2.1.2.4 OBSERVATION

Assumptions O1–O4 connect the picture of physical reality painted by the previous sets of assumptions (canvas, composition, motion) to that which can be observed. Throughout, the notion of *observation* is taken as primitive. Operationally, observation can be treated as an abstraction of the experiences of an idealized experimenter. Ontologically, it could be viewed in several ways, for example as reflecting the action of an abstract mind or consciousness of unspecified nature capable of becoming *aware* of (and thus *observing*) outcomes of measurement devices, or as reflecting the actualizing of events (outcomes of measurement devices) that are themselves taken as primitive (and as distinct from material objects and their properties).

Assumption O1 posits that the ideal process of observation is *passive* (or sight-like) in the sense that an ideal measurement of a particle's properties *registers* those properties without bringing about any change in their values. This assumption ensures that one can regard a given *appearance* as a direct manifestation of a pre-existing property—for example, the appearance of a particle at a location in space as underpinned by the body *possessing* (immediately prior to the observation) the property of being located at that location in space.

Assumption O2 posits that particle properties, which are quantifiable (P5), are *exactly* observable. Assumption O3 allows for the existence of idealized rods and (together with M3) idealized clocks, thereby furnishing observational access to the posited space and time structure insofar as spatial *intervals* and temporal *durations* are concerned. An idealized clock is one that indefinitely beats at a constant rate (irrespective of its location or state of motion), while an idealized rod is one that indefinitely possesses one and the same length (irrespective of its position, orientation, or state of motion). Note that the homogeneity (C2) and isotropy (C3) of space, together with inertial-frame equivalence (O4), precludes measurement of *absolute* location, direction, and velocity, while temporal homogeneity (C4) precludes measurement of absolute time. Assumptions O2 and O3 jointly imply that the measurement of all particle properties can be accomplished via suitable position measurements.

Finally, assumption O4 is a specific assumption, connected to M3, which asserts that all inertial frames (namely those in which an isolated body is observed to travel rectilinearly at constant speed) are 'physically equivalent' in the sense that the laws of motion apply equally to any such frame.

11.2.1.2.5 OBSERVATION VS. MANIPULATION

The notion of passive observation (O1) is in tension with an experimenter's capability—which he necessarily possesses—of *manipulating* physical bodies (such as experimental equipment) *at will*. Here, 'will' is taken as a primitive and is modelled within the mechanical conception of causation as that which can *initiate* causal chains.

Given that the experimenter is able to manipulate matter at will, questions concerning the nature of this will cannot be entirely separated from a description of the nature of the physical world. In particular, what constraints apply to 'at will' manipulation? Does the will act only through the body of an experimenter? Do other kinds of physical systems manifest a will? Is the will of the experimenter influenced by the physical system of interest (for example by its state)? If the will is not wholly independent of the physical state of the system of interest (or the physical world *in toto*), as one would expect from experience, then the existence of the experimenter in the physical world generates a complex feedback between the otherwise-deterministic physical world and the experimenter, blurring the line between the physical and 'non-physical'.

The classical conceptual framework provides no fundamental answers to these questions, in particular providing no explanatory grounds for our experience as causally efficacious *actors* in the physical world. Nevertheless, the framework posits that a physical system subject to observation *in the absence of manipulation* evolves deterministically and reversibly. Hence the Newtonian metaphor of the solar system as a clockwork mechanism holds good as long as one excludes such agential manipulations as the deflection of asteroids by explosive detonations.

11.2.1.3 Relation between Overarching Goal of Physics and the Classical Conceptual Framework

The experimental and theoretical ideals described above (Section 11.2.1.1), in combination with the assumption of knowability, are supported by the conceptual framework of classical physics (Section 11.2.1.2). Below, the experimental and theoretical ideals, and the fundamental assumptions that underpin the conceptual framework, are referenced by their labels (C1, etc.).

1. The ideal of *isolability* (E1) is supported by:
   i. a particular view of causation, namely that the behaviour of a system may be understood via efficient cause, without recourse to final cause. Accordingly, the future trajectory (or any aspect thereof) of the system is irrelevant to the present, provided that sufficient cognizance is taken of its past. Moreover, only its *immediate* past is relevant—its more distant history is rendered irrelevant by sufficient knowledge of its immediate past (M1a).
   ii. the assumption that a system is disproportionately influenced by bodies in close physical proximity (M6), or that influences due to more distant bodies can be regarded as providing a stable background for experimentation.
2. The ideal of *observability* (E2) is supported by the assumption that the idealized process of observation has negligible impact on the system under observation, and moreover that the system and its properties *exist*

and are well-defined quite independently of the process of observation. That is, idealized observation is *sight-like* rather that *touch-like* (O1).

3. Due to *generality* (T1), one aims at a theory that describes a 'bottom' layer of physical reality that is *maximally symmetric*, and in which there are a minimum of primitive entities. In particular, one aims to:

    i. treat all places, times, frames of observation, and objects—as far as possible—in the *same way*. For example, space is viewed as homogeneous and isotropic (C2, C3); time as homogeneous (C4); all inertial frames as equivalent (O4); and all objects—whatever their visible form—as describable in terms of the same small number of basic properties (P3, P4) whose values can gleaned through position measurements (O3).
    ii. treat 'nonfundamental' things as simple composites of 'fundamental' things. For example, regard macroscopic bodies as simple composites of *particles* (P1) that are drawn from a small number of possible types (an expression of Greek atomism). Here, simple composition precludes strong emergence.

4. The ideals that a theory be expressed *mathematically* (T2) and that the behaviour of a system can be *precisely* observed (E2) are supported by the assumption that particle properties are quantifiable (P5) and can be exactly observed (O2). Since our visual sense is the most readily and unambiguously quantifiable, that in turn favours the view that all properties can be mapped (using suitable experimental and theoretical tools) to spatial extension (O3), and that spatial intervals and temporal durations are observable (O3, M3). Mathematical expressibility also favours:

    i. *domain-specific principles:* uniformity assumptions such as the indefinite persistence (eternality) of the fundamental constituents of matter (P2) and dynamical uniformity principles (such as uniform motion of isolated bodies (M3) and conservation principles (M4)), as well as other dynamical principles (for instance extremization principles like the principle of least action) that can readily be expressed in mathematical terms; and
    ii. *technical assumptions:* such as continuity of trajectories (M2), and the existence of symmetries (such as associativity and commutativity of composition).

5. The ideal of *testability* is supported by the assumptions of determinism and reversibility (M1), since both promise the most perfect conceivable level of prediction. This ideal also disfavours theories that posit quantities (such as 'hidden variables') or detailed structure which is not directly amenable to experimental test.

11.2.1.4 *Theories of Classical Physics*

Newtonian mechanics, the first comprehensive theory of classical physics, is formulated within the foregoing classical conceptual framework. In brief, the theory introduces a crucial distinction between time-independent and time-dependent properties, positing that particles have a time-independent mass, together with two time-dependent properties—position and velocity. On this basis, conservation (M4) can be mathematically expressed as the conservation of total momentum in collective motion. The interactions of two separated bodies is handled through Newton's concept of *force* together with a specific law connecting force and change in velocity, with the posited inverse-square gravitational force between two bodies satisfying the influence fall-off assumption (M6). Differential composability (M5) is then formalized via the parallelogram rule for the composition of forces, enabling the model of two-body systems to be 'scaled up' to systems of any number of bodies[5].

Other fundamental theories of classical physics—chiefly electromagnetism and Einstein's relativity theories—proceed analogously, introducing additional domain-specific assumptions that posit specific quantities of interest and that enable the construction of a set of differential equations of motion describing how these quantities change over time. As mentioned above, each such theory has challenged the Newtonian-era classical conception, which has either stretched to accommodate or has been modified appropriately. Nevertheless, broadly speaking, the fundamental theories of classical physics—Newtonian mechanics, Faraday–Maxwell electrodynamics, and Einstein's relativity theories—are all recognizably *classical* insofar as they share most of the same basic convictions as to the nature of the physical world and the nature of observation.

11.2.2 *Why Is Classical Physics Intelligible?*

It is frequently suggested or supposed that the intelligibility of classical physics is a rather trivial consequence of the fact that it reflects 'common sense'—that its conceptual framework is intelligible primarily *because* is a refinement and abstraction of our mental model of an everyday physical world, namely a world populated by persistent objects immersed in space which possess various properties (like location, shape, colour), move smoothly from place to place, and which we observe (chiefly through the sense of sight) largely as they really are.

While it is certainly true that classical physics inherits many of these commonsense notions, it is also the case that, in many respects, it departs from everyday notions and intuitions. For example, whereas the canvas assumptions posit an isotropic space (C3), our lived experience reflects an asymmetry of 'up' and 'down'. Similarly, the posit of homogeneous time (C4) together with determinism and reversibility of motion (M1) erase any theoretical distinction between past, present and future, and makes it difficult to understand, in *fundamental* terms (i) why we do not experience the entire history of the universe in one go but only the *present moment* and (ii) what is the nature of genuine creativity (or genuine novelty) or a sense of purpose that is future-directed. And the assumption of simple mereology (P1) reduces all objects to mere arrangements of particles, depriving a 'whole' with any fundamental meaning other than the arrangement of its parts, for example erasing any *fundamental* distinction between the living and nonliving.

Ideally, the project of physics would wish to *contingently recover* these intuitions as far as possible, *viz.* to explain by reference to specific contingent facts just how it can be the case that we have these intuitions in spite of the validity of the classical conceptual framework. In a few cases, this can be said to have been achieved. For example, it is now widely accepted that our sense of the asymmetry of up and down is a consequence of our contingent Earth-bound existence, and that this asymmetry would effectively disappear if we were to live 'on the float' on board a non-rotating space station far from gravitating bodies. However, in most other cases—such as in the distinction between past, present and future; between the living and nonliving—no such robust reconciliation can be said to have been achieved, notwithstanding the enormous insights that have in some instances been gained by proceeding in the hope or belief that such a reconciliation might be possible.

For these reasons, I assert that the intelligibility of classical physics primarily reflects the fact that it abstracts *some particular aspects* of human experience in such a way as to create a tripartite structure which tightly weaves a clearly articulated project, a well-defined conceptual framework, and empirically successful mathematical theories of specific physical phenomena. From this perspective, classical physics is the product of what may well be a necessary *trade-off:* between the desideratum of creating general, mathematically precise, predictively capable physical theories, on the one hand; and the desideratum of taking full account of the vast richness of the human experience of that which we call reality, on the other. In this sense, classical physics is a spotlight that brightly illuminates—and brings into sharp relief—some aspects of our experience of reality, while leaving the substantial remainder in shadow.

11.3 Intelligibility of Quantum Theory

11.3.1 Development of Quantum Theory

What is commonly referred to as *quantum theory*—the von Neumann–Dirac abstract quantum formalism at its core—is the end-point of a period of development that spanned some twenty-five years. From the first to the last, this development was marked by bold conjectures that were made, above all, in the quest to construct viable models of specific groups of microphysical phenomena that, until that point, had entirely eluded models which squarely obeyed the strictures of classical physics. Almost without exception, these conjectures lacked any compelling *a priori* physical justification or oftentimes any justification at all. Sometimes they were directly at odds with classical physics. But, by and large, they were taken seriously because they *worked*.

For example, at the start of this period, in 1900, Planck noticed that if one posits that the energy of certain abstract oscillators is *quantized*, one can use then-standard statistical arguments to derive the blackbody radiation curve. This assumption was essentially an *ad hoc* mathematical device, whose sole justification was that it produced a result consistent with known experimental data, something which classical models had completely failed to do. No *a priori* physical rationale for such quantization was offered, nor was one on the horizon—classical physics embraced continuity. Similarly, in order to make theoretical contact with the discrete pattern of spectral lines exhibited by hydrogen, Bohr freely combined Newtonian mechanics and electrostatics with a novel assumption—the quantization of angular momentum—inspired by Planck's, which gave rise to a discrete set of electronic orbits. In this case, the novel assumption was not only a freestanding assumption (*viz.* apparently not derivable from classical physics and not provided with a deeper rationale) but was at odds with an established theory of classical physics—electromagnetism predicts that an accelerating charge radiates away its energy, so on this account no stable electronic orbits should exist.

This pattern continued up to and including the formulation of a general mathematical formalism of quantum theory in the mid-1920s. For example, Schrödinger's wave equation for a single particle (such as an electron), itself based on de Broglie's bold conjecture of wave–particle duality, was originally conceived as analogous to a classical wave equation. But, although Schrödinger initially proposed to consider the wave as describing a spatial charge density, it soon became clear that, in order to make proper connection between the wave equation and the brute fact of particle-like detections in the laboratory, one had to supplement the wave equation with a separate rule—the Born rule—that gives the *probability* that a particle will be detected within a region of space. Similarly, in order to handle systems of identical particles, such as multielectronic atoms, Dirac and Heisenberg independently introduced novel mathematical rules (the symmetrization postulate, in particular) which they rationalized by claiming that identical particles are *indistinguishable* from one another. Both of these ideas were at odds with basic tenets of the classical conceptual framework—probabilistic measurement outcomes in conflict with the sight-like nature of classical measurement and with the knowability assumption; indistinguishability in conflict with classical reidentifiability.

By the end of this period of development, the broad consensus in the physics community, particularly amongst most of the founders of quantum theory, seems to have been that the new theory departed from classical physics to such an extent that it was neither possible nor fruitful to try to stretch or modify the classical conceptual framework to accommodate it. Rather, it was necessary to somehow develop a new conceptual scheme that would render the mathematical formalism intelligible. However, given the huge cognitive cost of setting aside the classical framework—a framework that was widely regarded as *intelligible* and moreover had sustained the development of physics for more than two hundred and fifty years—there was understandably considerable dissent, including from such key figures as Einstein and Schrödinger.

Despite the substantial, multifaceted efforts made to elucidate quantum theory in the intervening period of roughly one hundred years, quantum theory is still broadly regarded as mysterious and counterintuitive, amongst both physicists and other academicians engaged in its use or conceptual investigation (such as other natural scientists, philosophers of physics, and philosophers of science), as well as amongst those members of the wider academy and general public interested in the broader implications of natural science.

11.3.2 Rendering Quantum Theory Intelligible

It is commonly asserted (or supposed) that quantum theory is unintelligible because it violates so many of our everyday intuitions, particularly those that seem to be conditioned by our engagement with the macroscopic physical world. There is undoubtedly some truth in this statement. But, from the perspective of the analysis of the intelligibility of classical physics given above, the unintelligibility of quantum physics does not *primarily* originate in its departure from 'common sense'. Rather, the lack of intelligibility is a symptom of quantum theory not being embedded in a conceptual framework—a *quantum conceptual framework*, if you will—that makes explicit *what* aspects of human experience it abstracts from and quite *how* it abstracts from them.

However, it is evident that the efforts made thus far to elucidate quantum theory have somehow fallen short—notwithstanding the valuable insights they have undoubtedly provided, they have not sufficed for the construction of a comprehensive quantum conception of physical reality. I contend that, in order to make decisive further progress, we need a better *methodology* for elucidating quantum theory, and that a suitable methodology is one that leverages the *quantum reconstruction program*, a program that has attracted considerable attention in the quantum foundations community over the past twenty or so years.

Below, we first briefly review the main approaches that have traditionally been used to elucidate quantum theory, and identify the nature of their limitations. We then turn to the methodology of reconstruction, and examine its potential for helping us develop a thoroughgoing understanding of quantum theory.

11.3.3 Traditional Approaches to Elucidate Quantum Theory

Broadly speaking, the most prominent and influential approaches traditionally employed to elucidate quantum theory are as follows:

1. *New concepts.* Formulate new concepts that encapsulate fundamental ways in which quantum theory seems to be at odds with the classical conception of physical reality, and then investigate quantum theory quantitatively through these new conceptual lenses. Examples include complementarity and entanglement.
2. *No-go theorems.* Show that certain predictions of quantum theory are inconsistent with a minimal classical model. Notable examples include Bell's theorem (nonlocality) (Bell, 1964) and the Kochen–Specker theorem (noncontextuality) (Kochen and Specker, 1967).
3. *Reformulations.* Re-write some or all of the quantum formalism in an alternative mathematical form. This often establishes an illuminating parallel between (parts of) quantum theory and an existing mathematical or physical theory. Examples include Feynman's path-based reformulation (which establishes a parallel to probability theory) (Feynman, 1948), and Bohm's Hamilton-Jacobi formulation of the Schrödinger equation (which establishes a parallel to the Hamilton-Jacobi formalism of classical mechanics) (Bohm, 1952).
4. *Interpretations.* Formulate a conceptual framework which at least partially accounts for some of the nonclassical features of the quantum formalism. Examples include Bohr's complementarity-based interpretation (Bohr, 1928), Heisenberg's potentiality/actuality-based interpretation (Heelan, 2016), the Stapp–Schwarz mind-based interpretation (Schwarz et al., 2005), the transactional interpretation (Cramer, 1986), the many worlds interpretation (Everett, 1957), the de Broglie–Bohm interpretation (Bohm, 1952), and the QBist interpretation (Fuchs, 2017).

A brief remark: It is important to bear in mind a distinction between a *research program* and a *component* of it. For example, understood as a research program, Bohmian mechanics contains both a reformulative and interpretative component (both of which components are mentioned above)[6]. But it also involves other components, such as ongoing attempts to derive the quantum equations of motion from more basic assumptions, or attempts to test the interpretation[7].

Similarly, the many worlds research program has not only an interpretative component, but also a reconstructive component (which one could view as part of its attempt to *test* or bolster the interpretation) consisting of a derivation of the Born rule from decision-theoretic axioms. And the QBist research program has—in addition to its interpretative component—reformulative (re-expressing the quantum formalism as far as possible in terms of SICs and associated probabilities) and reconstructive components.

However, it is usually the case that a research program is 'known' mostly for one of its components. Some of these components may be relatively stable, while others may be in active development. Research programs also differ in the degree of cohesion between their components. These nuances would need to be taken into account by an in-depth discussion of any of the research program components mentioned above. Here, I am focussed on the relative merits and limitations of the methodologies that have been traditionally employed for elucidating quantum theory, rather than on entire research programs *per se*. So, the focus is on components. I shall mention the connections to other components in a research program only if directly relevant.

11.3.3.1 *Illuminative Value of Traditional Approaches*

Each of these approaches has proven itself capable of illuminating quantum theory in its own distinct way:

1. The coining of new concepts is a crucial first step to elucidation, insofar as such concepts *point* to distinct non-classical features of quantum theory that appear to be of fundamental importance. Bohr's concept of *complementarity* sought to provide de Broglie's bold conjecture of wave-particle duality with a philosophical underpinning, arguing that it reflected an irreducible trade-off between space-time coordination (specifying *when* and *where* an object is located) and causality (the possibility of connecting the future behaviour of an object with its past behaviour). The concept of complementarity has since been formalized in many different ways, such as in the context of POVMs, information trade-off relations, and mutually unbiased bases.

2. No-go theorems, such as Bell's theorem, establish that, in light of quantum theory, we must abandon one or more classical assumptions about the nature of physical reality or our access to it. As such, these theorems

force us to recognize that certain patterns of thinking are not viable, and spur efforts to find theoretical or experimental loopholes, and to develop alternative patterns of thinking that are in tune with quantum theory.

3. The aQF is standardly expressed in the language of complex vector spaces, with physical states represented as complex vectors (or rays), dynamics as unitary operators, and projective measurements as Hermitian operators. Re-expression of this formalism, or of specific quantum models, in mathematical garb that establishes parallels with an existing mathematical or physical theory can allow us to appreciate quantum theory from another angle, and often suggests new ways of understanding certain nonclassical features, as well as sometimes suggesting alternative computational techniques. For example, Feynman re-expressed the aQF in a manner that parallels the Lagrangian formulation of classical mechanics, which results in dramatic mathematical simplification, and establishes a close parallel to the formalism of probability theory. Similarly, Bohm expressed Schrödinger's equation in the form of a classical Hamilton-Jacobi equation supplemented by a so-called quantum potential. The Bohmian model for a system of two particles renders nonlocal influence explicit via the quantum potential, which directly inspired Bell's theorem.

4. Each interpretation of quantum theory tends to focus on making sense of some specific nonclassical features of the quantum formalism. Most are concerned with making sense of the quantum formalism's distinction between dynamics and measurement, or with the limited access that measurement provides to the degrees of freedom in a quantum state, but some (such as the transactional interpretation) are more focussed on the puzzle of nonlocality. Some interpretations seek to 'defuse' these nonclassical features, somehow stretching the classical framework to accommodate them; while others seek to provide some kind of deeper understanding of these features. In so doing, these interpretations bring attention to specific noteworthy aspects of the formalism, and provide a certain way of thinking about them, which can be creatively inspiring[8].

11.3.3.2 *Limitations of Traditional Approaches*

However, these traditional approaches are limited in fairly obvious ways.

11.3.3.2.1 LIMITATIONS OF NO-GO THEOREMS

Although no-go theorems are powerful in that they force us to recognize that certain sets of obvious-seeming assumptions are logically at odds with certain quantum theoretic predictions, they are essentially *negative* results—they block certain patterns of thinking, but provide little positive guidance on how we *could* think about physical reality so as make sense of these predictions. In addition, as these no-go theorems make use of *specific* quantum theoretic predictions rather than the theory *as a whole*, the insight that they provide is necessarily fragmentary—each theorem focusses on some specific nonclassical feature—so one is left with the daunting challenge of integrating these fragmentary insights.

11.3.3.2.2 LIMITATIONS OF TRADITIONAL INTERPRETATIONS

Interpretations of quantum theory are, in contrast, relatively *positive* in the sense that they offer a specific way to make sense of some nonclassical features of the theory. However, they only harness a small fraction of the physical content of the quantum formalism: they focus only on making sense of some specific part of the quantum formalism—typically the Born rule—and offer no explanation of the remainder, which they take as a given.

For example, the many worlds interpretation takes the unitary part of the aQF (that is, the aQF apart from the Born rule) as a given. As a result, the interpretation cannot, due to its starting point, offer an explanation of the unitary part of the quantum formalism—for example, why quantum states are represented by complex vectors and dynamics by unitary operators. Conversely, the interpretation is not *constrained* by these mathematical structures—as far as the interpretation is concerned, quantum states and dynamics could be differently represented. Similarly, the de Broglie–Bohm interpretation takes the Schrödinger equation as a given, which means that its so-called quantum potential (which effectively encodes all nonclassical behaviour) must be taken as a given—as a brute fact—within the confines of this interpretation.

A further serious difficulty with traditional interpretations is that they are difficult to subject to scientific test, whether theoretical or experimental. The de Broglie–Bohm interpretation fares better than most in that it is based on a *reformulation* of quantum theory which casts parts of the quantum formalism in a mathematical form that more closely parallels classical mechanics. Consequently, one way to build confidence in this type of reformulation is to extend it to other parts of the formalism, such as relativistic quantum theory or quantum field theory. However, in other cases, the possibility for such theoretical test is much more limited. For example, the many worlds interpretation offers only one obvious test, namely the derivation of the Born rule[9]. But, again, as the unitary part of the aQF is taken as a given, one cannot test the interpretation by asking for a derivation of that. More seriously, *experimental* tests of current interpretations seem almost entirely out of reach.

In summary, the plethora of interpretations—which cover a wide range from those that hew closely to the classical conceptual framework, to those which draw upon metaphysical or psychological notions—show that it is possible to clothe quantum theory in many different conceptual outfits in a manner that might be attractive (or even creatively inspiring) to some. However, it has proven difficult to devise compelling tests (theoretical or experimental) of these interpretations. Consequently, one has little evidential basis to choose between them. Moreover, as traditional interpretations take most of the quantum formalism as a given, they make very limited use of the actual physical content of that formalism.

11.3.4 *A New Approach: Reconstruction-based Interpretation of Quantum Theory*

11.3.4.1 *The Bottleneck: Taking the Quantum Formalism as the Starting Point*

As noted above, almost all traditional methods for elucidating quantum theory *start* with the mathematical formalism of quantum theory. They then reflect upon it, derive critical predictions from it, reformulate it, or interpret it. But, as summarized in Section 11.3.1, the quantum formalism was the end-point of a rather convoluted and complex process involving many novel physical ideas and a good deal of mathematical guesswork. As a result, quite what any part of the mathematical formalism really *means* is often unclear. Moreover, even in those cases where the meaning seems 'obvious', it is perilously easy to be misled.

For example, it is common to talk about *quantum states* and *particles*, but it is far from clear that quantum states are anything like the states of classical systems, in the sense of being a description of their *objective* physical states. And it is far from clear whether so-called quantum particles have *any* of the fundamental characteristics (such as continuous localization, persistence, and reidentifiability) that we attribute to classical Newtonian particles. Yet, it is all too easy for one to be unconsciously influenced by the classical associations conjured up by such language, in part because it is yet to be established which associations *do* carry over and which do not.

There are also more technical examples of considerable importance where one is easily mislead by classical associations. For example, in the interpretation of the quantum rules for handling a system of identical particles, a great deal turns on how one reads the indices in the (anti-)symmetrized states that describe such a system[10]. These indices have almost universally been read as *particle labels*, which is a carry-over from classical physics. Yet it has recently become increasingly apparent that this 'common sense' reading (recently dubbed 'factorism'; Caulton, 2014) may well be incorrect. But, in that case, quite what these indices *do* mean remains controversial, yet a great deal (such as the nature of entanglement in systems of identical particles) turns on this issue.

To be clear: at the instrumental level, the formalism is sufficiently well defined that it enables physicists to build models of *most* specific quantum systems of interest[11]. It is this capacity, and the empirical success of the resultant models, which justly confers such prestige upon quantum theory. However, extracting any reliable information from that formalism as to the nature of the underlying physical reality is fraught with difficulty. As mentioned above, the no-go theorems are the most reliable way of doing so, but these necessarily suffer from fragmentariness due to the fact that they only make use of specific quantum theoretic predictions rather than the quantum formalism as a whole.

11.3.4.2 A Remedy: Reconstruction Followed by Interpretation

The methodology of quantum reconstruction seeks to remove the interpretative bottleneck by systematically deriving the quantum formalism in an operational framework from postulates that are, ideally, physically well-motivated, thereby *distilling* the full mathematical content of the theory into precise natural-language statements that—unlike the abstract mathematical postulates of quantum theory—are amenable to philosophical reflection (Berghofer et al., 2021; Fuchs, 2002; Grinbaum, 2007b).

Such a reconstruction can then serve as a stepping stone to interpretation: with a reconstruction in hand, one can *interpret quantum theory by reflecting on the postulates of the reconstruction, rather than attempting to decipher the inscrutable abstract mathematical postulates in which quantum theory is standardly cast.* In contrast to traditional interpretations, such an interpretation would likely be more *reliable* since the postulates of the reconstruction would likely be couched in language that is closer to basic experimental operations and be expressed in simpler mathematical terms. And it would likely be more *comprehensive* in that, owing to the greater digestibility of these postulates, it would likely be easier to simultaneously take into account a larger number of the nonclassical features of quantum theory.

In summary, then, we propose a two-step approach to interpretation:

1. *Reconstruct the quantum formalism.* Systematically derive the mathematical formalism of quantum theory in an operational framework from postulates whose physical meaning is as clear as possible.
2. *Interpret the reconstruction.* Philosophically reflect upon the postulates of the reconstruction, bringing to bear whatever philosophical traditions or notions seems fruitful.

A few remarks are in order:

11.3.4.2.1 OPERATIONAL FRAMEWORK

As illustrated above, traditional interpretation of the quantum formalism suffers from considerable ambiguity in that it is far from clear to what extent concepts (such as 'state', 'particle') and mathematical features (such as indices in symmetrized states) have the meaning which they are usually ascribed in classical physics. One way to ameliorate such difficulties is to reconstruct the quantum formalism within an *operational framework*—an idealized representation of the experiments carried out on a laboratory workbench. As elaborated in Section 11.5.2, concepts such as physical system, measurement, and interaction, are taken as primitive, and understood in terms of concrete macroscopic devices and their observable outcomes. Proceeding in this manner, the additional concepts that one introduces in the process of reconstruction can be directly related to elementary experimental operations, which strongly constrains their meaning.

11.3.4.2.2 INTUITIVELY GRASPABLE POSTULATES

For a reconstruction to be *suitable* for interpretation, it is essential that its postulates be formulated with a view to interpretation—as far as that is possible given the severe challenge of devising a viable reconstruction (namely one that leads to some part of the quantum formalism). This speaks against postulates that are rather abstract or mathematical, and speaks in favour of postulates that express intuitively graspable ideas. In this regard, the principles that underlie classical mechanics—such as the principles of relativity and conservation, and Newton's action-and-reaction principle—are exemplars, with the caveat that the postulates needed to reconstruct quantum theory will undoubtedly be quite different in character and may severely challenge our customary patterns of thought.

11.3.4.2.3 RECONSTRUCTION OF FULL QUANTUM FORMALISM

What is ordinarily referred to as 'quantum theory' in fact consists of several distinct components (see Figure 11.2), all of which are necessary to create explicit quantum models of physical systems of interest. Since it is only these explicit models that have been subjected to experimental test, it is essential for the construction of a *full* picture of quantum reality that *all* of these components be reconstructed and subjected to interpretation.

11.3.4.2.4 FORM OF IDEAL INTERPRETATION

What should an ideal 'interpretation' of a quantum reconstruction look like? I contend that, to ensure maximal intelligibility (along the lines of

Abstract quantum formalism (aQF)

- ***Single systems.*** Mathematical representations of the physical states and temporal evolution of a physical system, and of measurements performed upon it. Representation of symmetry transformations of the frame of reference.
- ***Composite systems.*** Tensor product rule for a system composed of nonidentical subsystems in the case where the subsystems are in pure states.

Quantum symmetrization algorithm (QSA)

- Rules for constructing the states (the symmetrization postulate) and measurement operators for a system composed of identical subsystems (i.e. subsystems with the same time-independent properties).

Spin-statistics connection (SSC)

- Rule specifying the connection between an identical particle's spin and the applicable 'statistics'.

Quantization rules (QRs)

- ***Heisenberg's equation.*** General form of time-evolution unitary operator in terms of the quantum Hamiltonian operator (*viz.* the operator that corresponds to what we classically understand as a "measurement of energy").
- ***Operators for specific measurements.*** Operators representing measurements classically described as measurements of function of classical observables, given the operators corresponding to the latter observables. Canonical commutation relations. Explicit representation of fundamental measurement operators.

Quantum wave equations (QWEs)

- Single-particle wave equations, especially the Schrödinger, Dirac and Klein-Gordon equations.

Figure 11.2 Main components of the standard quantum formalism. The abstract quantum formalism (aQF) is an abstract mathematical *shell* in which quantum models of specific systems of interest can be built. To construct a model of the electron in a hydrogen atom, one must employ Heisenberg's equation, and then appeal to the operator rules to establish the explicit form of the operators that represent measurements of position and momentum (this requires that one assume the classical Hamiltonian for a charged particle in an electromagnetic field). For a system composed of more than one identical particle—such as the helium atom or the conduction electrons in a metal—one also requires the quantum symmetrization algorithm (QSA) and the spin-statistics connection (SSC). Finally, to adequately describe a single particle in the relativistic regime, one requires specific wave equations, most importantly the Dirac equation for the electron. Excluded here, for brevity, are the rules associated with quantum field theory (QFT).

classical physics, as discussed in Section 11.2.2), such an interpretation must provide:

(i) *A quantum conceptual framework.* This framework should be articulated in the form of a set of fundamental assumptions which one can put alongside those that underpin the classical conceptual framework. This will enable a point-by-point comparison of the two conceptual frameworks.
(ii) *A revised notion of the project of physics*, in which the new conception is firmly rooted. For example, if quantum measurement is taken to be touch-like, so that observation is regarded fundamentally as an active process, then it is essential that this assumption be rationalized and rendered intelligible through a suitably revised notion of the overarching goal of the project of physics.

Over the past two decades, the quantum reconstruction program has generated intense interest in the quantum foundations community. Most of the attention thus far has focussed on the aQF, of which there are now several detailed reconstructions. The other parts of the formalism (QSA, SSC, QRs, QWEs)—see Figure 11.2—have, in contrast, received relatively little attention. Meanwhile, the *interpretation* of quantum reconstructions is in the early stages.

But, before going further, we will first consider in more detail reconstruction as a methodology *per se*. In particular, we examine how the reconstructive methodology has been used to elucidate the theories of classical physics as a way of better understanding just why reconstruction of quantum theory is both a natural part of the life-cycle of the theory and is prerequisite for its proper interpretation. We shall then return to the reconstruction of quantum theory (Section 11.5), and survey some of the recent interpretational insights that have been extracted from quantum reconstructions (Section 11.6).

11.4 The Methodology of Reconstruction

A physical theory must somehow balance two very different demands. It must not only allow us to better *grasp* some aspect of the workings of the physical world, but must yield a *workable* conceptual and mathematical tool that allows us to describe actual laboratory experiments and make precise predictions that conform to the brute facts of experience. In the development of a theory, if push comes to shove, the demand for workability tends to win out—rather as an individual tends to opt for safety over curiosity or self-actualization (as per Maslow's hierarchy of needs), a theory's survival in the scientific community normally depends far more on its capacity to successfully grapple with empirical regularities in the domain of interest and on the economy and usability of its mathematical formalism than on its intuitive graspability.

As a consequence, a freshly developed physical theory is inevitably a *compromise*. This tends to manifest in two main ways. First, there will be features of the theory's mathematical formalism that (i) have been compelled to some degree by specific phenomena rather than being an expression of a more general physical idea or principle; (ii) are a compromise between conflicting physical ideas or desiderata; and (iii) or are the product of somewhat non-physical rationales (such as 'mathematical simplicity'). Second, the theory may refer to physical entities that are not directly observed (and are not directly observable according to the theory), but which form an essential part of the theory's conceptual scaffolding.

Accordingly, once a new physical theory has been developed (the *developmental* phase) and its empirical power has been sufficiently demonstrated (the *proving* phase), there typically follows a *reflective phase* in which attempts are made to rectify these perceived weaknesses. In this reflective phase, *reconstruction* refers to the methodology whereby one elucidates the physical meaning and origin of physically obscure aspects of the mathematical formalism by deriving these from more fundamental or intuitively graspable physical ideas or assumptions[12]. In the reflective phase, attempts are also made to eliminate the need for any unobservable entities by tracing the relevant parts of the formalism back to what can be directly observed, and such efforts are often part of—or directly or indirectly lead to—reconstructive work.

11.4.1 *Reconstruction in Classical Mechanics*

Although Newtonian mechanics is generally regarded as a near-ideal theory (especially in contrast to quantum theory), its reconstructive phase witnessed a series of efforts to address its perceived formal and conceptual defects. For example, Newton's framework brings up the following questions:

1. *Momentum as quantity of motion.* Why is the quantity of motion associated with a body $m\mathbf{v}$? In particular, why not a scalar quantity such as Descartes' mv, or some more general vector $\mathbf{f}(m, \mathbf{v})$ or scalar quantity $g(m, \mathbf{v})$?
2. *Equation of motion.* Why is $\mathbf{F} = d\mathbf{p} / dt$ the equation of motion?
3. *Parallelogram of forces.* Why do the forces that act on a single body combine vectorially?

In each case, reconstructive work—some of which continues to this day—provides answers to these questions, and thereby grounds these crucial aspects of the Newtonian framework in more fundamental physical ideas and principles. The first and third of these are considered in some detail in Sections 11.4.1.1 and 11.4.1.2. For the second, see Darrigol (2020) and Goyal (2020, §2.3).

The Newtonian framework also posits entities that are not directly observable, namely absolute space, absolute time, and the notion of force. Reconstructive work has not been particularly successful in obviating the need for these, but has spurred important advances. Mach argued that the notions of absolute rotation and absolute acceleration relative to an unobservable absolute space should be abandoned in favour of the idea of rotation and acceleration relative to the (observable) 'fixed stars'. This notion—dubbed 'Mach's principle' by Einstein—was a key input to Einstein's general theory of relativity, but as that theory recovers Newtonian mechanics in a limiting case in an empty universe, it cannot be regarded as successfully implementing Mach's principle. Poincaré noted the disconnect between the notion of absolute time and the conventionality inherent in any experimental determination of the simultaneity of distant events, which was likely a key input to Einstein's special theory of relativity (Darrigol, 2004, 2005). Finally, the notion of force was criticized by Hertz (Lützen, 2005, Ch. 4), who attempted to reconstruct mechanics without making use of this notion (Hertz, 1899). However, his reconstruction is unsuccessful insofar as it introduces new and rather abstract unobservables (hidden cyclic systems) of its own (Lützen, 2005, Ch. 6).

11.4.1.1 Momentum as Quantity of Motion

Descartes' notion that the motion of a system conserves its total scalar 'quantity of motion' was a key input to the development of classical mechanics, in large part owing to its intuitive graspability[13]. Yet, in the subsequent development of the laws of collinear collisions, this notion quickly came into conflict with two physical desiderata, namely that:

i. the total quantity of motion be conserved not only asymptotically (before and after the collision) but also at the moment of collision itself (i.e. *continuous* conservation, rather than merely asymptotic conservation) and
ii. the total quantity of motion be conserved during an inelastic collision.

As a result, Descartes' principle morphed[14] into a new conservation principle, the conservation of momentum, with momentum being a *vectorial* quantity of motion.

Thus, in retrospect, one can regard the conservation of momentum as a *compromise* between an intuitively attractive idea (Descartes' conservation principle) and other physically motivated desiderata (applicability to inelastic collisions; continuous conservation). The principle of momentum conservation is mathematically expressible and workable, but is not faithful to Descartes' original idea—a universe can 'wind down' and yet conserve total momentum[15]. In Newton's framework, it is rationalized via the idea that forces occur in opposed pairs ('action and reaction is equal and opposite'), an idea markedly different from Descartes'.

Descartes originally posited that a body's scalar quantity of motion had the mathematical form mv, which Newton (and others) vectorized to give momentum as $m\mathbf{v}$. The only motivation for these expressions appears to have been mathematical simplicity, leaving open the question of whether other mathematical expressions were physically viable.

Systematic derivations of the quantities of motion began to appear in the early twentieth century, apparently spurred by revisions to dynamics forced by special relativity, and have continued to appear ever since[16]. For example, in one recent derivation, the mathematical form of kinetic energy is derived from (i) a specific highly symmetric elastic collision, (ii) the requirement that some total (additive) scalar quantity of motion is asymptotically conserved; and (iii) Galileo's principle of relativity (Goyal, 2020, §2). This derivation also generalizes to the special relativistic case, yielding the correct relativistic energy (Goyal, 2020, §3). Moreover, an earlier argument by Schütz shows that asymptotic conservation of total kinetic energy and the principle of relativity together imply that total *momentum* is also asymptotically conserved (Schütz, 1897).

Thus, reconstruction reveals the intimate relations between the scalar and vectorial conservation principles, and shows that the mathematical forms of the corresponding quantities of motion follow from the notion of conservation once supplemented by a fundamental *kinematical* symmetry principle (Galileo's principle of relativity, O4).

11.4.1.2 *Newton's Parallelogram of Forces*

In his *Principia*, Newton obtained his so-called parallelogram law for combining forces that act on a single body by considering the changes in the body's motion that results in the special case where the forces act impulsively at separate moments of time, and then combining the resultant changes. The argument thus depends upon Newton's second law that describes the effect that a force has upon the motion of a body, a law whose deeper origin is itself in question. And since the argument considers the case where the forces are unbalanced (and so cause a change in motion), it does not cover the case where the forces are in static equilibrium.

Following Newton's formulation, numerous attempts were made to place the parallelogram of forces on a sounder conceptual footing (Lange, 2011). One of the most incisive arguments is due to d'Alembert (and subsequently refined by many others), which shows that the parallelogram law can be derived largely from elementary symmetry assumptions[17]:

1. The resultant of two parallel forces has magnitude equal to the sum of the magnitudes of these forces, and points in the same direction.
2. The resultant of a number of forces is commutative and associative.
3. The resultant of two forces is rotationally covariant.
4. The resultant of two equal forces varies continuously with the angle between those forces.

The first assumption is a particular case: if two forces are parallel, their magnitudes simply add. As it happens, one can simplify this further: additivity of magnitudes follows from the requirement that parallel-force composition is associative and continuous at a point[18]. The second and third assumptions are *symmetry* requirements, the second appealing to compositional symmetries, the third implicitly appealing to the isotropy of space (C3). Finally, the fourth assumption appeals to the requirement of continuity—the resultant of two forces should change *gradually* as the forces gradually change.

As this argument does not depend upon any relation between force and motion, it applies equally in the static and dynamic cases, and is independent of any law connecting force and motion. Moreover, it shows that the parallelogram law follows very generally from compositional and spatial symmetries, together with the primitive idea that the magnitude of a force can be quantified by a real number. As one can see, very little of the intuitive notion of force is left: the assumptions essentially only require that force is something that has a direction and real-valued magnitude and can be composed with other forces in a manner that satisfies basic compositional symmetries.

11.4.2 Reconstruction in Electromagnetism

Reconstructive work in classical mechanics has generally served to more solidly ground aspects of the Newtonian mathematical framework in more elementary physical principles, but has not fundamentally challenged the classical conceptual framework in which it is embedded. However, in electromagnetism, the reconstructive method historically led to a profound transformation in the interpretation of the theory—and of the nature of space and time—which was wholly unexpected.

11.4.2.1 *Faraday's Field Conception*

Beginning in the early nineteenth century, theories of electromagnetic phenomena developed along two parallel streams. The major stream sought to embed electric and magnetic phenomena within the existing Newtonian framework by ascribing a new time-independent property—charge—to each particle, and by formulating new force laws (Coulomb, Ampère, Biot–Savart, Weber, etc.), patterned after Newton's law of gravitation, that govern the interaction between static and moving charges and between current elements. The minor stream, initiated and sustained by Faraday, sought to understand electromagnetic phenomena through a novel conception of space-filling electric and magnetic *fields* produced by—and influencing—charges and currents.

A rather crucial feature of both streams is that their key features were strongly shaped by the peculiarities of the phenomena of interest, rather than by some general *a priori* principles as was the case with classical mechanics. Thus, in Faraday's conception, the need for *two* distinct fields (electric, magnetic) as opposed to just one; the fact that these fields interact in the specific way that they do with charges and each other; and the fact that charges (electric monopoles) exist but magnetic monopoles do not—all were ultimately justified by the need to bring order to the phenomena of interest. In contrast, as noted previously, classical mechanics was largely based on general principles—the constant velocity of an isolated particle; the physical equivalence of all inertial frames; conservation principles—which seemed to spring more from the instinct to bring an ideal mathematical order to the physical realm rather than from an attempt to make sense of any specific regularities in the phenomena of interest.

11.4.2.2 *Maxwell's Equations and Their Interpretation*

Maxwell's equations, the heart of electrodynamics, are in essence a mathematical clothing of Faraday's field conception (although Maxwellian electrodynamics incorporates important elements from the force-law stream, for example in the form of Lorentz's force law). Due to the predicted existence of electromagnetic waves, and the belief that such waves—like sound waves—require a *medium* for their propagation, it was almost universally assumed until at least the end of the century that Maxwell's equations apply to a *particular* inertial frame, namely the frame that carries the requisite medium ('aether'). This naturally accounted for the fact that, unlike Newton's equations of motion, Maxwell's equations are not invariant under Galilean transformations. However, this 'natural' interpretation generated a clear tension with Galileo's principle of relativity (O4) and with the Newtonian mechanics that incorporated that principle.

Meanwhile, Maxwell's equations possess a rich mathematical structure, which was gradually brought to light. In particular, as discovered by Lorentz in 1892 (although to some degree anticipated by Voigt in 1887), these equations are invariant under the so-called Lorentz transformations, which suggests that these equations *are* valid in non-aether inertial frames provided that one introduce new abstract frame-dependent space and time coordinates. However, the physical meaning of these transformations and these abstract coordinates was unclear. Moreover, the aether hypothesis also faced a mounting challenge due to (i) the difficulty of coming up with a single aether model capable of supporting the full range of physical phenomena (such as the apparently frictionless passage of planetary bodies through the aether, as well as the propagation of light at such high speeds) and (ii) the failure of experimental attempts to detect motion relative to the aether (in particular the 1887 Michelson–Morley null result).

11.4.2.3 *Einstein's Reconstruction and Its Interpretational Implications*

The meaning of the Lorentz transformations, and of the failure to detect motion relative to the aether, was dramatically elucidated in 1905 by Einstein, who *reconstructed* the Lorentz transformations in an operational framework. Astonishingly, he derived the transformations without any direct reference to Maxwellian electrodynamics, the field concept, or the aether hypothesis. Instead, the Lorentz transformations were traced back to elementary spatial and temporal measurements carried out using rods and clocks. The derivation was based on two key assumptions: (i) the one-way speed of light is independent of the speed of the source (a reasonable extrapolation of experimental facts to date) and (ii) Galileo's principle of relativity. Einstein reconciled these two seemingly contradictory ideas by positing a reasonable definition of light-based synchronization of distant clocks, and then showed how they led to the Lorentz transformations. He went on to build a new mechanics compatible with the Lorentz transformations, which led to specific testable predictions.

Einstein's reconstruction thus showed that the Lorentz transformations *were* compatible with Galileo's principle of relativity, but that there was a cost: observers in different inertial frames would, in general, disagree as to the spatial distance and temporal duration between two events. Thus, even if there were an absolute distance and duration between the events (as per the Newtonian classical conceptual framework—see Section 11.2.1.2), it would be experimentally inaccessible.

Einstein's reconstruction was so compelling, and the various solutions hitherto proposed to address the challenges faced by the aether hypothesis sufficiently unattractive, that the aether hypothesis was effectively abandoned within a few years, and a new interpretation of Maxwell's electrodynamics established. Moreover, the very notion of absolute space and time was brought into question by the reconstruction's implication that absolute distances and durations (if they exist) are experimentally inaccessible, and it became the norm to require that a theory or model be Lorentz invariant.

11.5 Reconstruction of Quantum Theory

The development of quantum theory bears many important similarities to that of electrodynamics. In the absence of general physical principles of sufficient power, both theories were strongly shaped by the peculiarities of—and observed regularities in—the specific phenomena of interest. And the mature formalisms of both theories possess mathematical features and mathematical structure whose deeper physical meaning was initially obscure. As we have seen above, the physical meaning of the principal symmetry of Maxwell's equations, namely the Lorentz transformations, was elucidated by Einstein's reconstruction.

Can the reconstructive method similarly elucidate the various striking mathematical features of the quantum formalism? For example, why does the formalism employ complex numbers and why are complex numbers so well suited to the expression of the theory? Why are dynamics represented by unitary transformations, rather than a broader class of transformations? Why does the Born rule (which connects complex-valued states and outcome probabilities) take the specific mathematical form that it does rather than a more general form? Why is the tensor product operation appropriate to construct the states of a composite system? Why are the states of identical particles subject to the symmetrization postulate? A reconstruction of the quantum formalism which traces these abstract mathematical features to clearly stated physical principles in an operational framework—analogous to Einstein's reconstruction of the Lorentz transformations—could yield deep, unexpected insights into the reality that is so astonishing well described by quantum theory.

Recognition of the importance of reconstruction for elucidating the quantum formalism was not lost on the founders. For example, Heisenberg recognized that it would be highly desirable if the quantum formalism could somehow be derived using his uncertainty principle as a key axiom. And, in his 1946 Nobel lecture (Pauli, 1998), Pauli expressed the view that his exclusion principle (even after incorporation into the symmetrization postulate) called for a deeper explanation in the form of a 'rigorous derivation' from more general assumptions, but that no such explanation had hitherto been forthcoming[19].

Broadly speaking, reconstructive attempts prior to the 1980s tended towards abstract, intricate systems of axioms, which made little impact

(Grinbaum, 2007a, 2007b). Since the 1980s, the program of reconstruction has gained fresh impetus and inspiration from several directions, and has gradually gained traction in the foundations of quantum physics community[20]. There now exist a number of fairly rigorous reconstructions of the aQF in finite dimensions, and a fewer number of reconstructions of many of the other parts of the formalism[21].

11.5.1 *Informational Perspective*

One of the major forces behind the renewed interest in reconstruction has been the *informational* perspective on physical theory[22], which is aptly summarized in Wheeler's slogan '*It from Bit*' (Wheeler, 1989, 1990):

> '*It from bit*' symbolizes the idea that every item of the physical world has at bottom—at a very deep bottom, in most instances—an immaterial source and explanation; that which we call reality arises in the last analysis from the posing of yes-no questions and the registering of equipment-evoked responses; in short, that all things physical are information-theoretic in origin, and this in a participatory universe.

and

> What we call reality consists of a few iron posts of observation between which we fill an elaborate papier-mâché of imagination and theory.

These views colourfully echo Mach's view of physical theory as first and foremost an economical representation of physical observations. But they go significantly further. First, the assertion that the 'physical' world has an 'immaterial source' places *observation* at the centre and decisively demotes the notion of matter to part of the elaborate conceptual papier-mâché that we build in order to make sense of these observations[23]. Second, Wheeler speaks of measurement not as a passive observation of that which exists, but rather as an *active* process whereby we *pose questions* to nature and *evoke a response*. This view of measurement arises from a particular reading of the quantum formalism, namely one that takes measurement as a primitive, and moreover one that posits that measurement is not a passive registration of pre-existing properties but rather that measurement outcomes are in some sense *co-created* through the act of measurement.

The fertility of the informational viewpoint has been strongly supported by the emergence of the fields of quantum information and quantum computation. These fields' many successes and striking discoveries (such as the possibility of secure information transfer and better-than-classical computation) have deepened the conviction that quantum theory rewards an informational perspective, in particular that it provides a new lens through which to look at the quantum realm which may well allow us to make new progress in understanding the nature of quantum reality.

11.5.2 *Operational Framework*

Most reconstructions take place in an *operational framework*. In essence, this is an idealized, abstract representation of the laboratory workbench, the place where theory comes into contact with physical reality.

11.5.2.1 *Experimental Set-Up*

The operational framework rests on the idea that a *physical system* is subject to an experiment, namely a sequence of *measurements* and *interactions*. The measurements and interactions are presumed to be implemented by objects that are, at least to some extent, classically describable (i.e. 'macroscopic objects'), and whose settings are under an experimenter's control. Measurements differ from interactions in that measurements yield macroscopically observable events (such as flashes on a scintillation screen), and it is through these events that the physical system comes to be indirectly known. Idealized measurements are typically assumed to be *repeatable* in the sense that immediate repetition of a measurement on a physical system yields the same outcome with certainty.

11.5.2.1.1 NOTION OF PHYSICAL SYSTEM

The idea that an experiment is performed upon a 'physical system' consists of two distinct notions. First, that there exists some entity that *persists* for the duration of the experiment. That is, in spite of interactions and measurements performed upon it, there is some meaningful sense in which there is something which retains its identity over the course of the experiment, so that one can say that a measurement at t_2 is performed on 'the same' system as the measurement at t_1. The notion of persistence thus provides a minimal yet crucial means to link outcomes obtained at different times.

Second, that there is some definite sense in which all of these interactions and measurements are probing the same *aspect* of the physical object. We accordingly say that the experiment is performed upon a *physical system* or abstract physical object. In experiments on physical objects deemed to be classically describable, this notion is usually implicit: we understand that experiments on real objects (such as billiard balls) must be carefully circumscribed in order to extract meaningful information. For example, consider an experiment which is designed to probe the centre-of-mass spatial behaviour of a classically describable billiard ball. A preparation would consist in fixing the initial position and velocity of the ball. Subsequent measurements would need to be restricted to those which only probe these degrees of freedom (position, velocity)—a measurement whose outcome depends upon the ball's rotational motion or, say, its material composition or colour, would need to be excluded. Similarly, interactions with the ball would need to be restricted to those that do not couple the spatial (position, velocity) degrees of freedom and its rotational or non-spatial degrees of freedom. For instance, interactions which change the ball's velocity in a way that depends upon its colour would need to be excluded. With these restrictions in place, the outcomes of measurements on the ball would be independent of pre-preparation interactions with it (such as re-painting it a different colour), and the experiment would be probing only the spatial sub-component of the actual physical object. One could then abstractly describe this sub-component as 'a particle', namely, an abstract object whose only time-dependent properties are position and velocity.

As described in Goyal (2008, §II A), these ideas can be generalized in a way that can be applied to non-classical systems—specifically systems to which *a priori* we (i) cannot attribute properties and (ii) cannot assume these properties can be passively observed—via the notion of *closure*. The closure condition can be used to operationally establish that two measurements are probing the same aspect of an object. One can then operationally define the *set* of all measurements that probe the same aspect of the object, and define a set of interactions that can be thought to act wholly on the abstract object thereby defined. In short, the 'physical system' refers to some *operationally defined aspect* of a physical object.

11.5.2.1.2 NATURE OF THE AGENT

In the operational framework, it is usually left implicit that there exists an entity—an *agent*—that is capable of *observing* or *registering* measurement outcomes, and that the entity possesses the capacity to make changes in device settings without being influenced by the system under study or by past or future measurement outcomes. These notions are usually implicit in experimental science, but the precision required in reconstructive work is such that they sometimes need to be made explicit.

11.5.2.2 Abstraction of Key Notions

The operational framework is an *abstraction* of the laboratory workbench, not a *representation* of it. Accordingly, a great deal that one tends to unconsciously associate with an 'experiment' is not an intrinsic part of the operational framework, and indeed turns out not to be needed in many reconstructions of the aQF.

For example, we tend to think of an agent as an embodied being localized in space; a physical system as an object that is spatially localized in our laboratory at all times; or a measurement as carried out by a chunk of equipment in one corner of a laboratory. But the operational framework abstracts away all of these *spatial* notions. So, a *physical system* is simply an entity that *persists*—it does not necessary exist *anywhere* in particular at a given moment in time. A *measurement* is an abstract parameterized process that acts on a physical system to generate an *outcome* and to output the same physical system—it is not a spatially localized piece of equipment. The agent is simply *an entity* that exists and persists over time, and is capable of *observing outcomes* and of *freely acting* to change *settings* associated with measurement and interaction devices—it is not a spatially localized human being.

On the other hand, the notion of *temporal order* is essential to the operational framework. It is assumed that the measurements and interactions occur in a well-defined temporal sequence. In particular, temporal order is essential to the notion of closure (which is essential to the definition of a physical system and to the set of measurements and interactions), and the notion of 'immediately afterwards' is required to ground the above-mentioned notion of repeatability.

11.5.2.3 Classical Component of Reality as a Portal

The operational framework makes essential reference to physical objects that an experimenter can observe and manipulate. In so doing, the framework presumes that there is a component of physical reality that is well described by our everyday object model of sensations and/or by classical physics. For example, it is assumed that the apparatus and the agent both persist over time; that the agent can passively observe measurement outcomes; that the apparatus has well-defined properties (in practice, such things as dial-positions and knob settings) that can be unproblematically adjusted by the agent.

This *classical component* of physical reality then effectively acts as a *portal* through which we probe some other part of physical reality, be it classically describable or not. For example, in an electron diffraction experiment, the experimenter has access to a device that generates the accelerating voltage and to devices that generate deflecting electric and magnetic fields; and can observe flashes on a scintillation screen which he interprets as evidence of electronic collisions. But the behaviour of the 'electron'—the microphysical object which is commonly supposed to underpin the scintillations—is not classically describable. But it is only *through* these classically describable bits of equipment that we learn about its behaviour.

To be clear: we cannot *directly* access the non-classical because we do not know if our classical notion of physical object (namely, an persistent entity that possesses properties) is valid, and we do not know if we can passively observe the properties (if they indeed exist) of these objects. But these notions are the basis for defining 'an experiment', and underpin the possibility of carrying out many trials of 'the same' experiment and accumulating statistically analysable data.

For instance, when we speak of a measurement device having a *setting*, we are evoking an object–property model, in particular attributing the setting to the object itself. And we also presume that an agent can passively observe that setting. These are all 'common sense' notions drawn from everyday experience with physical objects, assumptions that are enshrined in the classical conceptual framework. It is far from clear what is left of the notion of 'experiment' if such assumptions cannot be made.

In short, it would seem that access—at least *scientific* access—to the non-classical component of reality requires a classical portal. This is not, of course, to say that one cannot access the non-classical component of reality without, for example, the possibility of changing 'settings' in a reliable way, or without the possibility of carrying out multiple trials of an 'experiment'. Indeed, such *non-scientific* access to non-classical reality might well correspond to much of the so-called subjective experience.

11.5.3 *Physical Principles*

Once the operational framework is in place, one must posit *physical principles* which precisely articulate guesses or hunches about the nonclassical physical reality which is manifested in experiments. It is characteristic of most recent operational reconstructions that these physical principles refer primarily to the *data* gathered in experiments, rather than attempting to *directly* posit features of the physical system under scrutiny. For that reason, such reconstructions are often referred to as *informational* or *information-theoretic* reconstructions, with the latter particularly common if the machinery of Shannon's information theory or Bayesian inference is employed.

11.5.3.1 *Wootters' Derivation of Malus' Law*

An excellent illustration of the kinds of physical principle that are employed in recent reconstructive work is afforded by one of the earliest informational reconstructive results due to Wootters (1980). Consider an experiment on a physical system subject to measurements that yield one of *two* possible outcomes (which we label 1 and 2), where each measurement is parameterized by a single angle[24]. This is an abstraction of Stern–Gerlach measurements performed on spin-1/2 systems.

Consider a game in which Alice *prepares* one such two-outcome system by performing a measurement with θ, and then transmitting that system to Bob if the outcome happens to be 1 (she discards those systems that yield outcome 2). On receipt, Bob performs a measurement with setting $\theta' = 0$, and records its outcome (either 1 or 2). In total, Alice sends n identically prepared systems to Bob.

Now, let us suppose that reality is lawful in the minimal sense that the *probability* of Bob's outcome is determined. That is, the probability, p, of his obtaining outcome 1 is a function of θ. It follows that, if Bob were to know the function $p(\theta)$, then he could make a reasonable guess about the setting angle, θ, given the frequency with which he obtains outcome 1. That is, the string of outcomes that he obtains in the n measurements provides him with some information about θ. Accordingly, one can view this as a game in which Alice imperfectly transmits the angle θ to Bob by 'encoding' the angle in n two-outcome physical systems.

Wootters now posits that the laws of nature are such that the amount of information that is transmitted—quantified using the relative entropy—is *maximized* in the limit as $n \to \infty$. Wootters then effectively shows[25] that, if one assumes that the prior probability over θ, $\Pr(\theta \mid I)$, is uniform, then $p(\theta) = \cos^2(m(\theta - \theta_0)/2)$, where $m \in \mathbb{Z}$. This *agrees* with the predictions of quantum theory for a two-outcome system, and is known as Malus' law.

11.5.4 *Informational Reconstructions*

Wootters' derivation of Malus' law is minimalistic in its assumptions, and is a powerful illustration of the potential of informational approaches to quantum theory. However, this minimalism does not survive the passage to reconstructions of the quantum formalism proper.

In particular, most reconstructions introduce the notion of *state* in the form of a mathematical object that is associated with the physical system at each moment in time, and whose role is defined operationally as that which allows the prediction of the outcome probabilities of any[26] measurement that could be performed on the system. In contrast, in the reconstruction of Feynman's formulation of quantum theory (Goyal et al., 2010), one considers a *transition* between given outcomes of successive measurements and associates a mathematical object—a pair of real numbers, which eventually becomes a complex-valued *amplitude*—to that transition, one role of which is to determine the transition probability.

The introduction of the notion of state or transition amplitude adds a layer of abstraction to the reconstruction, which is then reflected in the postulates to some degree. For example, Hardy (2001a, 2001b) formulates postulates that refer to the number of degrees of freedom associated with a state. Similarly, the derivation of Feynman's rules emerge as a pair-valued quantification of an experimental logic. The interpretation of this abstraction layer poses a considerable challenge.

Nonetheless, certain specific assumptions which are then made in order to give shape to the resulting mathematical structure can be more readily understood. One class of postulates concerns bipartite systems. For example, one operationally expressible postulate known as *tomographic locality* posits that the state of a bipartite system can always be determined from the statistics of a sufficient number of different joint measurements performed separately on the two sub-systems. This postulate is employed to good effect by Barrett (2007) to account for at least some of the structure of the quantum formalism, such as the tensor product rule for determining the state of a composite system when its subsystems are in known pure states. Hardy interprets one of his postulates as an expression of tomographic locality (Hardy, 2013).

Another class of postulates concern the behaviour of individual systems. For example, it is possible to reconstruct the aQF via the Feynman rules of quantum theory by suitably formalizing the notion of complementarity and by introducing a *no-disturbance* postulate which posits that certain measurements which yield no useful information about a physical system also do not disturb its state in any detectable way (Goyal, 2014; Goyal and Knuth, 2011; Goyal et al., 2010). Here, the no-disturbance postulate is an expression of the idea that measurement is an *active* process. In particular, that although the acquisition of information about a system in general forces a change in the state of the system, in the limiting case that the measurement yields *no* new information about the system (it simply tells us that the system exists, which we already knew), no such change occurs.

11.6 Interpretation of Quantum Reconstructions

As described in Section 11.3.4.2, an ideal interpretation of quantum reconstructions would yield (i) a quantum conceptual framework (precisely articulated in a form analogous to the classical conceptual framework as in Figure 11.1) and (ii) a revised notion of the project of physics that ties together the framework's assumptions.

The interpretation of quantum reconstructions is presently in its infancy. Be that as it may, some general interpretative implications of reconstructive work are already clearly visible, which I shall sketch in Section 11.6.1. More specific implications are also starting to come into view. For example, over the past few years, I have developed an interpretation of the

QSA based on a reconstruction of the same (Goyal, 2015), which gives rise to a new understanding of the nature of identical quantum particles (Goyal, 2019, 2022). The key ideas are summarized in Section 11.6.2. An interpretation of the reconstruction of Feynman's rules (Goyal et al., 2010) is also underway, which I expect will yield a metaphysically sharp formulation of Bohr's principle of complementarity.

11.6.1 *Interpretative Implications from Reconstruction of the Abstract Quantum Formalism*

The aQF is the core of quantum theory. Although an abstract shell rather than an explicit model of a particular physical system, it is highly contentful. For example, the striking protocols in quantum information (such as quantum teleportation or quantum cryptography) or schemes for quantum computation only make use of the aQF. Key no-go theorems, such as Bell's theorem and the Kochen–Specker theorem, can likewise be formulated wholly within the aQF. And more broadly, most of the familiar nonclassical features of quantum theory, such as the existence of entangled states, the change of quantum states due to measurement, and the notion of complementarity (in a variety of forms), can be articulated entirely within the aQF.

For these reasons, the aQF is the prime reconstructive target, and its interpretation is of the greatest importance. Below we make some preliminary remarks on some of the general interpretative insights that can be drawn from many of the operational reconstructions of the aQF (such as Chiribella et al., 2011; Goyal, 2014; Goyal et al., 2010; Hardy, 2001a).

11.6.1.1 *Notion of Space*

Operational reconstructions of the aQF typically make no explicit reference to the notion of space, either in the operational framework or in their postulates. In particular, the postulates make no use of the idea that objects are localized in space, or that space has a certain dimension, topology, or metric.

As mentioned previously, the operational framework presumes that an observer has some ground for saying that *this* measurement is performed on *that* system, and *this* outcome (rather than some other) is obtained. In practice, an experimenter grounds such assertions on spatial perception, and relies upon the assumed persistence and continuous localization of macrophysical objects. But reconstruction of the aQF makes clear that we do not *need* the notion of space (along with the rich set of ideas that accompany it) *per se*—the observer only needs *some* means to ground such assertions.

11.6.1.2 *Notion of Measurement*

As indicated in Section 11.5.2.1, whereas one ordinarily thinks of a measurement as implemented by a physical device localized in space, the operational framework abstracts from the laboratory workbench to the extent that all that is left of the notion of *measurement* is that it is an abstract parameterized physical process that acts on the physical system and yields an *outcome*. This more abstract notion of measurement is indeed in keeping with the above remarks on space.

This more abstract notion of measurement is, in fact, indispensable in the usual applications of the quantum formalism. For example, one commonly regards the outcomes of several localized measurements as a single outcome of a *joint* measurement, and that joint measurement is not located anywhere in particular. At a more fundamental level, a more abstract notion of measurement frees one's imagination to consider processes of actualization—such as Penrose's gravitationally induced objective reduction (OR) or GRW's collapse model—which are not tied to macroscopic measurement devices.

11.6.1.3 *Notion of Time*

In contrast to space, the notion of *temporal order* is central to the operational framework. In particular, the observer *must* be able to say '*this* measurement was performed, yielding such-and-such outcome; *then* that measurement was performed …', and so on. Thus, the aQF depends on the notion of time, at least in the very specific sense that it is assumed that an observer has the means to temporally order their experiences. In addition, the notion of 'immediately afterwards' is required to ground the notion that a measurement is repeatable.

11.6.1.4 *Primacy of Time Over Space*

Based on the above remarks about space and time in the operational framework, the aQF—the core of quantum theory—does not 'know' about space in the Newtonian sense. It does, however, 'know' about time in the limited sense of temporal order and immediate succession. This suggests an obvious interpretation, namely that the aQF describes a reality that *precedes* space[27]. This, coupled with the fact that the aQF captures so much of what distinguishes quantum physics from classical physics, is intriguing on a number of levels.

First, the aQF was historically obtained via a process of abstraction from concrete quantum models designed to account for such phenomena as the spectral lines generated by excited atoms. Moreover, these models were—in the work of, say, Bohr, Heisenberg, and Schrödinger—typically arrived at by subverting classical models of those same systems by introducing new ideas (such as quantization of angular momentum or wave-particle duality). But the aQF makes no reference to a gamut of fundamental classical notions such as space, energy, and momentum. So, the fact that the aQF can be reconstructed without explicit reference to these classical notions is remarkable, and shows that the aQF is a free-standing structure that does not need to lean on the classical conceptual framework or on any particular classical physical theory (such as classical mechanics or electromagnetism).

Second, the idea of a richly textured reality that precedes space may be scientifically useful. The idea is, in fact, quite prevalent in certain approaches to quantum gravity in the form of the posit that there exists some structure prior to space ('pre-space'), of which space is only an effective, approximate description. At a more metaphysical level, the flexibility that this notion provides in conceiving of physical reality opens up new possibilities for conceptualizing identical particles (see Section 11.6.2).

11.6.1.5 *State Concept*

In reconstructions of the aQF such as Hardy (2001a), the notion of state is introduced as a means of connecting together measurement outcome data. In the Feynman pathway to the aQF (Goyal 2014; Goyal et al., 2010), amplitudes (initially pairs of real numbers) are introduced in order to connect together pairs of measurement outcomes obtained at different times, and states are then built up as collections of these amplitudes. Thus, the mathematical state object reflects the chosen measurements as well as the physical system in question.

This contrasts sharply with the state concept in the classical framework, which is thought to describe the actual physical state of a system (such as the position and velocity of a particle) quite independently of whether or not it is observed, or indeed independently of whether there are any observers at all.

11.6.1.6 *Nature of the Agent*

As mentioned in Section 11.5.2.1, it is usually implicit in the operational framework that there exists an agent capable of (i) passively observing measurement outcomes and (ii) changing measurement and interaction device settings 'at will'. As described in Section 11.2.1.2, the agent also appears in this dual role in classical physics. However, the classical framework posits that, absent agential action, physical reality can be described and modelled without including the agent. The key interpretational question is then whether that remains the case in quantum theory, or whether there is a compelling reason to include the agent explicitly as part of the description of physical reality.

As described above, the notion of a quantum state is introduced as a means of connecting measurement outcomes in an operational context. That context is shaped by the actor-as-agent—the agent chooses the measurements to which to subject a physical system, and the set of quantum states that can be ascribed to the system is a function of that choice. As such, the quantum description of physical reality is—from the reconstructive point of view—inextricably context-dependent, and that context is shaped by agential action. This contrasts sharply with the situation in classical physics, wherein one is permitted to speak of the physical state of a system—or the universe as a whole—without regard to agential action.

We could then choose to say that quantum theory differs from classical theories in that it is a *context-dependent* formalism, namely that its formalism presupposes a classically describable experimental context. In such a case, the agent (understood in the dual sense as above) does not play a role that fundamentally differs from its role in classical physics. Alternatively, one could assimilate the context to the agent (as appears to be favoured in QBism—see Fuchs, 2017, especially Figure 11.1), so that the formalism is inextricably bound up with agential experiences and choices. My own view is that such an assimilation is unnecessary and is liable to lead to intractable difficulties, and that the first view—bearing in mind that the experimental context is agentially shaped—is the more fruitful.

Two final remarks. First, it is important to note that, insofar as the reconstruction of the aQF is concerned, the nature of the agent beyond that which is capable of serving a passive observer or an 'at will' actor are not specified. In particular, the notion of a bodily extended, spatially bounded agent is unnecessary—one could conceive of the agent as an incorporeal observer and actor. One need not even suppose that there are many distinct actors—one could conceive of a single (universal) actor.

Second, the agent-as-actor—namely the entity capable of actively changing experimental settings 'at will' (i.e. uninfluenced by the physical system or by future measurement outcomes)—is essential to the *carrying out* of actual experiments and its capabilities are implicit in the interpretation of experimental results. Although this is also true in classical physics, it is intriguing that certain key results in quantum foundations require that this be made explicit. For example, in the statement of Bell's theorem, it must be made explicit that experimenters in each wing are 'free' in their choice of measurement settings, in particular that they are not in any way influenced by the present or previous state of the system upon which the measurements are being performed. It also must be made explicit that the state of the system at an earlier time is not influenced by what measurement settings will be chosen at a later time. This is sometimes referred to as the 'no-conspiracy' assumption, which is in fact implicit in virtually all experimental design.

11.6.2 *The Nature of Identical Quantum Particles*

Atomism seeks to account for our everyday experience of persistent objects by positing the existence of elementary entities that have continuous transtemporal existence. As described in Section 11.2.1.2, this metaphysical conception is incorporated into the conceptual framework of classical particle mechanics: macroscopic bodies are assumed to be composed of eternal (indefinitely persistent) point particles, each of which possesses time-independent properties (such as mass and charge) and moves continuously through space. Each such particle can in principle be reidentified, either by measurement of its distinct time-independent properties (if it is unique), or by tracking it precisely over time.

According to present day particle physics, ordinary matter is composed of elementary entities such as electrons and quarks (which, in turn, compose protons and neutrons). But, quantum theory raises the question as to whether—and to what extent—these 'particles' possess the various characteristics ascribed to them by classical physics.

In particular, the quantum formalism contains a specific algorithm—the QSA—which must be employed to model systems composed of *identical* particles, such as the two electrons in a helium atom. According to this algorithm, the set of allowable states of such a system are restricted to those that are either symmetric or antisymmetric, depending upon the type of particle in question. According to the standard interpretation of this restriction (known as the symmetrization postulate), there exists no measurement that can 'address' a particular particle in a system of identical particles. Consequently, particle reidentification is impossible. Such an implication, however, is directly at odds with the basic experimental data of particle physics: a bubble chamber images is said to show *particle tracks*, and the geometry of these tracks is used to determine these particles' intrinsic properties. The question is, then, how to reconcile these two (theoretical and experimental) perspectives.

Over the past few years, I have developed an interpretation of the QSA based on a reconstruction of the same (Goyal, 2015), which reconciles these two perspectives (Goyal, 2019, 2022). The essential idea that emerges is that, *contra* the atomistic conception, particles cannot be said to simply *exist*. Rather, the particle notion is simply an element of a model of detection events, and that model is only strictly valid in limiting cases. In general—as, for example, in a helium atom—a duality of object-models is needed to conceptualize the underlying microphysical reality. In one of these models (the so-called persistence model), there exist two distinct persistent entities. However, in the other model (the nonpersistence model), there exists but a single persistent entity—a holistic object, if you will—which manifests as multiple point-like events at a single moment in time.

In general, *both* models must be mathematically synthesized in order to generate empirically accurate models.

This radical conclusion can be further illuminated through connection to metaphysical debates on such issues as event ontology, strong emergence, and monism *vs.* pluralism. For example, I argue that microphysical events have primacy over microphysical objects (a restricted form of event ontology); and that one can regard the formation of microphysical composites of identical particles (such as in the formation of a helium atom) as the emergence of a holistic object, and, conversely, the decomposition of such composites as the emergence of particles (Goyal, 2022).

11.7 Concluding Remarks

Quantum theory is our most successful physical theory, both in the sheer *range* of physical phenomena it is capable of describing and the *precision* with which is does so; and in its capacity to repeatedly make *unexpected novel predictions* that are subsequently borne out[28].

Quantum theory is also by far our most enigmatic physical theory. Despite a century of efforts to peer through the quantum veil, we are still left in the strange predicament of being in possession of an astonishingly powerful theoretical tool that we scarcely comprehend, *viz.* a tool that we do not know how to speak about in natural language in a coherent, comprehensive, and precise manner—let alone with any logical-philosophical precision and systematicity—without resorting to theoretical jargon or leaning on mathematical crutches.

The elucidation of quantum theory is of vital importance for the development of physics, in particular for development of theories of quantum gravity, and—as the development of the fields of quantum information and computation attests—for the full-bodied exploration and technological harnessing of the quantum realm. But the importance of such elucidation goes far deeper: quantum theory forces reconsideration of a world-view that has influenced or guided the development of Western society for three centuries. Hence, a vivid understanding of quantum theory, one precise and comprehensive enough to be put alongside the mechanical conception of physical reality, would likely have widespread consequences for many compartments of human knowledge and human culture.

I believe that it is entirely reasonable to expect to be able to understand quantum theory at the same level of clarity as classical physics. In particular, I have argued that the intelligibility of classical physics rests on it possessing a coherent tripartite structure (which spans from an overarching conception of physics and physical reality—the mechano-geometric conception—all the way to specific mathematical theories) rather than it comporting with our experience of the everyday physical world. Accordingly,

I contend that quantum theory will achieve a comparable level of intelligibility once it is embedded within an analogous tripartite structure.

In this chapter, I have described a reconstruction-based strategy for elucidating quantum theory that, unlike most traditional elucidative methodologies, has the potential of yielding such a tripartite structure. As I have argued, reconstruction is part of the natural life-cycle of physical theories, and is likely to yield the most radical insights when used to investigate theories whose formalism was largely shaped by regularities in the phenomena of interest rather than by *a priori* general principles. As quantum theory is such a theory *par excellence*, its reconstruction is particularly apt. Yet, for the first eighty or so years of its existence, no compelling reconstruction was available. It is thus a blessing that, at this moment in time, we have access to numerous detailed reconstructions of many key parts of the quantum formalism, a bonanza that is largely (although by no means exclusively) due to the informational perspective on physical theories.

The next step in the elucidative strategy is the *interpretation* of suitably chosen reconstructions. The *goal* of such interpretation is clear: we wish for a precisely articulated quantum conceptual framework, analogous to the classical conceptual framework, together with an overarching conception of physical reality which renders intelligible the particular assumptions within that framework.

Although the interpretation of quantum reconstructions is in its early stages, certain insights have already been obtained. In particular, I have sketched some general implications of reconstructions of the aQF, and have summarized some rather striking implications drawn from a recent reconstruction of the QSA.

The prospects for further rapid progress in the interpretation of reconstructions is unclear. The fundamental limiting factor is the sheer cognitive difficulty of bringing to bear a reflective or philosophical mindset onto reconstructions that are often articulated in a distinctive mathematical framework, often employ unfamiliar mathematical machinery (such as functional equations or the geometry of convex sets), and posit numerous physical principles whose primary raison d'être is often their sufficiency for reconstruction rather than their perspicuity. Social factors pose an additional barrier. For example, very few workers in the foundations of physics who develop reconstructions have gone on to interpret them in a deliberate manner. That may, in part, reflect a lack of philosophical training or the modern-day gulf between the physics and philosophy communities. In addition, from my outsider's perspective, the philosophical landscape itself appears rather balkanized, and only certain very specific traditions appear to have maintained a strong connection to mainstream physics. Furthermore, in the mainstream philosophy of physics, reconstructive work does not seem to be taken seriously as an alternative pathway to the interpretation of quantum theory. Instead, traditional interpretations (particularly those that seek to preserve as much as possible of the classical conceptual framework) continue to hold sway.

My recent efforts to interpret my previous reconstructive work have convinced me that many disparate areas of philosophy—including metaphysics (both analytical and other forms), continental philosophy, and non-western philosophy (such as Mādhyamika philosophy)—have a great deal to offer. For example, Husserl's phenomenology seems to strongly resonate with the operational reconstructive strategy for elucidation of physical theories: the emphasis on operational procedures reflects the centrality of moment-by-moment perceptual experience (Berghofer and Wiltsche, 2019), while the importance of reconstruction echoes Husserl's genetic phenomenology (Berghofer and Wiltsche, 2020, §1.2.8), *viz.* the importance of becoming fully aware of the metaphysical assumptions implicit in the highly sedimented practices which we unwittingly absorb through inculturation (Berghofer et al., 2021).

In this regard, it is encouraging to witness 'metaphysics of physics' sessions at a recent physics conference, conferences on the metaphysics of science, and the vibrant 'Phenomenological approaches to Physics' conference series. It is my hope that a confluence of physicists and philosophers from diverse traditions, acting in a concerted manner to interpret the fruits of the quantum reconstruction program, will finally succeed in lifting the quantum veil.

Acknowledgments

I am most grateful to Philipp Berghofer and Harald Wiltsche for their enthusiastic resonance with the reconstruction program, and for the invitation to contribute to this volume. I would also like to thank Michel Bitbol and Florian Boge for very helpful comments on a draft version of this chapter.

Notes

1 The degree to which these demands are in fact insisted upon, or indeed are deemed appropriate, varies with time and with the physics sub-community (or indeed the physicist) in question. For example, Mach's view of a theory as merely (or primarily) an economic codification of sense data seems to de-emphasize the first demand, while certain modern-day research programs (such as the string theory program) appear to de-emphasize the second.

2 This echoes Wigner's view of symmetry principles, *viz.* just as physical theories formalize regularities in sensory data, symmetry principles formalize regularities in the laws posited by those theories (Wigner, 1960).

3 For example, reconstruction of the classical quantities of motion (momentum and kinetic energy) shows that these quantities are a direct consequence of Galilean relativity and the general desideratum of conservation, thereby

establishing a clear connection between the *dynamical* and *kinematic* aspects of the theory, and also makes clear the necessity of forms of energy not bound to massive bodies (Goyal, 2020). See Section 11.4.1 for details.

4 The strategy is refined in due course—see Section 11.3.4.2.

5 In Goyal (2020), Newtonian mechanics is reconstructed on the basis of these and other special assumptions. A categorization of the assumptions employed is also given.

6 To be clear: the reformulative component consists in a re-expression of parts of the quantum formalism, for example the reformulation of the Schrödinger equation in the form of a Hamilton-Jacobi equation. Minimally, the interpretative component consists in the posit of the existence of particles obeying a guidance condition, together with an interpretation of the wave function. These two components are tightly connected, but can be separated—the reformulation is a purely mathematical rewriting, without any additional ontological commitment. And the reformulative component is being actively developed, for example being extended to relativistic quantum mechanics and quantum field theories.

7 For example, through establishing whether a $|\psi|^2$-distribution of particles can be viewed as an 'equilibrium' distribution by showing that a 'non-equilibrium' distribution—i.e. one not conforming to $|\psi|^2$—will generically 'relax' to an equilibrium distribution under unitary dynamics.

8 For example, David Deutsch, one of the founders of the field of quantum computation, credits the many worlds interpretation as providing a way of thinking which inspired the idea that it might be possible to *use* 'parallel universes' to carry out certain computational tasks more effectively than a classical computer. Similarly, John Bell has cited the de Broglie–Bohm interpretation as an inspiration for his eponymous theorem. These examples illustrate the powerful heuristic value of attempts—however inadequate or incomplete—to penetrate the quantum veil.

9 A decision-theoretic argument for the Born rule has been offered (Wallace, 2010), but as the many worlds interpretation does away with the idea that an agent has one and only one future successor, the applicability of decision-theoretic axioms to such a scenario is questionable. In particular, in a branching universe, the notion of probability has to be reconceived.

10 Under the standard interpretation, if a system of two identical particles is in a symmetrized state such as $\psi(x_1, x_2) = [\alpha(x_1)\beta(x_2) \pm \beta(x_1)\alpha(x_2)]/\sqrt{2}$, then index i refers to particle i.

11 But, even at this level, it should be noted that the standard quantum formalism is sometimes ambiguous. For example, given the standard interpretation of the quantum symmetrization algorithm for handling systems of identical particles, it is unclear how to determine whether or not a given state is entangled (Ghirardi et al., 2002).

12 A wide-ranging survey of the reconstructive methodology in physics is given in Darrigol (2015a).

13 Descartes' rationalization: 'It is obvious that when God first created the world, He not only moved its parts in various ways, but also simultaneously caused some of the parts to push others and to transfer their motion to these others. So in now maintaining the world by the same action and with the same laws with which He created it, He conserves motion; not always contained in the same parts of matter, but transferred from some parts to others depending on the ways in which they come in contact' (Descartes, 1982, II.42).

14 For more historical detail on this process of transformation, see Goyal (2020, §5.3.1).
15 Newton (amongst others) asserted that atoms were hard bodies that collide completely *inelastically* (Scott, 1970, pp. 4–5). Hence the fundamental importance of formulating laws applicable to inelastic collisions.
16 For a sample of such derivations, see Goyal (2020, §4).
17 For mathematical details, see, for instance, Aczél and Dhombres (1989, Ch. 1).
18 See, for instance Aczél (1966, §6.2).
19 'Already in my original paper I stressed the circumstance that I was unable to give a logical reason for the exclusion principle or to deduce it from more general assumptions. I had always the feeling and I still have it today, that this is a deficiency. Of course in the beginning I hoped that the new quantum mechanics, with the help of which it was possible to deduce so many half-empirical formal rules in use at that time, will also rigorously deduce the exclusion principle. Instead of it there was for electrons still an exclusion: not of particular states any longer, but of whole classes of states, namely the exclusion of all classes different from the antisymmetrical one' (Pauli, 1998, p. 32).
20 A comprehensive overview of the reconstruction program, which details the full range of approaches and their inspirations, assumptions, and techniques, has yet to been written. This is not surprising: the diversity in the approaches' conceptual starting points and mathematical techniques is immense, and the program is still in flow. Nevertheless, various perspectives—some written by researchers in the field and some by philosophers of physics—do exist. For example, in Hardy (2013, §2), Hardy—one of the first to present a compelling reconstruction of the aQF—offers brief but insightful remarks on the history of reconstruction. Grinbaum (2007a, b), a philosopher of physics, provides another interesting perspective. See also Dickson (2015), Stairs (2015) and Felline (2016) for philosophical perspectives on the reconstruction program. A detailed discussion of Hardy's reconstruction (and related reconstructions) is given in Darrigol (2015b).
21 A range of talks from leading researchers in the field presented at the 'Reconstructing Quantum Theory' workshop held at Perimeter Institute in 2009 are available online at https://pirsa.org/C09016.
22 The informational perspective in physics precedes quantum theory, and can be traced at least as far back as the development of statistical mechanics. For a more detailed treatment of its development, see Goyal (2012).
23 In contrast, in his *Science of Mechanics*, Mach recognizes that observations are central and casts doubt on the Newtonian notion of absolute space due to its lack of direct observability, but does not go so far as to question the notion of matter as primitive (Mach, 1919).
24 It is understood that these measurements satisfy the conditions of the operational framework. In particular, these measurements are repeatable, and satisfy the closure condition.
25 For more mathematical details, see Goyal (2012, §4.1) and Wootters (2013).
26 As per the operational framework (Section 11.5.2.1.1), the set of possible measurements needs to be circumscribed by a specific procedure.
27 In this connection, it is useful to compare the aQF with its analogue in classical physics, to which one could refer as the abstract classical formalism (aCF). In the aCF, one speaks of an abstract system in state that evolves continuously, deterministically, and reversibly; of the existence of an ideal measurement whose outcome determines the state without disturbing the state; and of

a system's subsystems always being in well-defined states which determine the state of the system. In the aQF, dynamics is—as in the aCF—reversible and deterministic, but the ideal measurement and compositional axioms differ.

28 These include the prediction of Bell-violating correlations, which are generally interpreted as a manifestation of nonlocality. According to Lakatos, novel predictions that are unexpected relative to previously existing theories, and which are subsequently experimentally verified, are a hallmark of progressive research programs (Lakatos, 1978). In that respect, it is astonishing that quantum theory has continued to form the heart of a progressive research program for a century, and shows no signs of flailing. In comparison, so-called regressive research programs typically play catch-up, merely accommodating new findings (whether they be experimentally discovered or anticipated by rival research programs).

References

Aczél, J. (1966). *Lectures on Functional Equations and Their Application*. Academic Press.

Aczél, J., and Dhombres, J. (1989). *Functional Equations in Several Variables*. Cambridge Univ. Press.

Barrett, J. (2007). Information processing in generalized probabilistic theories. *Physical Review A*, 75(032304). https://journals.aps.org/pra/abstract/10.1103/PhysRevA.75.032304

Bell, J. S. (1964). On the Einstein Podolsky Rosen paradox. *Physics*, 1:195–200.

Berghofer, P., Goyal, P., and Wiltsche, H. A. (2021). Husserl, the mathematization of nature, and the informational reconstruction of quantum theory. *Continental Philosophy Review*, 54:413–436.

Berghofer, P., and Wiltsche, H. A. (2019). The co-presentational character of perception. In Limbeck, C. and Stadler, F., editors, *The Philosophy of Perception and Observation*. De Gruyter.

Berghofer, P., and Wiltsche, H. A. (2020). Phenomenological approaches to physics. Mapping the field. In Wiltsche, H. A. and Berghofer, P., editors, *Phenomenological Approaches to Physics.*, pages 1–47.

Bohm, D. (1952). A suggested interpretation of the quantum theory in terms of "hidden" variables, I and II. *Physical Review*, 85:166–193.

Bohr, N. (1928). The quantum postulate and the recent development of atomic theory. *Nature*, 121:580–590.

Caulton, A. (2014). Qualitative individuation in permutation-invariant quantum mechanics. https://arxiv.org/abs/1409.0247.

Chiribella, G., Perinotti, P., and D'Ariano, G. M. (2011). Informational derivation of quantum theory. *Physical Review A*, 84:012311.

Cramer, J. G. (1986). The transactional interpretation of quantum mechanics. *Rev. Mod. Phys.*, 58:647–688.

Darrigol, O. (2004). The mystery of the Einstein–Poincaré connection. *Isis*, 95:614–626.

Darrigol, O. (2005). The genesis of the theory of relativity. *Séminaire Poincaré*, 1:1–22.

Darrigol, O. (2015a). *Physics and Necessity: Rationalist Pursuits from the Cartesian Past to the Quantum Present*. Oxford University Press.

Darrigol, O. (2015b). 'Shut up and contemplate!': Lucien Hardy's reasonable axioms for quantum theory. *Studies in History and Philosophy of Modern Physics*, 52:328–342.

Darrigol, O. (2020). Deducing Newton's second law from relativity principles: A forgotten history. *Archive for History of Exact Sciences*, 74:1–43.

Descartes, R. (1982). *Principles of Philosophy*. Kluwer.

Dickson, M. (2015). Reconstruction and reinvention in quantum theory. *Foundations of Physics*, 45:1330–1340.

Everett, H. (1957). Relative state formulation of quantum mechanics. *Reviews of Modern Physics*, 29(3):452–462.

Felline, L. (2016). It's a matter of principle. Scientific explanation in information-theoretic reconstructions of quantum theory. *Dialectica*, 70(4):549–575.

Feynman, R. P. (1948). Space-time approach to non-relativistic quantum mechanics. *Reviews of Modern Physics*, 20:367.

Fuchs, C. A. (2002). Quantum mechanics as quantum information. https://arxiv.org/abs/quant-ph/0205039.

Fuchs, C. A. (2017). Notwithstanding Bohr, the reasons for QBism. *Mind and Matter*, 15(2):245–300.

Ghirardi, G., Marinatto, L., and Weber, T. (2002). Entanglement and properties of composite quantum systems: A conceptual and mathematical analysis. *Journal of Statistical Physics*, 108:9–122

Goyal, P. (2008). Information-geometric reconstruction of quantum theory. *Physical Review A*, 78(5):052120.

Goyal, P. (2012). Information physics—Towards a new conception of physical reality. *Information*, 3(4):567.

Goyal, P. (2014). Derivation of quantum theory from Feynman's rules. *Physical Review A*, 89:032120.

Goyal, P. (2015). Informational approach to the quantum symmetrization postulate. *New Journal of Physics*, 17:013043.

Goyal, P. (2019). Persistence and nonpersistence as complementary models of identical quantum particles. *New Journal of Physics*, 21:063031.

Goyal, P. (2020). Derivation of classical mechanics in an energetic framework via conservation and relativity. *Foundations of Physics*, 50:1426–1479.

Goyal, P. (2022). Persistence and Reidentification in systems of identical quantum particles: Towards a post-atomistic conception of matter. Preprint: https://www.philsci-archive.pitt.edu/21849/

Goyal, P., and Knuth, K. H. (2011). Quantum theory and probability theory: Their relationship and origin in symmetry. *Symmetry*, 3:171–206.

Goyal, P., Knuth, K. H., and Skilling, J. (2010). Origin of complex quantum amplitudes and Feynman's rules. *Physical Review A*, 81:022109.

Grinbaum, A. (2007a). Reconstructing instead of interpreting quantum theory. *Philosophy of Science*, 74:761–774.

Grinbaum, A. (2007b). Reconstruction of quantum theory. *British Journal for the Philosophy of Science*, 58:387–408.

Hardy, L. (2001a). Quantum theory from five reasonable axioms. https://arxiv.org/abs/quant-ph/0101012.

Hardy, L. (2001b). Why quantum theory? Contribution to NATO Advanced Research Workshop "Modality, Probability, and Bell's Theorem", Cracow, Poland, 2001.

Hardy, L. (2013). Reconstructing quantum theory. https://arxiv.org/abs/1303.1538.

Heelan, P. A. (2016). *The Observable. Heisenberg's Philosophy of Quantum Mechanics*. Peter Lang.

Hertz, H. (1899). *The Principles of Mechanics Presented in a New Form*. Macmillan & Co. English translation.

Kochen, S., and Specker, E. P. (1967). The problem of hidden variables in quantum mechanics. *Journal of Mathematics and Mechanics*, 17:59.

Lakatos, I. (1978). Falsification and the methodology of scientific research programmes. In Worrall, J. and Currie, G., editors, *The Methodology of Scientific Research Programmes*, pages 8–101. Cambridge Univ. Press.

Lange, M. (2011). Why do forces add vectorially? A forgotten controversy in the foundations of classical mechanics. *American Journal of Physics*, 79:380–388.

Lützen, J. (2005). *Mechanistic Images in Geometric Form: Heinrich Hertz's 'Principles of Mechanics'*. Oxford University Press.

Mach, E. (1919). *The Science of Mechanics: A Critical and Historical Account of Its Development*. Open Court Pub. Co., 4th edition.

Pauli, W. (1998). Exclusion principle and quantum mechanics. In *Nobel Lectures in Physics (1942–1962)*. World Scientific.

Schütz, J. R. (1897). Prinzip der absoluten Erhaltung der Energie. *Nachr. v. d. Gesellschaft der Wissenschaften zu Göttingen, Math.-Phy. Kl.*, 2:110–123.

Schwarz, J. M., Stapp, H. P., and Beauregard, M. (2005). Quantum physics in neuroscience and psychology: A neurophysical model of mind-brain interaction. *Philosophical Transactions of the Royal Society B*, 360:1309–1327.

Scott, W. L. (1970). *The Conflict between Atomism and Conservation Theory 1644 to 1960*. Macdonald (London), Elsevier (NY).

Stairs, A. (2015). Quantum logic and quantum reconstruction. *Foundations of Physics*, 45:1351–1361.

Wallace, D. (2010). How to prove the Born rule. In Saunders, S., Barrett, J., Kent, A., and Wallace, D., editors, *Many Worlds? Everett, Quantum Theory, and Reality*. Oxford University Press.

Wheeler, J. A. (1989). It from bit. In *Proceedings of the 3rd International Symposium on the Foundations of Quantum Mechanics, Tokyo*.

Wheeler, J. A. (1990). Information, physics, quantum: The search for links. In Zurek, W. H., editor, *Complexity, Entropy, and the Physics of Information*. Addison-Wesley.

Wigner, E. P. (1960). The unreasonable effectiveness of mathematics in the natural sciences. *Communications on Pure and Applied Mathematics*, 13:1–14.

Wootters, W. K. (1980). *The acquisition of information from quantum measurements*. PhD thesis, University of Texas at Austin.

Wootters, W. K. (2013). Communicating through probabilities: Does quantum theory optimize the transfer of information? *Entropy*, 15:3130–3147.

Index

Note: Page references in *italics* denote figures, and with "n" endnotes.

For Product Safety Concerns and Information please contact our EU representative GPSR@taylorandfrancis.com
Taylor & Francis Verlag GmbH, Kaufingerstraße 24, 80331 München, Germany

www.ingramcontent.com/pod-product-compliance
Lightning Source LLC
Chambersburg PA
CBHW061615220326
41598CB00026BA/3773